How To Build WOODFRAME Houses

DRAKE PUBLISHERS INC. NEW YORK

Acknowledgment is made to the following individuals
for their contributions to this work:

The U. S. Government Printing Office
L. O. Anderson, Engineer, Forest Products Laboratory
Forest Service
U. S. Dept. of Agriculture

John M. Black
Theodore C. Scheffer
Herbert W. Eickner
Otto C. Heyer

Published in 1973 by
Drake Publishers Inc.
381 Park Avenue South
New York, N.Y. 10016

Library of Congress Cataloging in Publication Data
Main entry under title.

How to build wood-frame houses.

1. Wooden-frame houses.
TH4818.W6H68 694 73-4301
ISBN 0-87749-493-2

Printed in the United States of America

CONTENTS

WOOD-FRAME HOUSE CONSTRUCTION

INTRODUCTION

This publication presents sound principles for wood-frame house contruction and suggestions for selecting suitable materials that will greatly assist in the construction of a good house. It is also meant as a guide and handbook for those without this type of construction experience.

Many wood houses are in existence today that were built more than 200 years ago when early settlers arrived. The modern conventional wood-frame house, with wood or wood product covering materials, is economical, long lasting, and can be constructed in any location. The United States is well supplied with timber and has a diversified industry that manufactures lumber and other wood products used in the house. Few, if any, materials can compete with wood framing in the construction of houses. However, to provide this efficient wood house, good construction details are important as well as the selection of materials for each specific use.

Three essentials to be considered in building a satisfactory house are: (1) An efficient plan, (2) suitable materials, and (3) sound construction. The house may be large or small, elaborate or unpretentious, modern or traditional, yet without all three of these essentials it may be neither permanent nor satisfactory.

While designing and planning are beyond the scope of this publication, the information on materials and building practices is intended to guide builders and prospective homeowners in erecting a good house with a minimum of maintenance. This Handbook can also be used as a training aid for apprentices or as a standard by which to judge the quality of house construction.

It sets forth what are considered to be acceptable practices in assembling and arranging the parts of a well-designed wood-frame house. While details of construction may vary in different localities, the fundamental principles are the same. This handbook deals essentially with established methods of construction, and does not attempt to show new ones that are used in various parts of the country.

Construction details for houses are given in a series of drawings with accompanying text, which show the methods used in assembling the various parts.

In general, the order of presentation conforms to the normal sequence of constructing the building—from foundation to finish work. The final chapters add information on painting, protecting wood from decay and fire, and maintenance. A glossary of housing terms is also included at the back of the handbook to aid with unfamiliar or specific word usage.

CHAPTER 1

LOCATION AND EXCAVATION

Condition at Site

Before excavating for the new home, determine the subsoil conditions by test borings or by checking existing houses constructed near the site. A rock ledge may be encountered, necessitating costly removal; or a high water table may require design changes from a full basement to crawl space or concrete slab construction. If the area has been filled, the *footings* [2] should always extend through to undisturbed soil. Any variation from standard construction practices will increase the cost of the *foundation* and footings. Thus

[1] Maintained at Madison, Wis., in cooperation with the University of Wisconsin.

[2] Key words in italics appear in the glossary, p. 210.

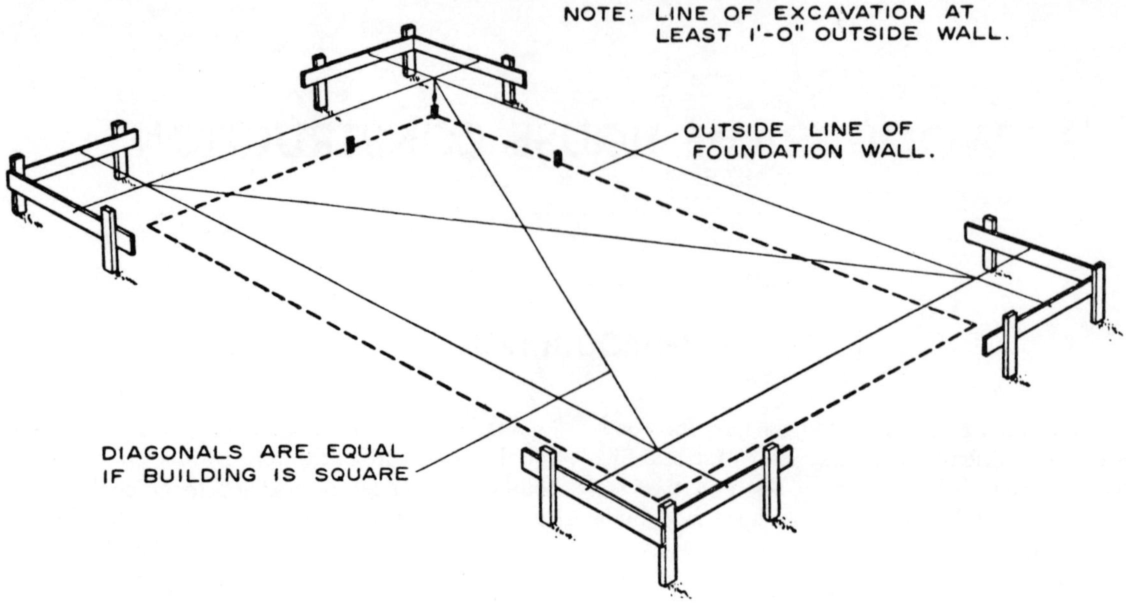

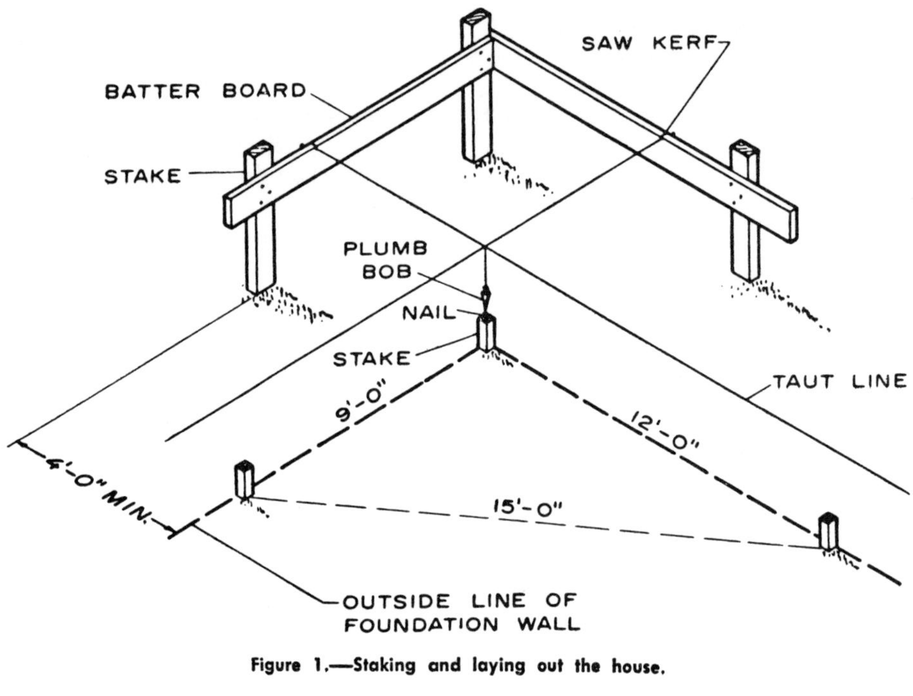

Figure 1.—Staking and laying out the house.

it is good practice to examine the type of foundations used in neighboring houses—this might influence the design of the new house.

Placement of the House

After the site is cleared, the location of the outer walls of the house is marked out. In general, the surveyor will mark the corners of the lot after making a survey of the plot of land. The corners of the proposed house also should be roughly marked by the surveyor.

Before the exact location of the house is determined, check local codes for minimum setback and side-yard requirements; the location of the house is usually determined by such codes. In some cases, the setback may be established by existing houses on adjacent property. Most city building regulations require that a plot plan be a part of the house plans so its location is determined beforehand.

The next step, after the corners of the house have been established, is to determine lines and grades as aids in keeping the work level and true. The *batter board* (fig. 1) is one of the methods used to locate and retain the outline of the house. The height of the boards is sometimes established to conform to the height of the foundation wall.

Small stakes are first located accurately at each corner of the house with nails driven in their tops to indicate the outside line of the foundation walls. To assure square corners, measure the diagonals to see if they are the same length. The corners can also be squared by measuring along one side a distance in 3-foot units such as 6, 9, and 12 and along the adjoining side the same number of 4-foot units as 8, 12, and 16. The diagonals will then measure the equal of 5-foot units such as 10, 15, and 20 when the unit is square. Thus, a 9-foot distance on one side and a 12-foot distance on the other should result in a 15-foot diagonal measurement for a true 90° corner.

After the corners have been located, three 2- by 4-inch or larger stakes of suitable length are driven at each location 4 feet (minimum) beyond the lines of the foundation; then 1- by 6- or 1- by 8-inch boards are nailed horizontally so the tops are all level at the same grade. Twine or stout string (carpenter chalkline) is next held across the top of opposite boards at two corners and adjusted so that it will be exactly over the nails in the corner stakes at either end; a *plumb* bob is handy for setting the lines. Saw kerfs at the outside edge are cut where the lines touch the boards so that they may be replaced if broken or disturbed. After similar cuts are located in all eight batter boards, the lines of the house will be established. Check the diagonals again to make sure the corners are square. An "L" shaped plan, for example, can be divided into rectangles, treating each separately or as an extension of one or more sides.

Height of Foundation Walls

It is common practice to establish the depth of the excavation, and consequently the height of the foundation, on ungraded or graded sites, by using the highest elevation of the excavation's perimeter as the control point (fig. 2). This method will insure good drainage if sufficient foundation height is allowed for the sloping of the final grade (fig. 3). Foundation walls at least 7 feet 4 inches high are desirable for full basements, but 8-foot walls are commonly used.

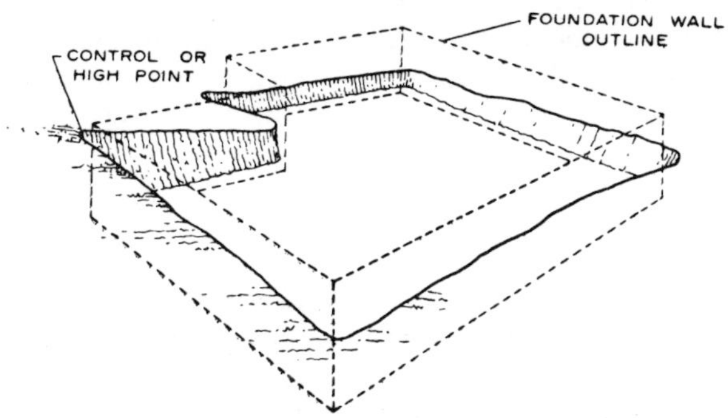

Figure 2.—Establishing depth of excavation.

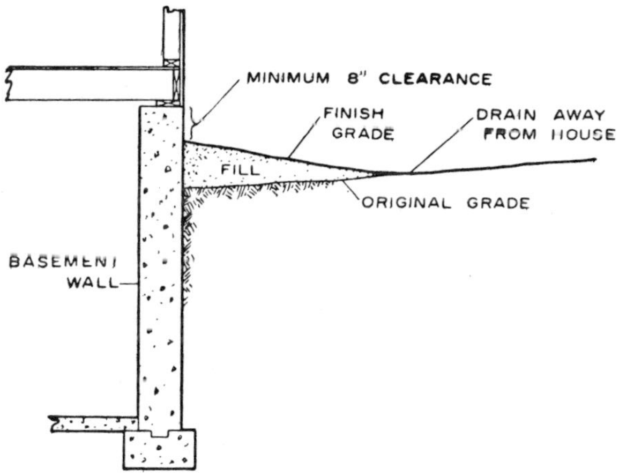

Figure 3.—Finish grade sloped for drainage.

Foundation walls should be extended above the finished grade around the outside of the house so that the wood finish and framing members will be adequately protected from soil moisture and be well above the grass line. Thus, in termite-infested areas, there will be an opportunity to observe any termite tubes between the soil and the wood and take protective measures before damage develops. Enough height should be provided in crawl spaces to permit periodic inspection for termites and for installation of soil covers to minimize the effects of ground moisture on framing members.

The top of the foundation wall should usually be at least 8 inches above the finish grade at the wall line. The finish grade at the building line might be 4 to 12 inches or more above the original ground level. In lots sloping upward from front to rear (fig. 3), this distance may amount to 12 inches or more. In very steeply sloped lots, a retaining wall to the rear of the wall line is often necessary.

For houses having crawl space, the distance between the ground level and underside of the *joist* should be at least 18 inches above the highest point within the area enclosed by the foundation wall. Where the interior ground level is excavated or otherwise below the outside finish grade, adequate precautionary measures should be made to assure positive drainage at all times.

Excavation

Excavation for basements may be accomplished with one of several types of earth-removing equipment. Top soil is often stockpiled by bulldozer or front-end loader for future use. Excavation of the basement area may be done with a front-end loader, power shovel, or similar equipment.

Power trenchers are often used in excavating for the walls of houses built on a slab or with a crawl space, if soil is stable enough to prevent caving. This eliminates the need for forming below grade when footings are not required.

Excavation is preferably carried only to the top of the footings or the bottom of the basement floor, because some soil becomes soft upon exposure to air or water. Thus it is advisable *not* to make the final excavation for footings until nearly time to pour the concrete unless formboards are to be used.

Excavation must be wide enough to provide space to work when constructing and waterproofing the wall and laying drain tile, if it is necessary in poor drainage areas (fig. 4). The steepness of the back slope of the excavation is determined by the subsoil encountered. With clay or other stable soil, the back slope can be nearly vertical. When sand is encountered, an inclined slope is required to prevent caving.

Some contractors, in excavating for basements, only roughstake the perimeter of the building for the removal of the earth. When the proper floor elevation has been reached, the footing layout is made and the earth removed. After the concrete is poured and set, the building wall outline is then established on the footings and marked for the formwork or concrete block wall.

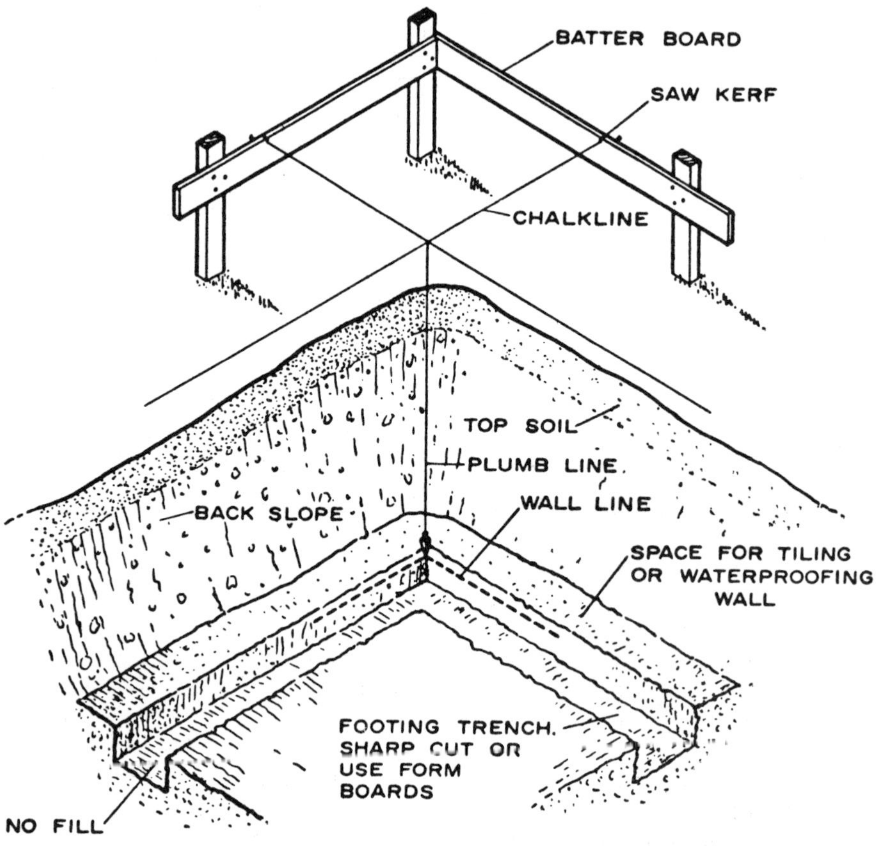

Figure 4.—Establishing corners for excavation and footings.

CHAPTER 2

CONCRETE AND MASONRY

Concrete and *masonry* units such as concrete block serve various purposes in most house designs, including concrete-slab and crawl-space houses which have poured concrete or concrete block foundation walls of some type. However, developments in treated wood foundation systems will permit all-weather construction and provide reliable foundations for crawl-space houses.

A great amount of concrete is supplied by ready-mix plants, even in rural areas. Concrete in this form is normally ordered by the number of bags per cubic yard, in addition to aggregate size and water-content requirements. Five-bag mix is considered minimum for most work, and where high strength or reinforcing is used, six-bag mix is commonly specified.

The size of gravel or crushed rock which can be obtained varies in different locations and it may be necessary to change the cement ratio normally recommended. Generally speaking, when gravel size is smaller than the normal 1½- to ¼-inch size, it is good practice to use a higher cement ratio. When gravel size is a maximum of 1 inch, add one-quarter sack of cement to the 5-bag mix; when gravel size is a maximum of ¾-inch, add one-half bag; and for ⅜-inch size add one bag.

Mixing and Pouring

Proportions of fine and coarse aggregate, amount of cement, and water content should follow the recommendations of the American Concrete Institute. Mixing plants are normally governed by these quantities. It is common practice to limit the amount of water to not more than 7½ gallons for each sack of cement, including that contained in the sand and gravel. Tables of quantities for field mixing on small jobs are available. For example, one combination utilizing a 1-inch maximum size of coarse aggregate uses: 5.8 sacks of cement per cubic yard, 5 gallons of water per sack of cement, and a cement to fine aggregate to coarse aggregate ratio of 1 to 2½ to 3½. Size of coarse aggregate is usually governed by the thickness of the wall and the spacing of reinforcing rods, when used. The use of 2-inch coarse aggregate, for example, is not recommended for slabs or other thin sections.

Concrete should be poured continuously wherever possible and kept practically level throughout the area being poured. All vertical joints should be keyed. *Rod* or *vibrate* the concrete to remove air pockets and force the concrete into all parts of the forms.

In hot weather, protect concrete from rapid drying. It should be kept moist for several days after pouring. Rapid drying lowers its strength and may injure the exposed surfaces of sidewalks and drives.

In very cold weather, keep the temperature of the concrete above freezing until it has set. The rate at which concrete sets is affected by temperature, being much slower at 40° F. and below than at higher temperatures. In cold weather, the use of heated water and aggregate during mixing is good practice. In severe weather, insulation or heat is used until the concrete has set.

Footings

The *footings* act as the base of the foundation and transmit the superimposed load to the soil. The type and size of footings should be suitable for the soil condition, and in cold climates the footings should be far enough below ground level to be protected from frost action. Local codes usually establish this depth, which is often 4 feet or more in northern sections of the United States.

Poured concrete footings are more dependable than those of other materials and are recommended for use in house foundations. Where fill has been used, the foundations should extend below the fill to undisturbed earth. In areas having adobe soil or where soil moisture may cause soil shrinkage, irregular settlement of the foundation and the building it supports may occur. Local practices that have been successful should be followed in such cases.

Wall Footings

Well-designed wall footings are important in preventing settling or cracks in the wall. One method of determining the size, often used with most normal soils, is based on the proposed wall thickness. The footing thickness or depth should be equal to the wall thickness (fig. 5,*A*). Footings should project beyond each side of the wall one-half the wall thickness. This is a general rule, of course, as the footing bearing area should be designed to the load capacity of the soil. Local regulations often relate to these needs. This also applies to column and fireplace footings.

If soil is of low load-bearing capacity, wider reinforced footings may be required.

A few rules that apply to footing design and construction are:

1. Footings must be at least 6 inches thick, with 8 inches or more preferable.
2. If footing excavation is too deep, fill with concrete—never replace dirt.

3. Use formboards for footings where soil conditions prevent sharply cut trenches.
4. Place footings *below* the frostline.
5. Reinforce footings with steel rods where they cross pipe trenches.
6. Use key slot for better resistance to water entry at wall location.
7. In freezing weather, cover with straw or supply heat.

Pier, Post, and Column Footings

Footings for *piers, posts,* or *columns* (fig. 5,*B*) should be square and include a *pedestal* on which the member will bear. A protruding steel pin is ordinarily set in the pedestal to anchor a wood post. Bolts for the bottom plate of steel posts are usually set when the pedestal is poured. At other times, steel posts are set directly on the footing and the concrete floor poured around them.

Footings vary in size depending on the allowable soil pressure and the spacing of the piers, posts, or columns. Common sizes are 24 by 24 by 12 inches and 30 by 30 by 12 inches. The pedestal is sometimes poured after the footing. The minimum height should be about 3 inches above the finish basement floor and 12 inches above finish grade in crawl-space areas.

Footings for fireplaces, furnaces, and chimneys should ordinarily be poured at the same time as other footings.

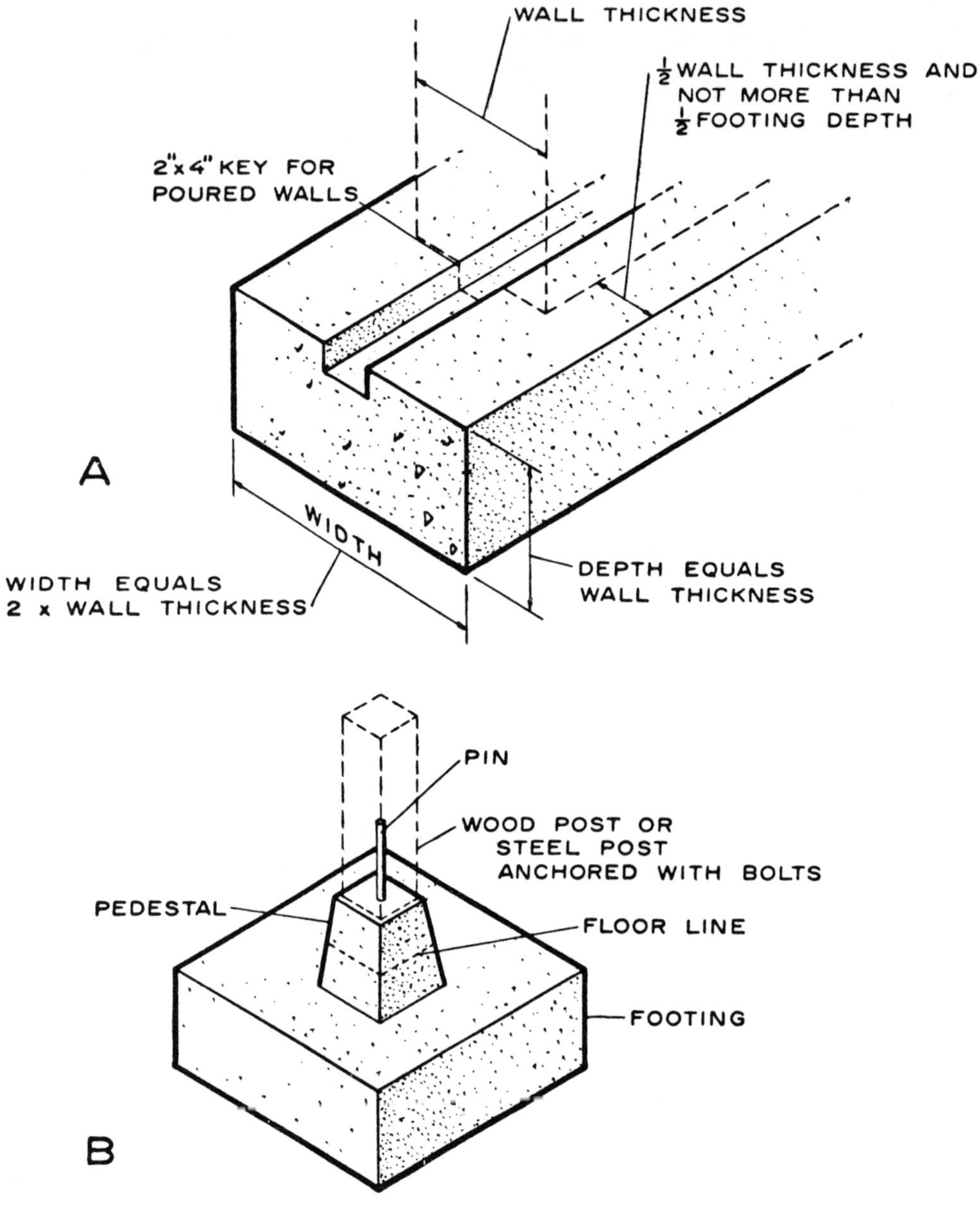

Figure 5.—Concrete footing: *A*, Wall footing; *B*, post footing.

Stepped Footings

Stepped footings are often used where the lot slopes to the front or rear and the garage or living areas are at basement level. The vertical part of the step should be poured at the same time as the footing. The bottom of the footing is always placed on undisturbed soil and located below the frostline. Each run of the footing should be level.

The vertical step between footings should be at least 6 inches thick and the same width as the footings (fig. 6). The height of the step should not be more than three-fourths of the adjacent horizontal footing. On steep slopes, more than one step may be required. It is good practices, when possible, to limit the vertical step to 2 feet. In very steep slopes, special footings may be required.

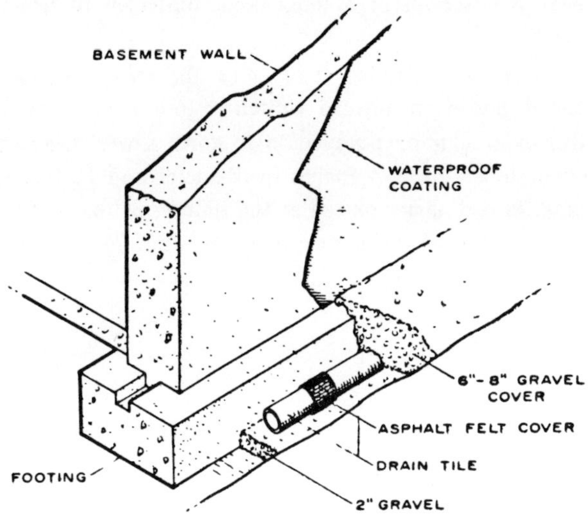

Figure 7.—Draintile for soil drainage at outer wall.

Draintile

Foundation or footing drains must often be used around foundations enclosing basements, or habitable spaces below the outside finish grade (fig. 7). This may be in sloping or low areas or any location where it is necessary to drain away subsurface water. This precaution will prevent damp basements and wet floors. Draintile is often necessary where habitable rooms are developed in the basement or where houses are located

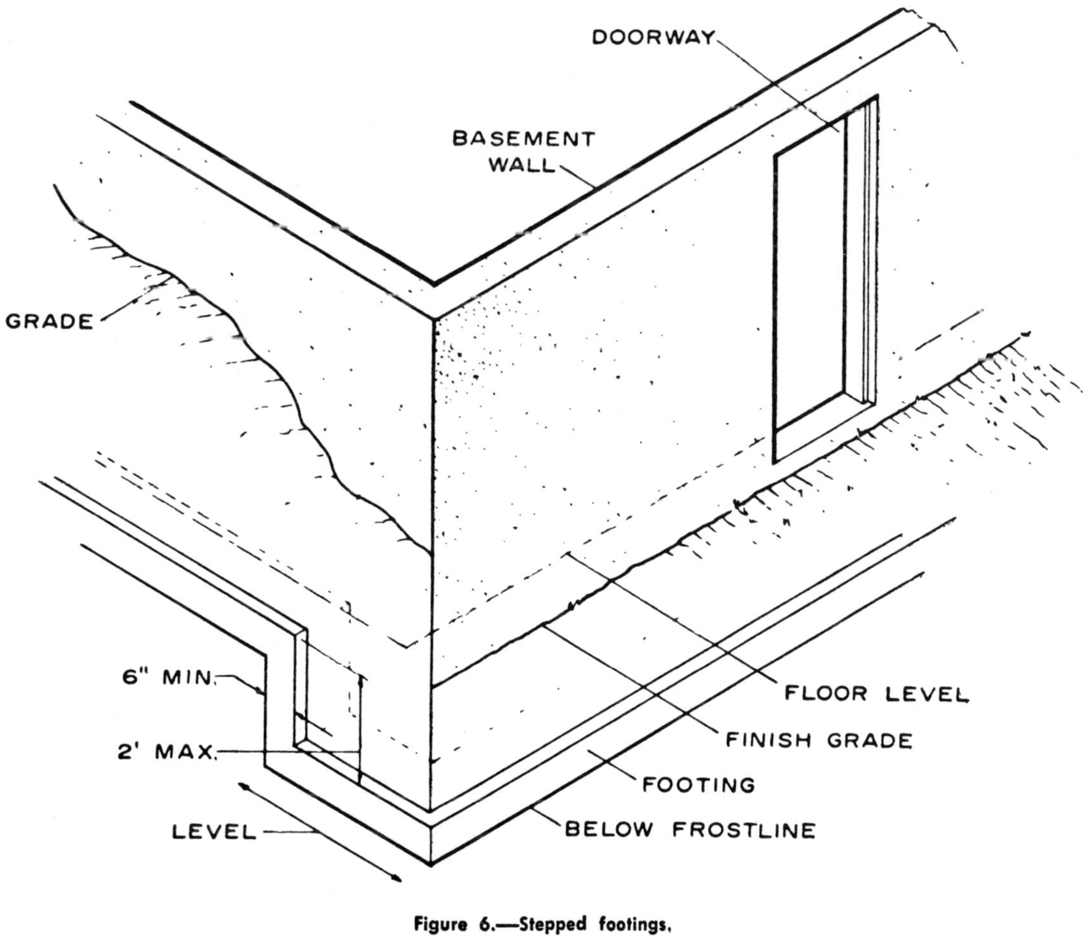

Figure 6.—Stepped footings.

near the bottom of a long slope subjected to heavy runoff.

Drains are installed at or below the area to be protected, and drain toward a ditch or into a sump where the water can be pumped to a storm sewer. Clay or concrete draintile, 4 inches in diameter and 12 inches long, is ordinarily placed at the bottom of the footing level on top of a 2-inch gravel bed (fig. 7). Tile are placed end to end and spaced about 1/8 inch apart. The top of the joint between the tile is covered with a strip of asphalt felt or similar paper; 6 to 8 inches of gravel is used over the tile. Drainage is toward the outfall or ditch. Dry wells for drainage water are used only when the soil conditions are favorable for this method of disposal. Local building regulations vary somewhat and should be consulted before construction of drainage system is started.

CHAPTER 3

FOUNDATION WALLS AND PIERS

Foundation walls form an enclosure for basements or crawl spaces and carry wall, floor, roof, and other building loads. The two types of walls most commonly used are *poured concrete* and *concrete block*. Treated wood foundations might also be used when accepted by local codes.

Preservative-treated posts and poles offer many possibilities for low-cost foundation systems and can also serve as a structural framework for the walls and roof (*6, 7,*).[3]

Wall thicknesses and types of construction are ordinarily controlled by local building regulations. Thicknesses of poured concrete basement walls may vary from 8 to 10 inches and concrete block walls from 8 to 12 inches, depending on story heights and length of unsupported walls.

Clear wall height should be no less than 7 feet from the top of the finish basement floor to the bottom of the joists; greater clearance is usually desirable to provide adequate headroom under girders, pipes, and ducts. Many contractors pour 8-foot-high walls above the footings, which provide a clearance of 7 feet 8 inches from the top of the finish concrete floor to the bottom of the joists. Concrete block walls, 11 courses above the footings with 4-inch solid cap-block, will produce about a 7-foot 4-inch height to the joists from the basement floor.

Poured Concrete Walls

Poured concrete walls (fig. 8) require forming that must be tight and also braced and tied to withstand the forces of the pouring operation and the fluid concrete.

Poured concrete walls should be double-formed (formwork constructed for each wall face). Reusable forms are used in the majority of poured walls. Panels may consist of wood framing with plywood facings and are fastened together with clips or other ties (fig. 8). Wood sheathing boards and studs with horizontal members and braces are sometimes used in the construction of forms in small communities. As in reusable forms, formwork should be plumb, straight, and braced sufficiently to withstand the pouring operations.

Frames for cellar windows, doors, and other openings are set in place as the forming is erected, along with forms for the beam pockets which are located to support the ends of the floor beam.

Reusable forms usually require little bracing other than horizontal members and sufficient blocking and bracing to keep them in place during pouring operations. Forms constructed with vertical studs and waterproof plywood or lumber sheathing require horizontal whalers and bracing.

Level marks of some type, such as nails along the form, should be used to assure a level foundation top. This will provide a good level sill plate and floor framing.

Concrete should be poured continuously without interruption and constantly puddled to remove air pockets and work the material under window frames and other blocking. If wood spacer blocks are used, they should be removed and not permitted to become buried in the concrete. Anchor bolts for the sill plate should be placed while the concrete is still plastic. Concrete should always be protected when temperatures are below freezing.

Forms should not be removed until the concrete has hardened and acquired sufficient strength to support loads imposed during early construction. At least 2 days (and preferably longer) are required when temperatures are well above freezing, and perhaps a week when outside temperatures are below freezing.

Poured concrete walls can be dampproofed with one heavy cold or hot coat of tar or asphalt. It should

[3] Numbers in parentheses refer to Literature Cited at the end of the Handbook.

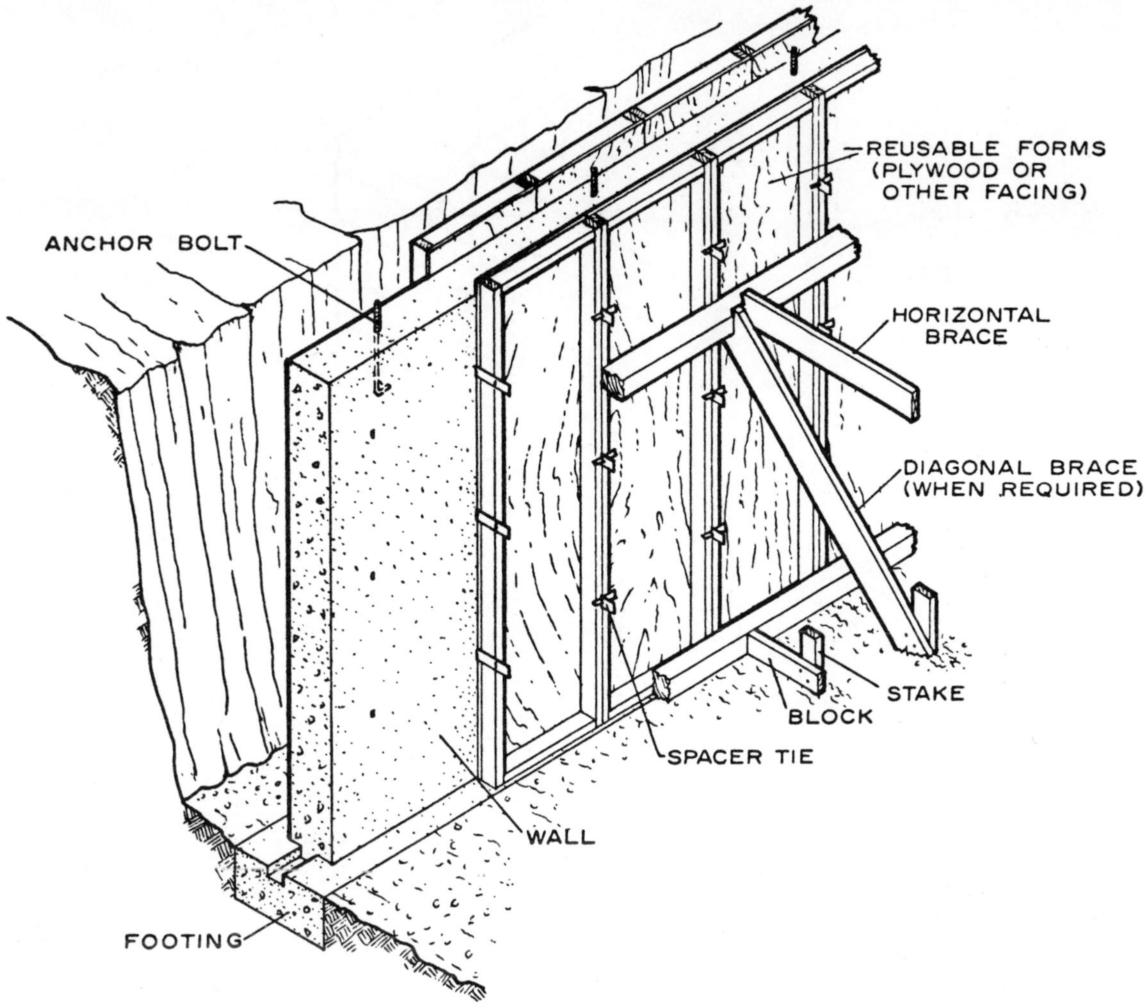

Figure 8.—Forming for poured concrete walls.

be applied to the outside from the footings to the finish gradeline. Such coatings are usually sufficient to make a wall watertight against ordinary seepage (such as may occur after a rainstorm), but should not be applied until the surface of the concrete has dried enough to assure good adhesion. In poorly drained soils, a membrane (such as described for concrete block walls) may be necessary.

Concrete Block Walls

Concrete blocks are available in various sizes and forms, but those generally used are 8, 10, and 12 inches wide. Modular blocks allow for the thickness and width of the mortar joint so are usually about 7⅝ inches high by 15⅝ inches long. This results in blocks which measure 8 inches high and 16 inches long from centerline to centerline of the mortar joints.

Concrete block walls require *no* formwork. Block courses start at the footing and are laid up with about ⅜-inch mortar joints, usually in a common bond (fig. 9). Joints should be tooled smooth to resist water seepage. Full bedding of mortar should be used on all contact surfaces of the block. When *pilasters* (column-like projections) are required by building codes or to strengthen a wall, they are placed on the interior side of the wall and terminated at the bottom of the beam or girder supported.

Basement door and window frames should be set with keys for rigidity and to prevent air leakage (fig. 9).

Block walls should be capped with 4 inches of solid masonry or concrete reinforced with wire mesh. *Anchor bolts* for sills are usually placed through the top two rows of blocks and the top cap. They should be anchored with a large plate washer at the bottom and the block openings filled solidly with mortar or concrete. (fig. 9).

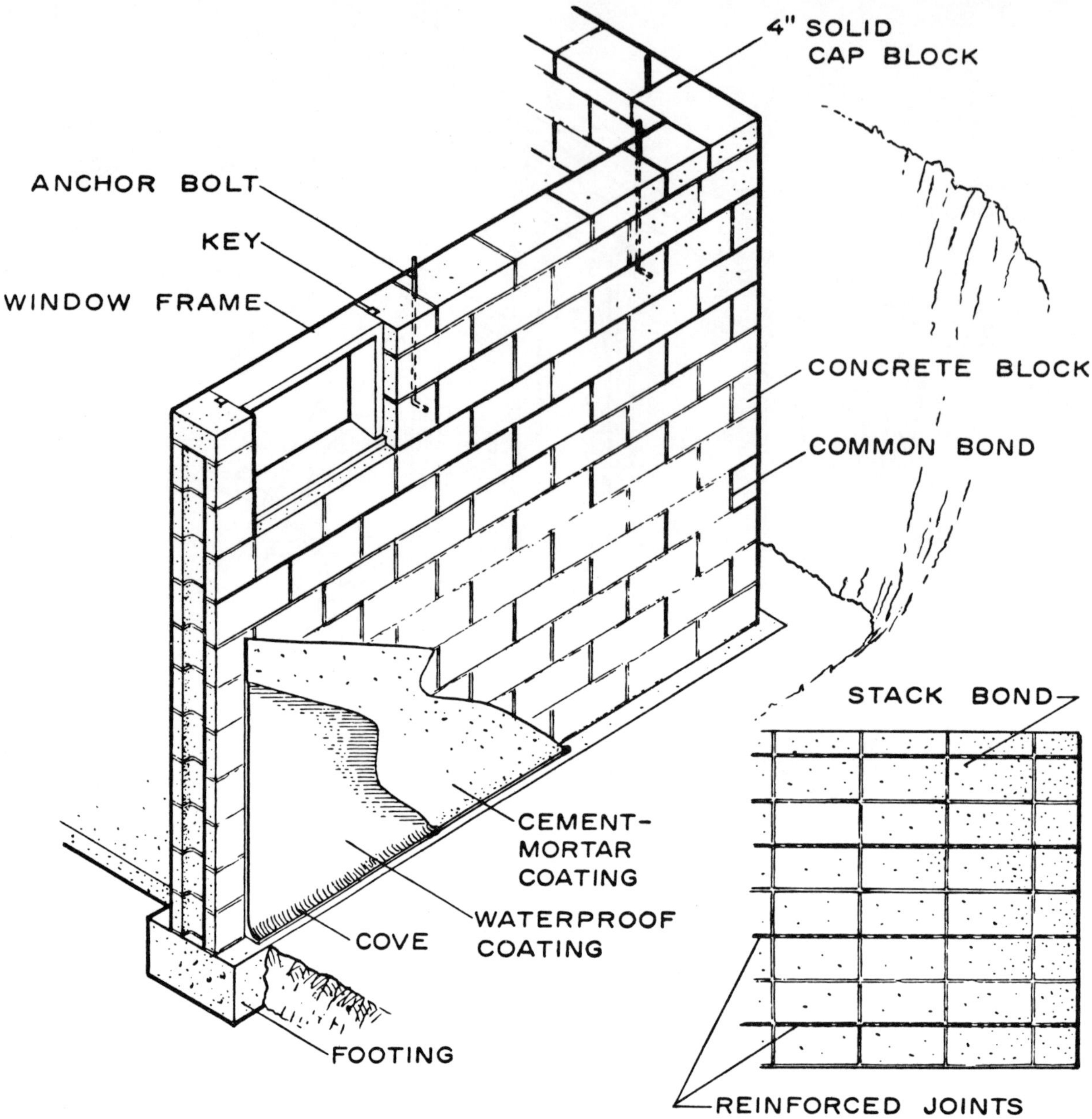

Figure 9.—Concrete block walls.

When an exposed block foundation is used as a finished wall for basement rooms, the *stack bond pattern* may be employed for a pleasing effect. This consists of placing blocks one above the other, resulting in continuous vertical mortar joints. However, when this system is used, it is necessary to incorporate some type of joint reinforcing every second course. This usually consists of small diameter steel longitudinal and cross rods arranged in a grid pattern. The common bond does not normally require this reinforcing, but when additional strength is desired, it is good practice to incorporate this bonding system into the wall.

Freshly laid block walls should be protected in temperatures below freezing. Freezing of the mortar before it has set will often result in low adhesion, low strength, and joint failure.

To provide a tight, waterproof joint between the footing and wall, an elastic calking compound is often used. The wall is waterproofed by applying a coating of cement-mortar over the block with a cove formed at the juncture with the footing (fig. 9). When the mortar is dry, a coating of asphalt or other waterproofing will normally assure a dry basement.

For added protection when wet soil conditions may be encountered, a waterproof membrane of roofing felt or other material can be mopped on, with shingle-style laps of 4 to 6 inches, over the cement-mortar coating. Hot tar or hot asphalt is commonly used over the membrane. This covering will prevent leaks if minor cracks develop in the blocks or joints between the blocks.

Masonry Construction for Crawl Spaces

In some areas of the country, the crawl-space house is often used in preference to those constructed over a basement or on a concrete slab. It is possible to construct a satisfactory house of this type by using (a) a good soil cover, (b) a small amount of ventilation, and (c) sufficient insulation to reduce heat loss. These details will be covered in later chapters.

One of the primary advantages of the crawl-space house over the full basement house is, of course, the reduced cost. Little or no excavation or grading is required except for the footings and walls. In mild climates, the footings are located only slightly below the finish grade. However, in the northern States where frost penerates deeply, the footing is often located 4 or more feet below the finish grade. This, of course, requires more masonry work and increases the cost. The footings should always be poured over undisturbed soil and never over fill unless special piers and grade beams are used.

The construction of a masonry wall for a crawl space is much the same as those required for a full basement (figs. 8 and 9), except that no excavation is required within the walls. Waterproofing and draintile are normally *not* required for this type of construction. The masonry pier replaces the wood or steel posts of the basement house used to support the center beam. Footing size and wall thicknesses vary somewhat by location and soil conditions. A common minimum thickness for walls in single-story frame houses is 8 inches for hollow concrete block and 6 inches for poured concrete. The minimum footing thickness is 6 inches and the width is 12 inches for concrete block and 10 inches for the poured foundation wall for crawl-space houses. However, in well constructed houses, it is common practice to use 8-inch walls and 16- by 8-inch footings.

Poured concrete or concrete block piers are often used to support floor beams in crawl-space houses. They should extend at least 12 inches above the groundline. The minimum size for a concrete block pier should be 8 by 16 inches with a 16- by 24- by 8-inch footing. A solid cap block is used as a top course. Poured concrete piers should be at least 10 by 10 inches in size with a 20- by 20- by 8-inch footing. Unreinforced concrete piers should be no greater in height than 10 times their least dimension. Concrete block piers should be no higher than four times the least dimension. The spacing of piers should not exceed 8 feet on center under exterior wall beams and interior girders set at right angles to the floor joists, and 12 feet on center under exterior wall beams set parallel to the floor joists. Exterior wall piers should not extend above grade more than four times their least dimension unless supported laterally by masonry or concrete walls. As for wall footing sizes, the size of the pier footings should be based on the load and the capacity of the soil.

Sill Plate Anchors

In wood-frame construction, the *sill plate* should be anchored to the foundation wall with ½-inch bolts hooked and spaced about 8 feet apart (fig. 10,*A*). In some areas, sill plates are fastened with masonry nails, but such nails do not have the uplift resistance of bolts. In high-wind and storm areas, well-anchored plates are very important. A *sill sealer* is often used under the sill plate on poured walls to take care of any irregularities which might have occured during curing of the concrete. Anchor bolts should be embedded 8 inches or more in poured concrete walls and 16 inches or more in block walls with the core filled with concrete. A large plate washer should be used at the head end of the bolt for the block wall. If termite shields are used, they should be installed under the plate and sill sealer.

Although not the best practice, some contractors construct wood-frame houses without the use of a sill plate. Anchorage of the floor system must then be provided by the use of steel strapping, which is placed during the pour or between the block joints. Strap is bent over the joist or the header joist and fastened by nailing (fig. 10,*B*). The use of a concrete or mortar beam fill provides resistance to air and insect entry.

Reinforcing in Poured Walls

Poured concrete walls normally do not require steel *reinforcing* except over window or door openings located below the top of the wall. This type of construction requires that a properly designed steel or reinforced-concrete *lintel* be built over the frame (fig. 11,*A*). In poured walls, the rods are laid in place while the concrete is being poured so that they are about 1½ inches above the opening. Frames should be prime painted or treated before installation. For concrete block walls, a similar reinforced poured concrete or a precast lintel is commonly used.

Where concrete work includes a connecting porch or garage wall not poured with the main basement wall, it is necessary to provide reinforcing-rod ties (fig. 11,*B*). These rods are placed during pouring of the main wall. Depending on the size and depth, at least three ½-inch deformed rods should be used at the intersection of each wall. Keyways may be used in addition to resist lateral movement. Such connecting walls should extend below normal frostline and be

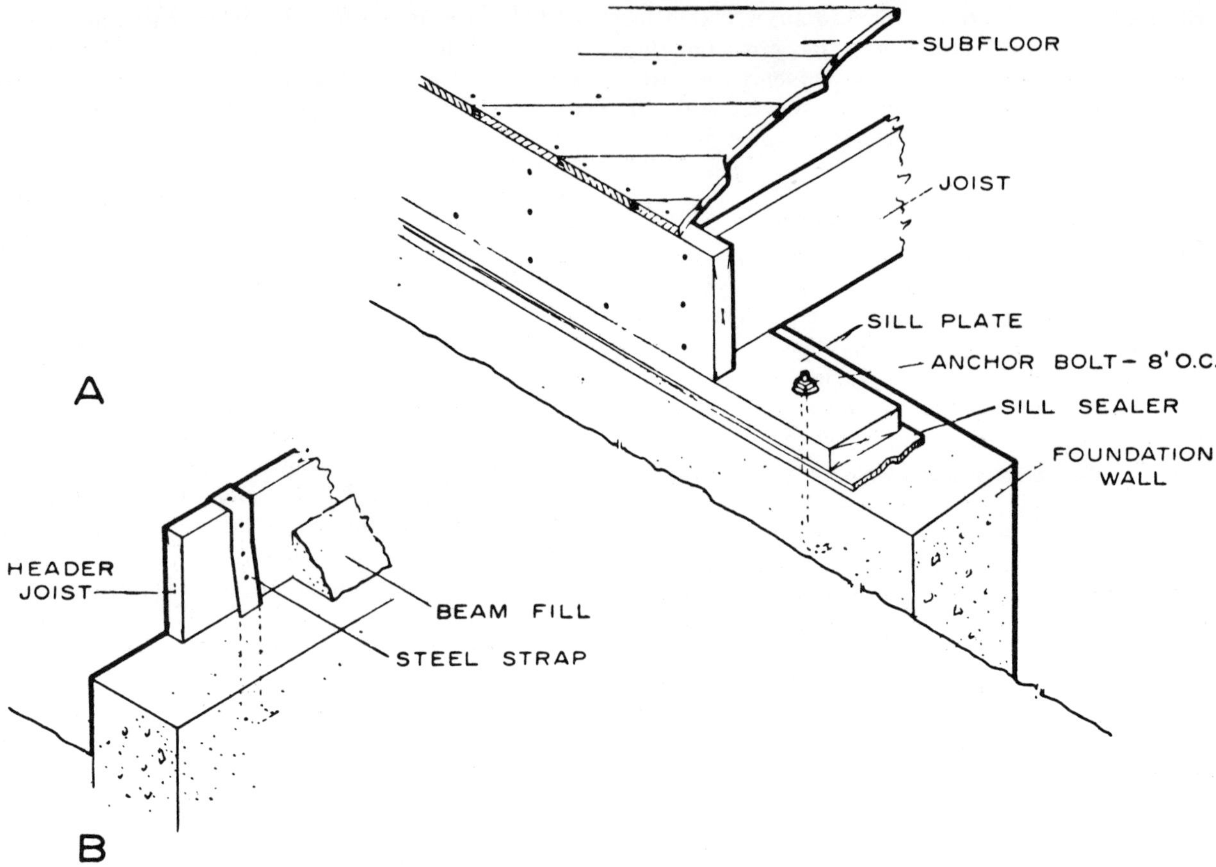

Figure 10.—Anchoring floor system to concrete or masonry walls: *A*, With sill plate; *B*, without sill plate.

supported by undisturbed ground. Wall extensions in concrete block walls are also of block and are constructed at the same time as the main walls over a footing placed below frostline.

Masonry Veneer Over Frame Walls

If *masonry veneer* is used for the outside finish over wood-frame walls, the foundation must include a supporting ledge or offset about 5 inches wide (fig. 12). This results in a space of about 1 inch between the masonry and the sheathing for ease in laying the brick. A base flashing is used at the brick course below the bottom of the sheathing and framing, and should be lapped with sheathing paper. Weep holes, to provide drainage, are also located at this course and are formed by eliminating the mortar in a vertical joint. Corrosion-resistant metal ties—spaced about 32 inches apart horizontally and 16 inches vertically—should be used to bond the brick veneer to the framework. Where other than wood sheathing is used, secure the ties to the studs.

Brick and stone should be laid in a full bed of mortar; avoid dropping mortar into the space between the veneer and sheathing. Outside joints should be tooled to a smooth finish to get the maximum resistance to water penetration.

Masonry laid during the cold weather should be protected from freezing until after the mortar has set.

Notch for Wood Beams

When basement beams or girders are wood, the wall notch or pocket for such members should be large enough to allow at least ½ inch of clearance at sides and ends of the beam for ventilation (fig. 13). Unless the wood is treated there is a decay hazard where beams and girders are so tightly set in wall notches that moisture cannot readily escape. A waterproof membrane, such as roll roofing, is commonly used under the end of the beam to minimize moisture absorption (fig. 13).

Protection Against Termites

Certain areas of the country, particularly the Atlantic Coast, Gulf States, Mississippi and Ohio Valleys, and southern California, are infested with wood-destroying termites. In such areas, wood construction

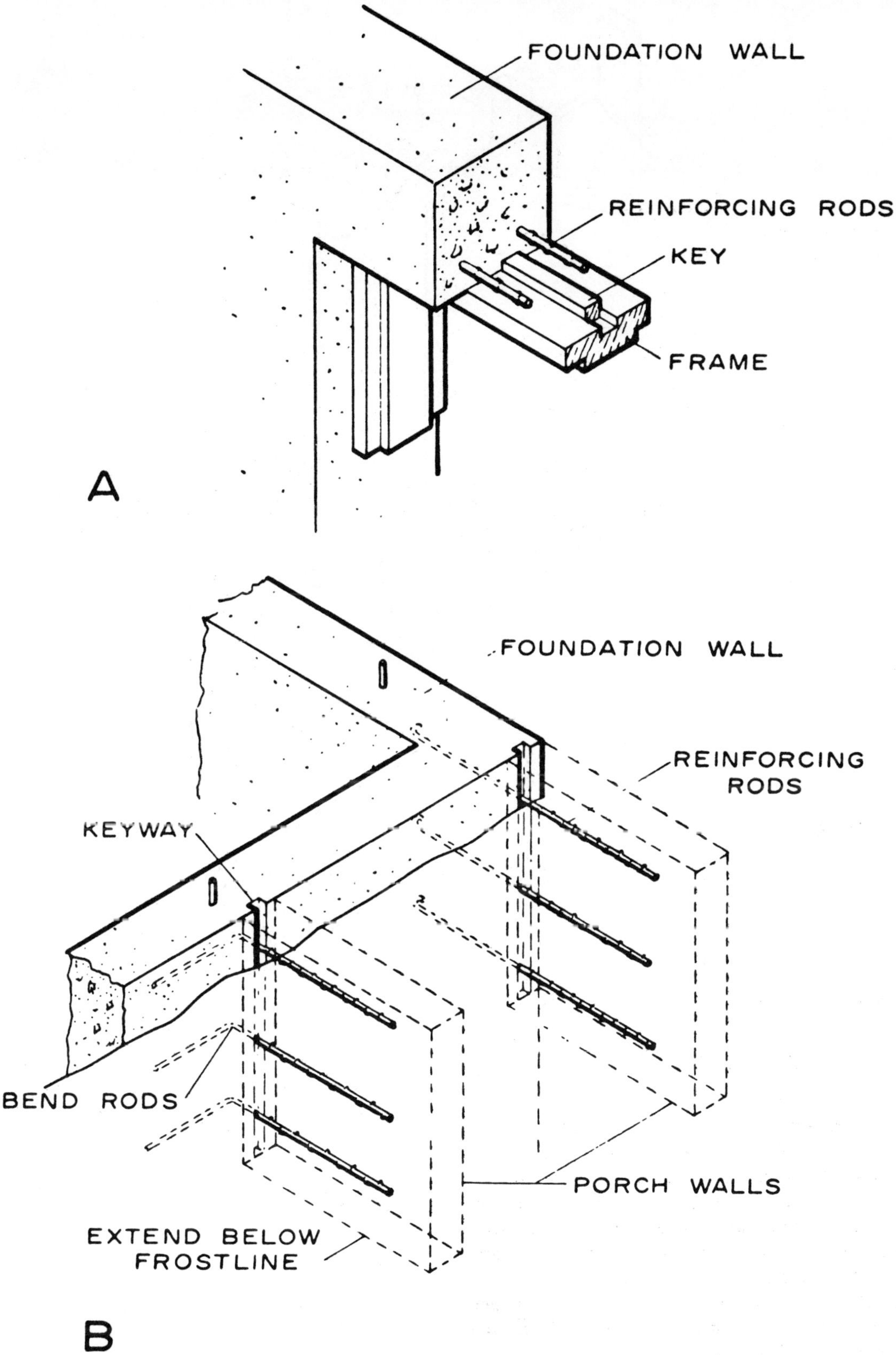

Figure 11.—Steel reinforcing rods in concrete walls: *A*, Rods used over window or doorframes; *B*, rod ties used for porch or garage walls.

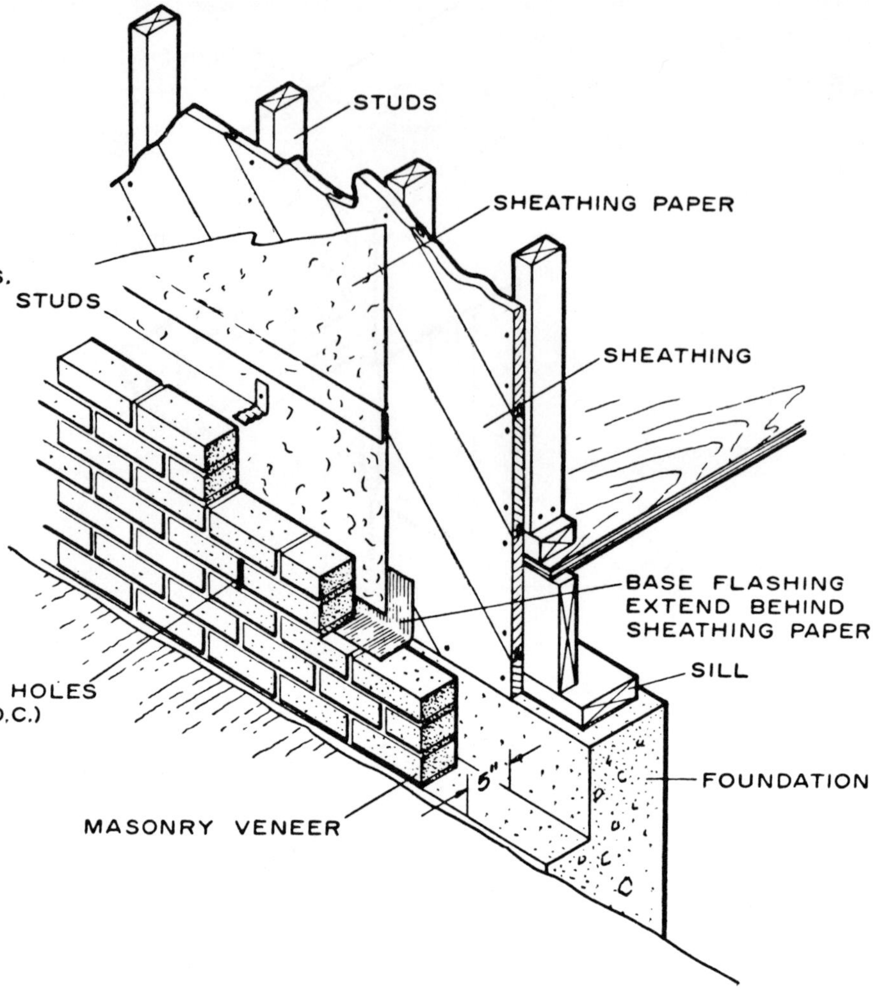

Figure 12.—Wood-frame wall with masonry veneer.

over a masonry foundation should be protected by one or more of the following methods:

1. Poured concrete foundation walls.
2. Masonry unit foundation walls capped with reinforced concrete.
3. Metal shields made of rust-resistant material. (Metal shields are effective only if they extend beyond the masonry walls and are continuous, with no gaps or loose joints. This shield is of primary importance under most conditions.)
4. Wood-preservative treatment. (This method protects only the members treated.)
5. Treatment of soil with soil poison. (This is perhaps one of the most common and effective means used presently.)

See Chapter 29 for further details on protection against termites.

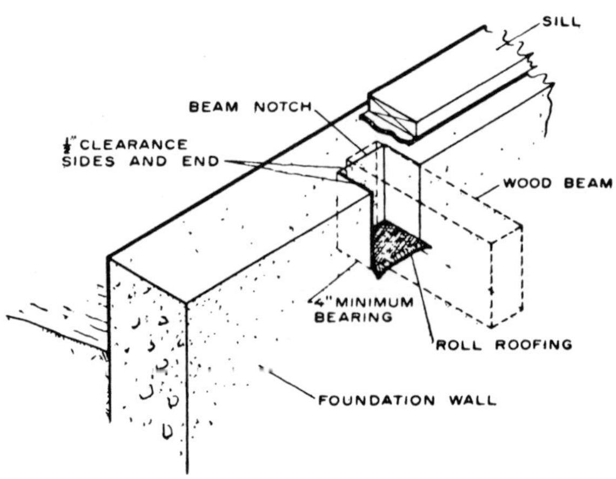

Figure 13.—Notch for wood beam.

CHAPTER 4

CONCRETE FLOOR SLABS ON GROUND

The number of new one-story houses with full basements has declined in recent years, particularly in the warmer areas of the United States. This is due in part to lower construction costs of houses without basements and an apparent decrease in need for the basement space.

The primary function of a basement in the past has been to provide space for a central heating plant and for the storage and handling of bulk fuel and ashes. It also houses laundry and utilities. With the wide use of liquid and gas fuels, however, the need for fuel and ash storage space has greatly diminished. Because space can be compactly provided on the ground-floor level for the heating plant, laundry, and utilities, the need for a basement often disappears.

Types of Floor Construction

One common type of floor construction for basementless houses is a concrete slab over a suitable foundation. Sloping ground or low areas are usually not ideal for slab-on-ground construction because structural and drainage problems would add to costs. Split-level houses often have a portion of the foundation designed for a grade slab. In such use, the slope of the lot is taken into account and the objectionable features of a sloping ground become an advantage.

The finish flooring for concrete floor slabs on the ground was initially asphalt tile laid in *mastic* directly on the slab. These concrete floors did not prove satisfactory in a number of instances, and considerable prejudice has been built up against this method of construction. The common complaints have been that the floors are cold and uncomfortable and that condensation sometimes collects on the floor, near the walls in cold weather, and elsewhere during warm, humid weather. Some of these undesirable features of concrete floors on the ground apply to both warm and cold climates, and others only to cold climates.

Improvements in methods of construction based on past experience and research have materially reduced the common faults of the slab floor but consequently increased their cost.

Floors are cold principally because of loss of heat through the floor and the foundation walls, with most loss occurring around the exterior walls. Suitable insulation around the perimeter of the house will help to reduce the heat loss. *Radiant floor heating* systems are effective in preventing cold floors and floor condensation problems. Peripheral warm-air heating ducts are also effective in this respect. Vapor barriers over a gravel fill under the floor slab prevent soil moisture from rising through the slab.

Basic Requirements

Certain basic requirements should be met in the construction of concrete floor slabs to provide a satisfactory floor. They are:

1. Establish finish floor level high enough above the natural ground level so that finish grade around the house can be sloped away for good drainage. Top of slab should be no less than 8 inches above the ground and the siding no less than 6 inches.

2. Top soil should be removed and sewer and water lines installed, then covered with 4 to 6 inches of gravel or crushed rock well-tamped in place.

3. A vapor barrier consisting of a heavy plastic film, such as 6-mil polyethylene, asphalt laminated duplex sheet, or 45-pound or heavier roofing, with minimum of ½-perm rating should be used under the concrete slab. Joints should be lapped at least 4 inches and sealed. The barrier should be strong enough to resist puncturing during placing of the concrete.

4. A permanent, waterproof, nonabsorptive type of rigid insulation should be installed around the perimeter of the wall. Insulation may extend down on the inside of the wall vertically or under the slab edge horizontally.

5. The slab should be reinforced with 6- by 6-inch No. 10 wire mesh or other effective reinforcing. The concrete slab should be at least 4 inches thick and should conform to information in Chapter 2, "Concrete and Masonry." A monolithic slab (fig. 14) is preferred in termite areas.

6. After leveling and screeding, the surface should be floated with wood or metal floats while concrete is still plastic. If a smooth dense surface is needed for the installation of wood or resilient tile with adhesives, the surface should be steel troweled.

Combined Slab and Foundation

The combined slab and foundation, sometimes referred to as the *thickened-edge slab,* is useful in warm climates where frost penetration is not a problem and where soil conditions are especially favorable. It consists of a shallow perimeter reinforced footing poured integrally with the slab over a vapor barrier (fig. 14). The bottom of the footing should be at least 1 foot below the natural gradeline and supported on solid, unfilled, and well-drained ground.

Independent Concrete Slab and Foundation Walls

When ground freezes to any appreciable depth during winter, the walls of the house must be supported

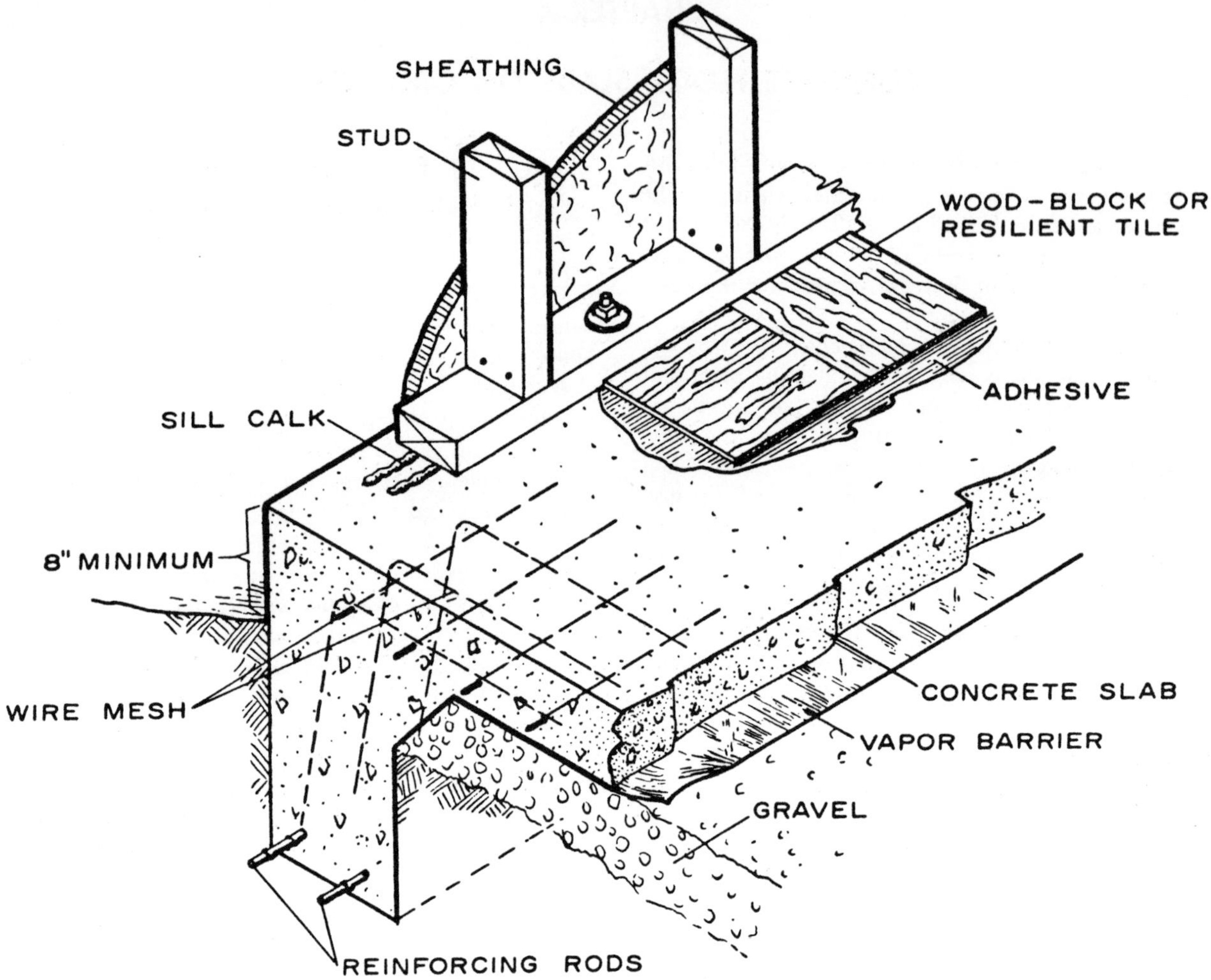

Figure 14.—Combined slab and foundation (thickened edge slab.)

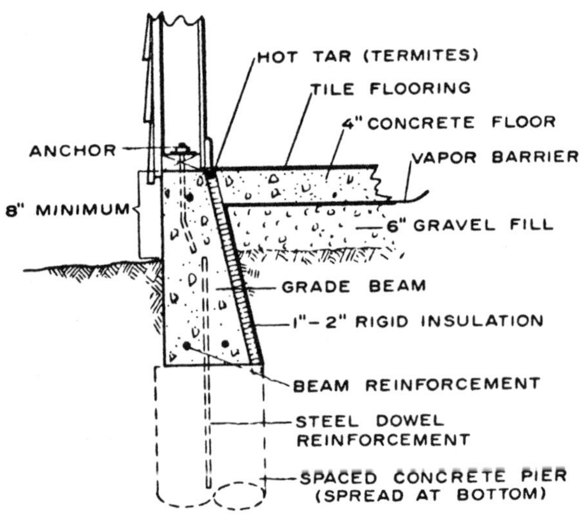

M 134 652

Figure 15.—Reinforced grade beam for concrete slab. Beam spans between concrete piers located below frostline.

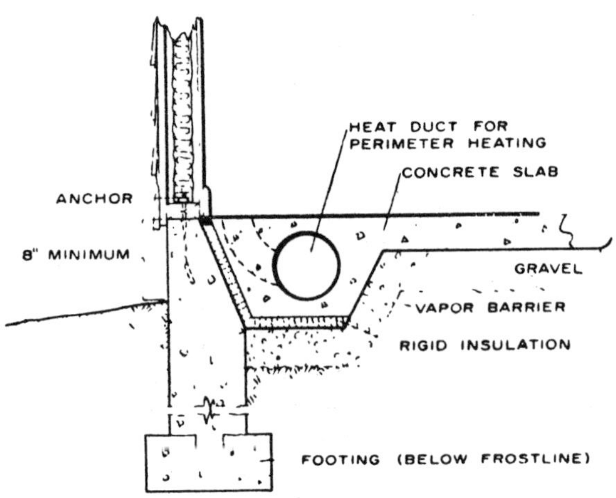

Figure 16.—Full foundation wall for cold climates. Perimeter heat duct insulated to reduce heat loss.

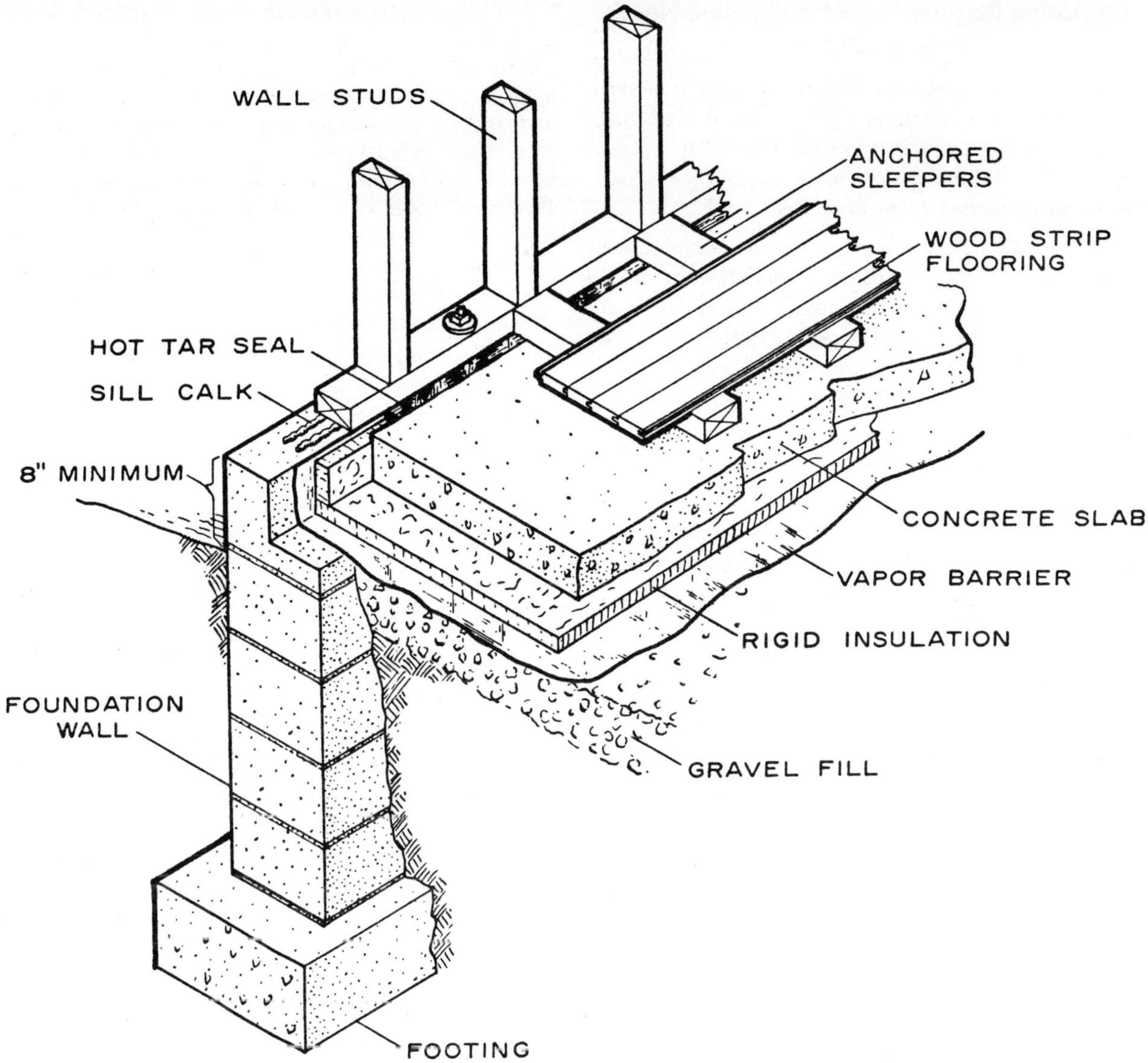

Figure 17.—Independent concrete floor slab and wall. Concrete block is used over poured footing which is below frostline. Rigid insulation may also be located along the inside of the block wall.

by foundations or piers which extend below the frostline to solid bearing on unfilled soil. In such construction, the concrete slab and foundation wall are usually separate. Three typical systems are suitable for such conditions (figs. 15, 16, and 17).

Vapor Barrier Under Concrete Slab

The most desirable properties in a vapor barrier to be used under a concrete slab are: (a) Good vapor-transmission rating (less than 0.5 perm); (b) resistance to damage by moisture and rot; and (c) ability to withstand normal usage during pouring operations. Such properties are included in the following types of materials:

1. 55-pound roll roofing or heavy asphalt laminated duplex barriers.
2. Heavy plastic film, such as 6-mil or heavier polyethylene, or similar plastic film laminated to a duplex treated paper.
3. Three layers of roofing felt mopped with hot asphalt.
4. Heavy asphalt impregnated and vapor-resistant rigid sheet material with sealed joints.

Insulation Requirements for Concrete Floor Slabs on Ground

The use of perimeter insulation for slabs is necessary to prevent heat loss and cold floors during the heating season, except in warm climates. The proper locations for this insulation under several conditions are shown in figures 15 to 17.

The thickness of the insulation will depend upon requirements of the climate and upon the materials used. Some insulations have more than twice the insulating value of others (see Chapter 15). The resistance (R) per inch of thickness, as well as the heating design temperature, should govern the amount required. Perhaps two good general rules to follow are:

1. For average winter low temperatures of 0° F. and higher (moderate climates), the total *R* should be about 2.0 and the depth of the insulation or the width under the slab not less than 1 foot.
2. For average winter low temperatures of −20° F. and lower (cold climates), the total *R* should be about 3.0 without floor heating and the depth or width of insulation not less than 2 feet.

Table 1 shows these factors in more detail. The values shown are minimum and any increase in insulation will result in lower heat losses.

TABLE 1.—*Resistance values used in determining minimum amount of edge insulation for concrete floors slabs on ground for various design temperatures.*

Low temperatures	Depth insulation extends below grade	Resistance (R) factor: No floor heating	Resistance (R) factor: Floor heating
°F.	Ft.		
−20	2	3.0	4.0
−10	1½	2.5	3.5
0	1	2.0	3.0
+10	1	2.0	3.0
+20	1	2.0	3.0

Insulation Types

The properties desired in insulation for floor slabs are: 1) High resistance to heat transmission, 2) permanent durability when exposed to dampness and frost, and 3) high resistance to crushing due to floor loads, weight of slab, or expansion forces. The slab should also be immune to fungus and insect attack, and should not absorb or retain moisture. Examples of materials considered to have these properties are:

1. *Cellular-glass insulation board,* available in slabs 2, 3, 4, and 5 inches thick. *R* factor, or resistivity, 1.8 to 2.2 per inch of thickness. Crushing strength, approximately 150 pounds per square inch. Easily cut and worked. The surface may spall (chip or crumble) away if subjected to moisture and freezing. It should be dipped in roofing pitch or asphalt for protection. Insulation should be located above or inside the vapor barrier for protection from moisture (figs. 15 to 17). This type of insulation has been replaced to a large extent by the newer foamed plastics such as polystyrene and polyurethane.

2. *Glass fibers with plastic binder,* coated or uncoated, available in thicknesses of ¾, 1, 1½, and 2 inches. *R* factor, 3.3 to 3.9 per inch of thickness. Crushing strength, about 12 pounds per square inch. Water penetration into coated board is slow and inconsequential unless the board is exposed to a constant head of water, in which case this water may disintegrate the binder. Use a coated board or apply coal-tar pitch or asphalt to uncoated board. Coat all edges. Follow manufacturer's instructions for cutting. Placement of the insulation inside the vapor barrier will afford some protection.

3. *Foamed plastic* (polystyrene, polyurethane, and others) insulation in sheet form, usually available in thicknesses of ½, 1, 1½, and 2 inches. At normal temperatures the *R* factor varies from 3.7 for polystyrenes to over 6.0 for polyurethane for a 1-inch thickness. These materials generally have low water-vapor transmission rates. Some are low in crushing strength and perhaps are best used in a vertical position (fig. 15) and not under the slab where crushing could occur.

4. *Insulating concrete.* Expanded mica aggregate, 1 part cement to 6 parts aggregate, thickness used as required. *R* factor, about 1.1 per inch of thickness. Crushing strength, adequate. It may take up moisture when subject to dampness, and consequently its use should be limited to locations where there will be no contact with moisture from any source.

5. *Concrete made with lightweight aggregate,* such as expanded slag, burned clay, or pumice, using 1 part cement to 4 parts aggregate; thickness used as required. *R* factor, about 0.40 per inch of thickness. Crushing strength, high. This lightweight aggregate may also be used for foundation walls in place of stone or gravel aggregate.

Under service conditions there are two sources of moisture that might affect insulating materials: (1) Vapor from inside the house and (2) moisture from soil. Vapor barriers and coatings may retard but not entirely prevent the penetration of moisture into the insulation. Dampness may reduce the crushing strength of insulation, which in turn may permit the edge of the slab to settle. Compression of the insulation, moreover, reduces its efficiency. Insulating materials should perform satisfactorily in any position if they do not change dimensions and if they are kept dry.

Protection Against Termites

In areas where termites are a problem, certain precautions are necessary for concrete slab floors on the ground. Leave a countersink-type opening 1-inch wide and 1-inch deep around plumbing pipes where they pass through the slab, and fill the opening with hot tar when the pipe is in place. Where insulation is used between the slab and the foundation wall, the insulation should be kept 1 inch below the top of the slab and the space should also be filled with hot tar (fig. 15). Further discussion of protection against termites, such as soil poisoning, is given in Chapter 29.

Finish Floors Over Concrete Slabs on the Ground

A natural concrete surface is sometimes used for the finish floor, but generally is not considered wholly satisfactory. Special dressings are required to prevent dusting. Moreover, such floors tend to feel cold. Asphalt or vinyl-asbestos tile laid in mastic in accordance with the manufacturer's recommendations is comparatively economical and easy to clean, but it also feels cold. Wood tile in various forms and wood parquet flooring may be used, also laid in mastic (fig. 14) in accordance with the manufacturer's recommendations. Tongued-and-grooved wood strip flooring 25/32 inch thick may be used but should be used over pressure-treated wood sleepers anchored to the slab (fig. 17). For existing concrete floors, the use of a vaporproof coating before installation of the treated sleepers is good practice.

CHAPTER 5

FLOOR FRAMING

The *floor framing* in a wood-frame house consists specifically of the posts, beams, sill plates, joists, and subfloor. When these are assembled properly on a foundation, they form a level anchored platform for the rest of the house. The posts and center beams of wood or steel, which support the inside ends of the joists, are sometimes replaced with a woodframe or masonry wall when the basement area is divided into rooms. Wood-frame houses may also be constructed upon a concrete floor slab or over a crawl-space area with floor framing similar to that used for a full basement.

Factors in Design

One of the important factors in the design of a wood floor system is to equalize shrinkage and expansion of the wood framing at the outside walls and at the center beam. This is usually accomplished by using approximately the same total depth of wood at the center beam as the outside framing. Thus, as beams and joists approach moisture equilibrium or the moisture content they reach in service, there are only small differences in the amount of shrinkage. This will minimize plaster cracks and prevent sticking doors and other inconveniences caused by uneven shrinkage. If there is a total of 12 inches of wood at the foundation wall (including joists and sill plate), this should be balanced with about 12 inches of wood at the center beam.

Moisture content of beams and joists used in floor framing should not exceed 19 percent. However, a moisture content of about 15 percent is much more desirable. Dimension material can be obtained at these moisture contents when so specified. When moisture contents are in the higher ranges, it is good practice to allow joists and beams to approach their moisture equilibrium before applying inside finish and trim, such as baseboard, base shoe, door jambs, and casings.

Grades of dimension lumber vary considerably by species. For specific uses in this publication, a sequence of first, second, third, fourth, and sometimes fifth grade material is used. In general, the first grade is for a high or special use, the second for better than average, the third for average, and the fourth and fifth for more economical construction. Joists and girders are usually second grade material of a species, while sills and posts are usually of third or fourth grade. Specific recommendations for each species are available (*5*).

Recommended Nailing Practices

Of primary consideration in the construction of a house is the method used to fasten the various wood members together. These connections are most commonly made with nails, but on occasions metal straps, lag screws, bolts, and adhesives may be used.

Proper fastening of frame members and covering materials provides the rigidity and strength to resist severe windstorms and other hazards. Good nailing is also important from the standpoint of normal performance of wood parts. For example, proper fastening of intersecting walls usually reduces plaster cracking at the inside corners.

The schedule in table 2 outlines good nailing practices for the framing and sheathing of a well-constructed wood-frame house. Sizes of common wire nails are shown in figure 18.

When houses are located in hurricane areas, they should be provided with supplemental fasteners. Details of these systems are outlined in "Houses Can Resist Hurricanes" (7).

Posts and Girders

Wood or steel posts are generally used in the basement to support wood girders or steel beams. Masonry piers might also be used for this purpose and are commonly employed in crawl-space houses.

The round steel post can be used to support both wood girders and steel beams and is normally supplied with a steel bearing plate at each end. Secure anchoring to the girder or beam is important (fig. 19).

Wood posts should be solid and not less than 6 by 6 inches in size for freestanding use in a basement. When combined with a framed wall, they may be 4 by 6

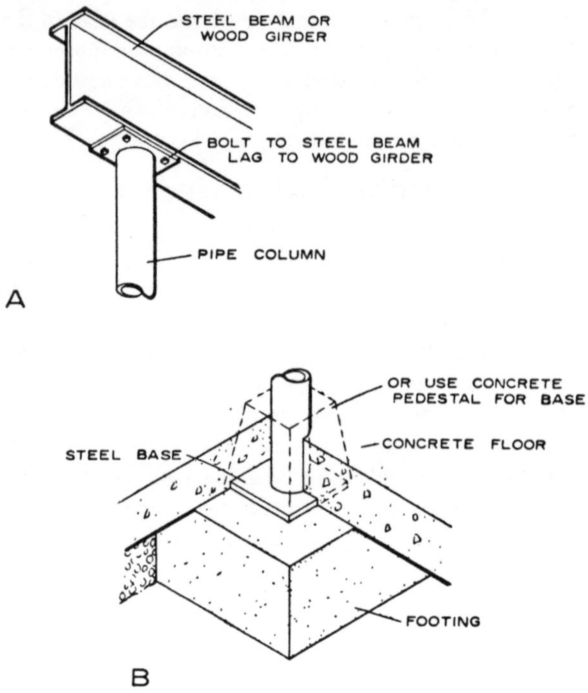

Figure 19.—Steel post for wood or steel girder: *A*, Connection to beam; *B*, base plate also may be mounted on and anchored to a concrete pedestal.

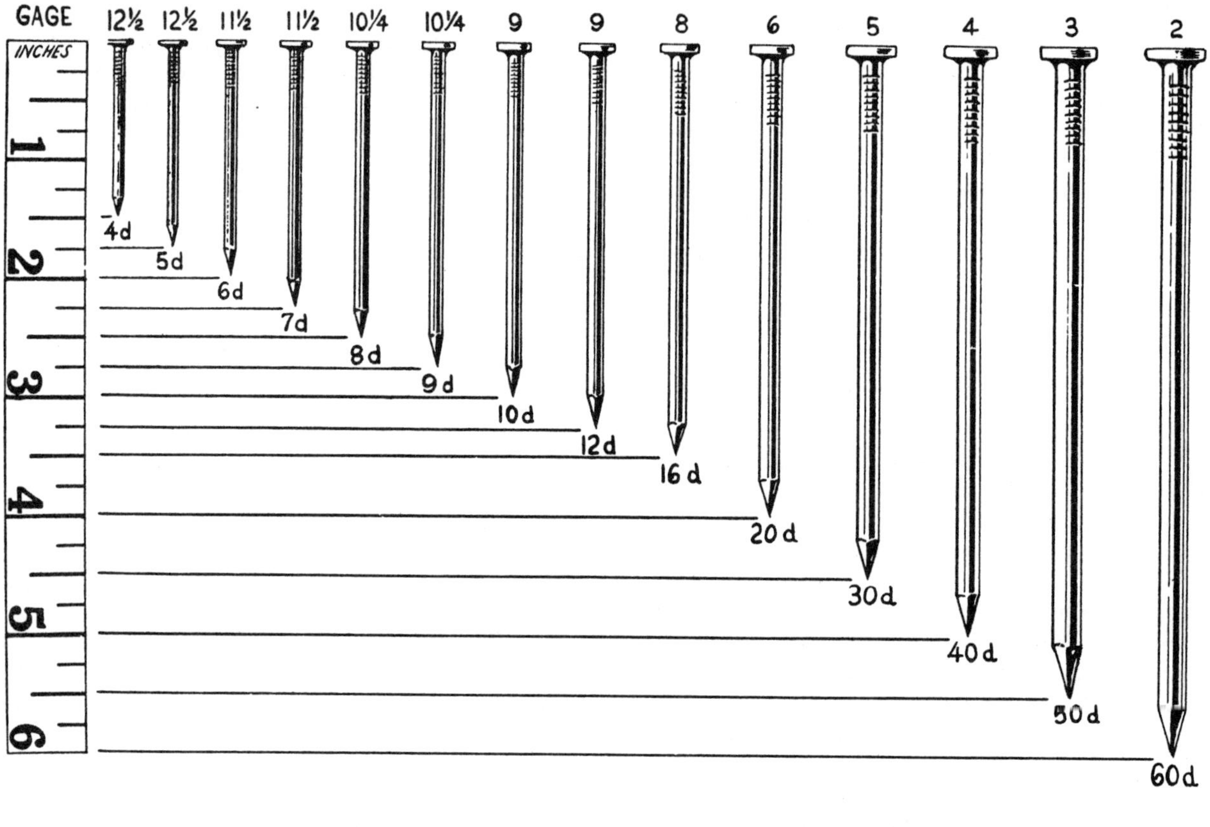

Figure 18.—Sizes of common wire nails.

TABLE 2.—*Recommended schedule for nailing the framing and sheathing of a well-constructed wood-frame house*

Joining	Nailing method	Nails: Number	Nails: Size	Nails: Placement
Header to joist	End-nail	3	16d	
Joist to sill or girder	Toenail	2 3	10d or 8d	
Header and stringer joist to sill	Toenail		10d	16 in. on center
Bridging to joist	Toenail each end	2	8d	
Ledger strip to beam, 2 in. thick		3	16d	At each joist
Subfloor, boards:				
1 by 6 in. and smaller		2	8d	To each joist
1 by 8 in.		3	8d	To each joist
Subfloor, plywood:				
At edges			8d	6 in. on center
At intermediate joists			8d	8 in. on center
Subfloor (2 by 6 in., T&G) to joist or girder	Blind-nail (casing) and face-nail	2	16d	
Soleplate to stud, horizontal assembly	End-nail	2	16d	At each stud
Top plate to stud	End-nail	2	16d	
Stud to soleplate	Toenail	4	8d	
Soleplate to joist or blocking	Face-nail		16d	16 in. on center
Doubled studs	Face-nail, stagger		10d	16 in. on center
End stud of intersecting wall to exterior wall stud	Face-nail		16d	16 in. on center
Upper top plate to lower top plate	Face-nail		16d	16 in. on center
Upper top plate, laps and intersections	Face-nail	2	16d	
Continuous header, two pieces, each edge			12d	12 in. on center
Ceiling joist to top wall plates	Toenail	3	8d	
Ceiling joist laps at partition	Face-nail	4	16d	
Rafter to top plate	Toenail	2	8d	
Rafter to ceiling joist	Face-nail	5	10d	
Rafter to valley or hip rafter	Toenail	3	10d	
Ridge board to rafter	End-nail	3	10d	
Rafter to rafter through ridge board	Toenail Edge-nail	4 1	8d 10d	
Collar beam to rafter:				
2 in. member	Face-nail	2	12d	
1 in. member	Face-nail	3	8d	
1-in. diagonal let-in brace to each stud and plate (4 nails at top)		2	8d	
Built-up corner studs:				
Studs to blocking	Face-nail	2	10d	Each side
Intersecting stud to corner studs	Face-nail		16d	12 in. on center
Built-up girders and beams, three or more members	Face-nail		20d	32 in. on center, each side
Wall sheathing:				
1 by 8 in. or less, horizontal	Face-nail	2	8d	At each stud
1 by 6 in. or greater, diagonal	Face-nail	3	8d	At each stud
Wall sheathing, vertically applied plywood:				
⅜ in. and less thick	Face-nail		6d	6 in. edge
½ in. and over thick	Face-nail		8d	12 in. intermediate
Wall sheathing, vertically applied fiberboard:				
½ in. thick	Face-nail			1½ in. roofing nail } 3 in. edge and
25/32 in. thick	Face-nail			1¾ in. roofing nail } 6 in. intermediate
Roof sheathing, boards, 4-, 6-, 8-in. width	Face-nail	2	8d	At each rafter
Roof sheathing, plywood:				
⅜ in. and less thick	Face-nail		6d }	6 in. edge and 12 in. intermediate
½ in. and over thick	Face-nail		8d }	

incnes to conform to the depth of the studs. Wood posts should be squared at both ends and securely fastened to the girder (fig. 20). The bottom of the post should rest on and be pinned to a masonry pedestal 2 to 3 inches above the finish floor. In moist or wet conditions it is good practice to treat the bottom end of the post or use a moisture-proof covering over the pedestal.

Both wood girders and steel beams are used in present-day house construction. The standard *I-beam* and wide *flange beam* are the most commonly used steel beam shapes. Wood girders are of two types—*solid* and *built up*. The built-up beam is preferred because it can be made up from drier dimension material and is more stable. Commercially available glue-laminated beams may be desirable where exposed in finished basement rooms.

The built-up girder (fig. 21) is usually made up of two or more pieces of 2-inch dimension lumber spiked together, the ends of the pieces joining over a supporting post. A two-piece girder may be nailed from one side with tenpenny nails, two at the end of each piece and others driven stagger fashion 16 inches apart. A three-piece girder is nailed from each side with twentypenny nails, two near each end of each piece and others driven stagger fashion 32 inches apart.

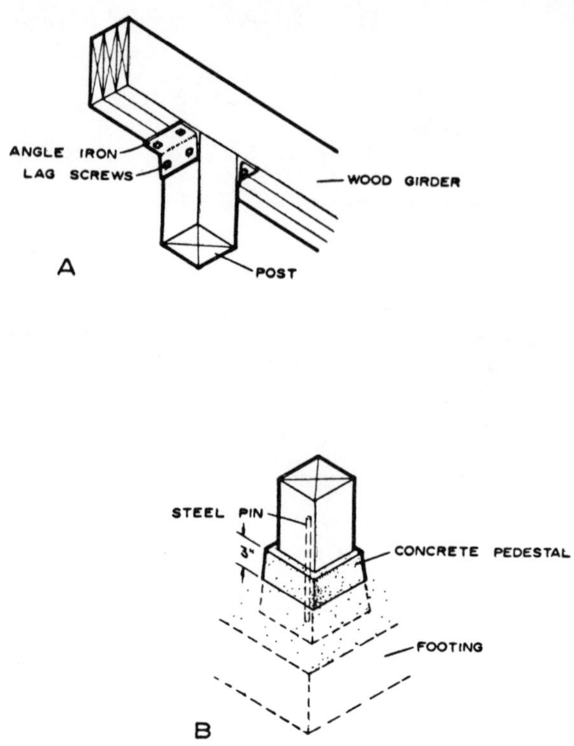

Figure 20.—Wood post for wood girder: *A*, Connection to girder; *B*, base.

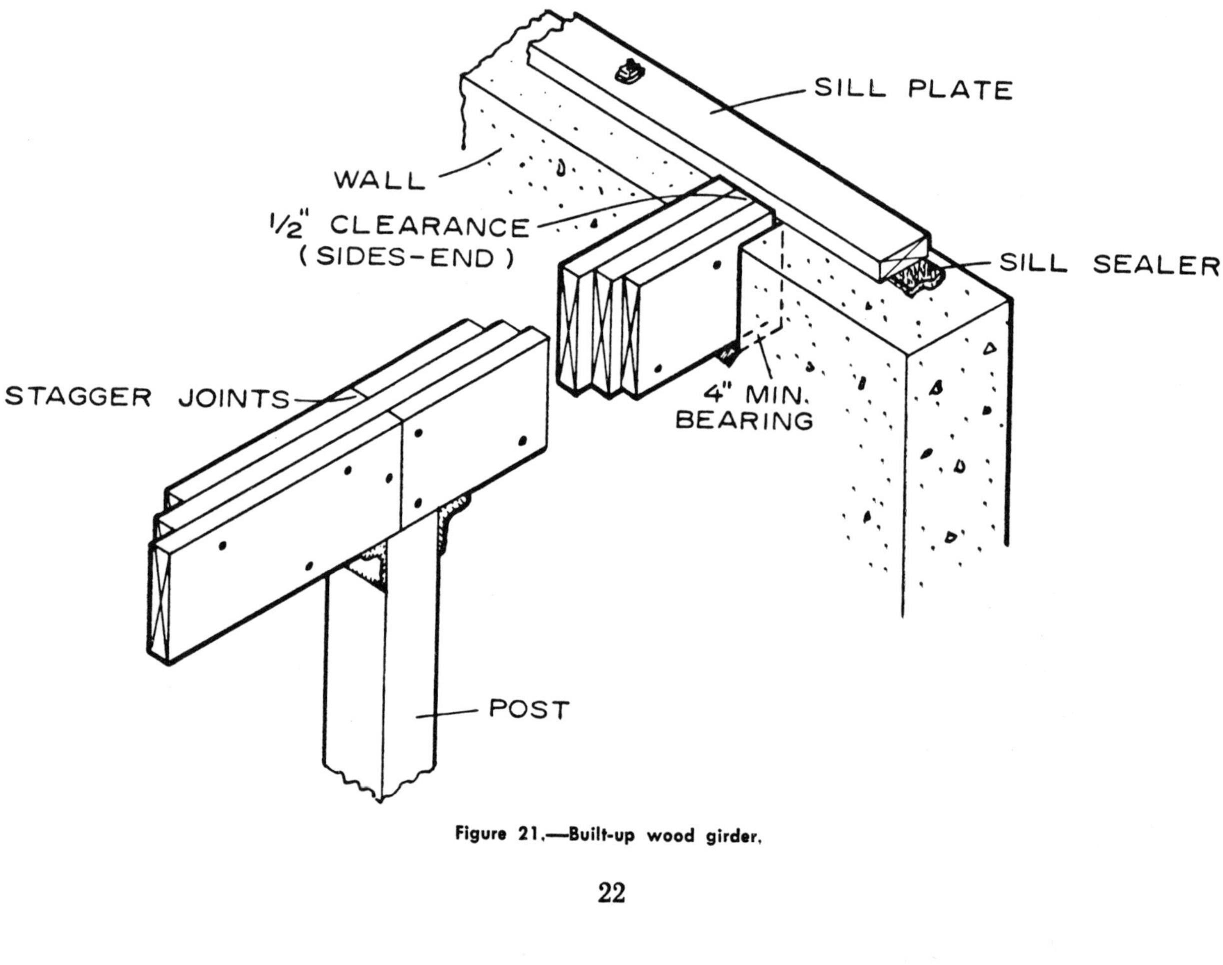

Figure 21.—Built-up wood girder.

Ends of wood girders should bear at least 4 inches on the masonry walls or pilasters. When wood is untreated, a ½-inch air space should be provided at each end and at each side of wood girders framing into masonry (fig. 21). In termite-infested areas, these pockets should be lined with metal. The top of the girder should be level with the top of the sill plates on the foundation walls, unless *ledger strips* are used. If steel plates are used under ends of girders, they should be of full bearing size.

Girder-joist Installation

Perhaps the simplest method of floor-joist framing is one where the joists bear directly on the wood girder or steel beam, in which case the top of the beam coincides with the top of the anchored sill (fig. 21). This method is used when basement heights provide adequate headroom below the girder. However, when wood girders are used in this manner, the main disadvantage is that shrinkage is usually greater at the girder than at the foundation.

For more uniform shrinkage at the inner beam and the outer wall and to provide greater headroom, joist hangers or a supporting ledger strip are commonly used. Depending on sizes of joists and wood girders, joists may be supported on the ledger strip in several ways (fig. 22). Each provides about the same depth of wood subject to shrinkage at the outer wall and at the center wood girder. A continuous horizontal tie between exterior walls is obtained by nailing notched joists together (fig. 22,*A*). Joists must always bear on the ledgers. In figure 22,*B*, the connecting scab at each pair of joists provides this tie and also a nailing area for the subfloor. A steel strap is used to tie the joists together when the tops of the beam and the joists are level (fig. 22,*C*). It is important that a small space be allowed above the beam to provide for shrinkage of the joists.

When a space is required for heat ducts in a partition supported on the girder, a *spaced wood girder* is sometimes necessary (fig. 23). Solid blocking is used at intervals between the two members. A single post support for a spaced girder usually requires a bolster, preferably metal, with sufficient span to support the two members.

Joists may be arranged with a steel beam generally the same way as illustrated for a wood beam. Perhaps the most common methods, depending on joist sizes, are:

1. The joists rest directly on the top of the beam.
2. Joists rest on a wood ledger or steel angle iron, which is bolted to the web (fig. 24,*A*).
3. Joists bear directly on the flange of the beam (fig. 24,*B*).

In the third method, wood blocking is required between the joists near the beam flange to prevent overturning.

Wood Sill Construction

The two general types of wood sill construction used over the foundation wall conform either to platform or balloon framing. The *box sill* is commonly used in platform construction. It consists of a 2-inch or thicker plate anchored to the foundation wall over a sill sealer which provides support and fastening for the joists and header at the ends of the joists (fig. 25). Some houses are constructed without benefit of an anchored sill plate although this is not entirely desirable. The floor framing should then be anchored with metal strapping installed during pouring operations (fig. 10,*B*).

Balloon-frame construction uses a nominal 2-inch or thicker wood sill upon which the joists rest. The studs also bear on this member and are nailed both to the floor joists and the sill. The subfloor is laid diagonally or at right angles to the joists and a firestop added between the studs at the floorline (fig. 26). When diagonal subfloor is used, a nailing member is normally required between joists and studs at the wall lines.

Because there is less potential shrinkage in exterior walls with balloon framing than in the platform type, balloon framing is usually preferred over the platform type in full two-story brick or stone veneer houses.

Floor Joists

Floor joists are selected primarily to meet strength and stiffness requirements. Strength requirements depend upon the loads to be carried. Stiffness requirements place an arbitrary control on deflection under load. Stiffness is also important in limiting vibrations from moving loads—often a cause of annoyance to occupants. Other desirable qualities for floor joists are good nail holding ability and freedom from warp.

Wood floor joists are generally of 2-inch (nominal) thickness and of 8-, 10-, or 12-inch (nominal) depth. The size depends upon the loading, length of span, spacing between joists, and the species and grade of lumber used. As previously mentioned, grades in species vary a great deal. For example, the grades generally used for joists are "Standard" for Douglas-fir, "No. 2 or No. 2KD" for southern pine, and comparable grades for other species.

Span tables for floor joists, which are published by the Federal Housing Administration (*11*) or local building codes can be used as guidelines. These sizes are of course often minimum, and it is sometimes the practice in medium- and higher priced houses to use the next larger size than those listed in the tables.

Joist Installation

After the sill plates have been anchored to the foundation walls or piers, the joists are located according to the house design. (Sixteen-inch center-to-center spacing is most commonly used.)

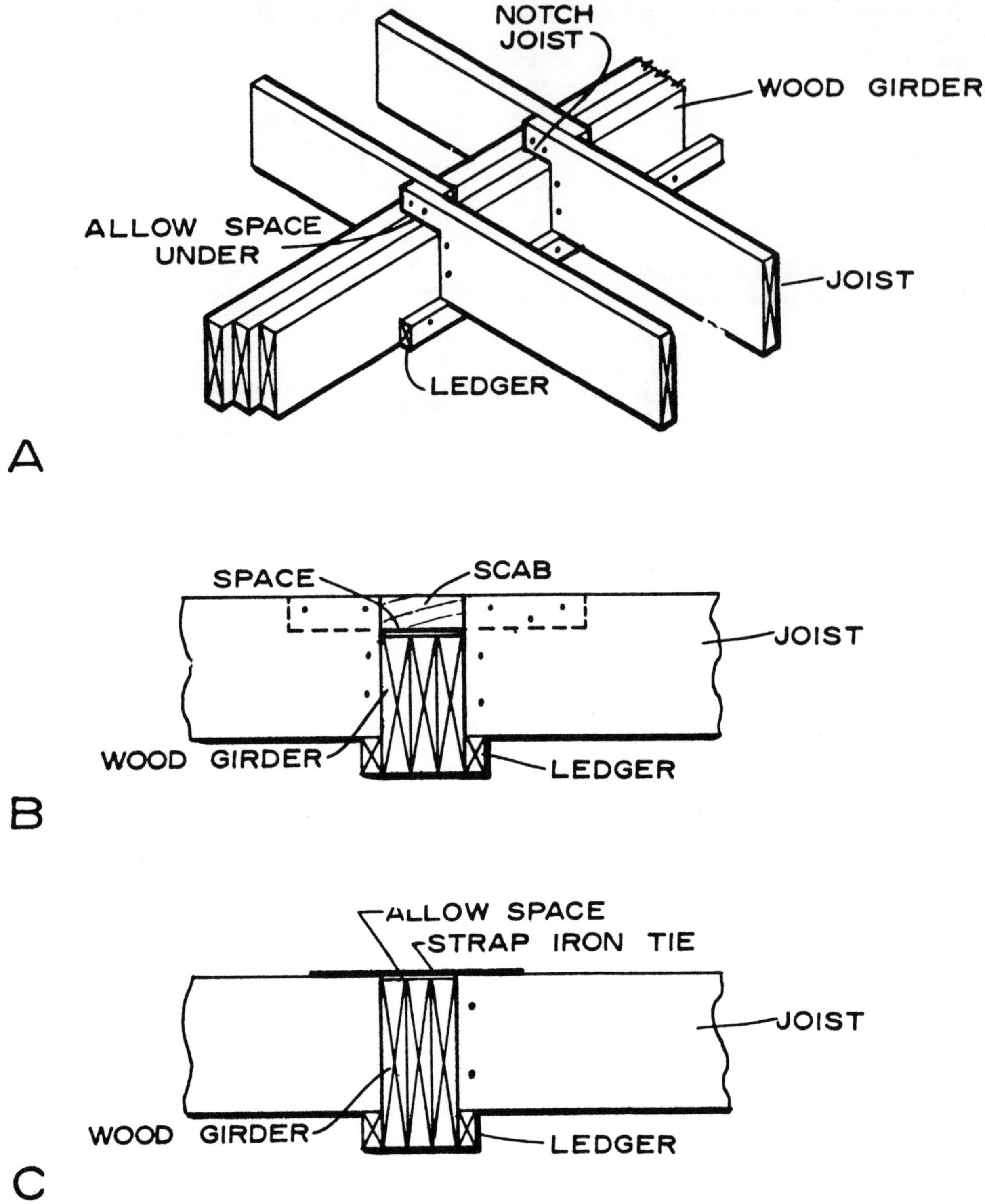

Figure 22.—Ledger on center wood girder: *A*, Notched joist; *B*, scab tie between joist; *C*, flush joist.

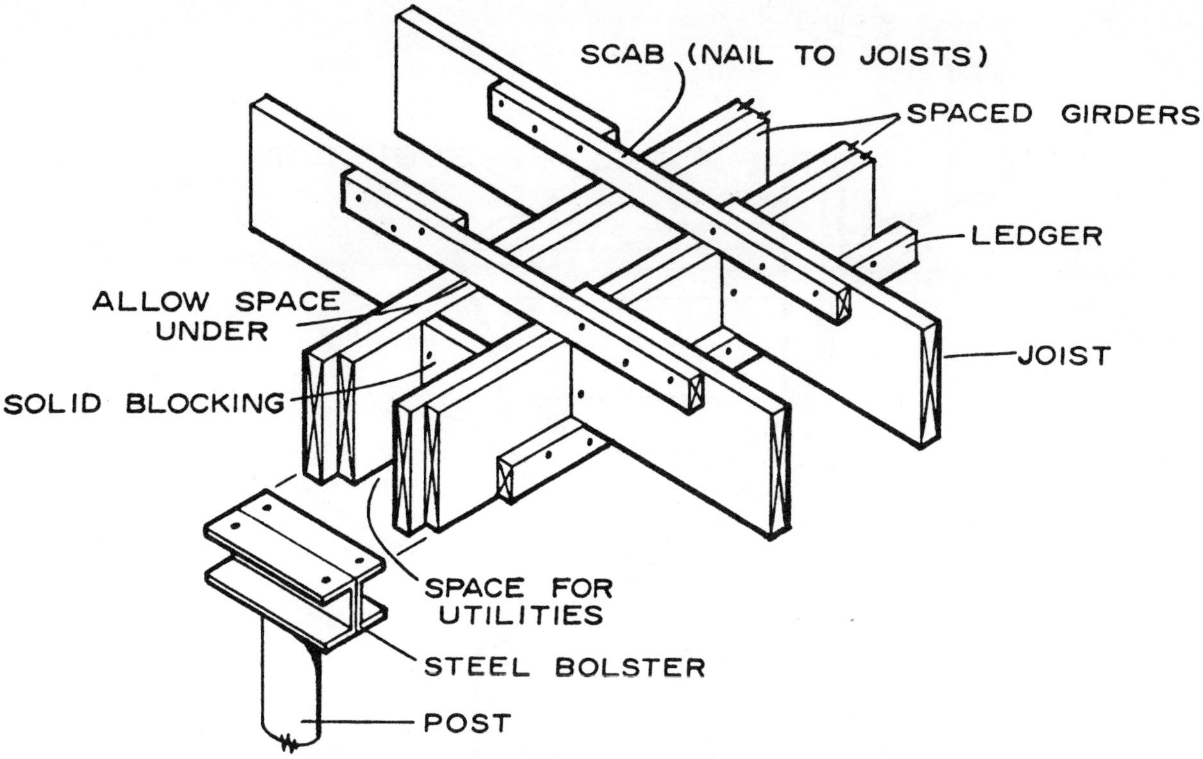

Figure 23.—Spaced wood girder.

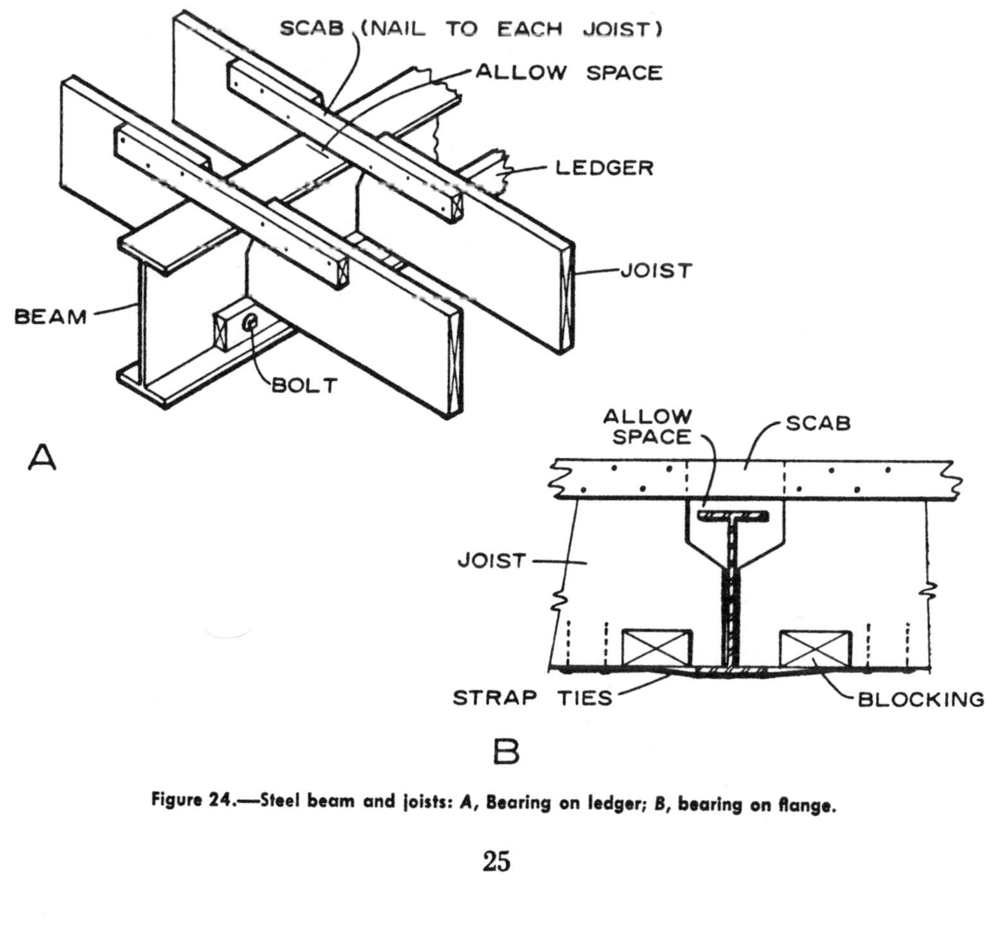

Figure 24.—Steel beam and joists: *A*, Bearing on ledger; *B*, bearing on flange.

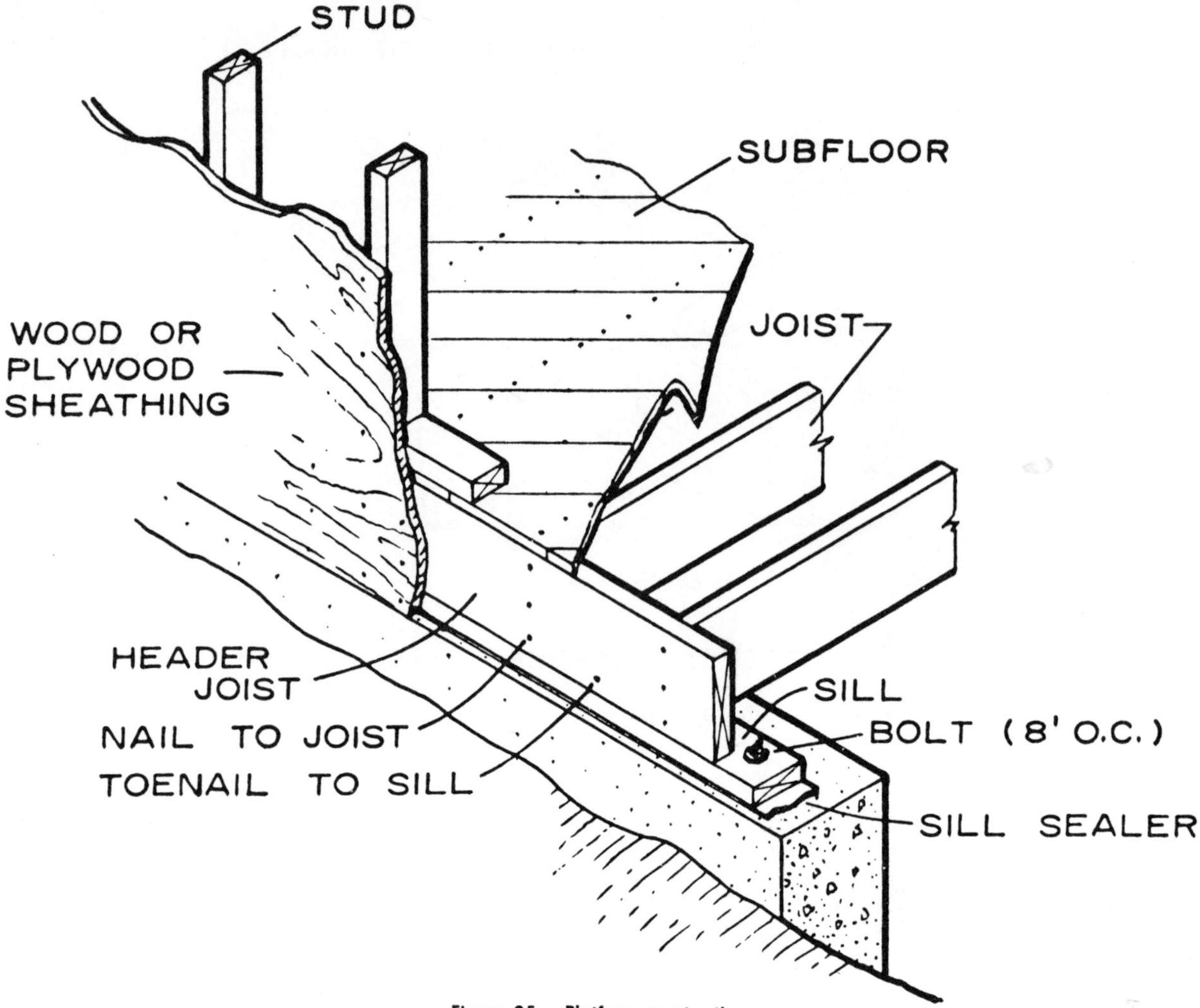

Figure 25.—Platform construction.

Any joists having a slight bow edgewise should be so placed that the crown is on top. A crowned joist will tend to straighten out when subfloor and normal floor loads are applied. The largest edge knots should be placed on top, since knots on the upper side of a joist are on the compression side of the member and will have less effect on strength.

The header joist is fastened by nailing into the end of each joist with three sixteenpenny nails. In addition, the header joist and the stringer joists parallel to the exterior walls in platform construction (fig. 27) are toenailed to the sill with tenpenny nails spaced 16 inches on center. Each joist should be toenailed to the sill and center beam with two tenpenny or three eightpenny nails; then nailed to each other with three or four sixteenpenny nails when they lap over the center beam. If a nominal 2-inch scab is used across butt-ended joists, it should be nailed to each joist with at least three sixteenpenny nails at each side of the joint. These and other nailing patterns and practices are outlined in table 2.

The "in-line" joist splice is sometimes used in framing for floor and ceiling joists. This system normally allows the use of one smaller joist size when center supports are present. Briefly, it consists of uneven length joists, the long overhanging joist is cantilevered over the center support, then spliced to the supported joist (fig. 28). Overhang joists are alternated. Depending on the span, species, and joist size, the overhang varies between about 1 foot 10 inches and 2 feet 10 inches. Plywood splice plates are used on each side of the end joints.[4]

It is good practice to double joists under all parallel bearing partition walls; if spacing is required for heat ducts, solid blocking is used between the joists (fig. 27).

[4] Details of this type of construction can be obtained from builders, lumber dealers, or architects, or by contacting the American Plywood Association, Tacoma, Wash. 98401.

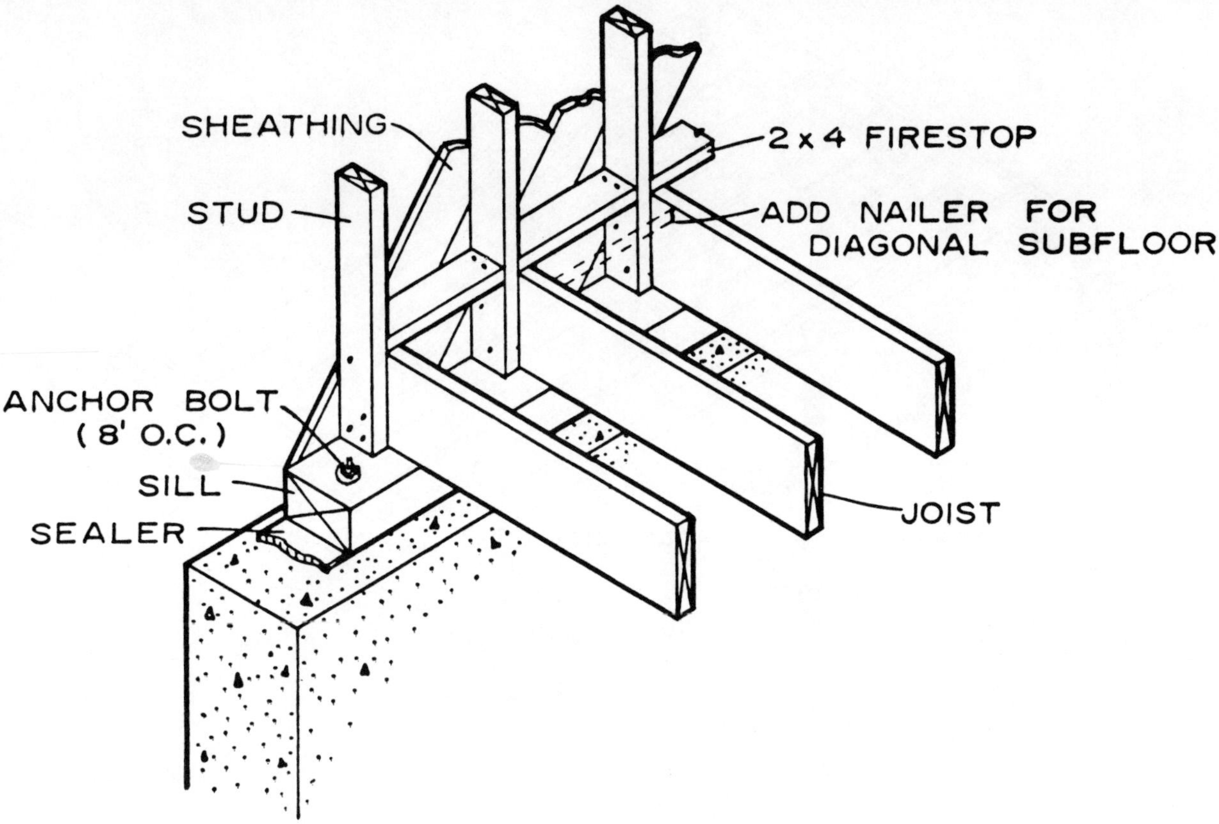

Figure 26.—Sill for balloon framing.

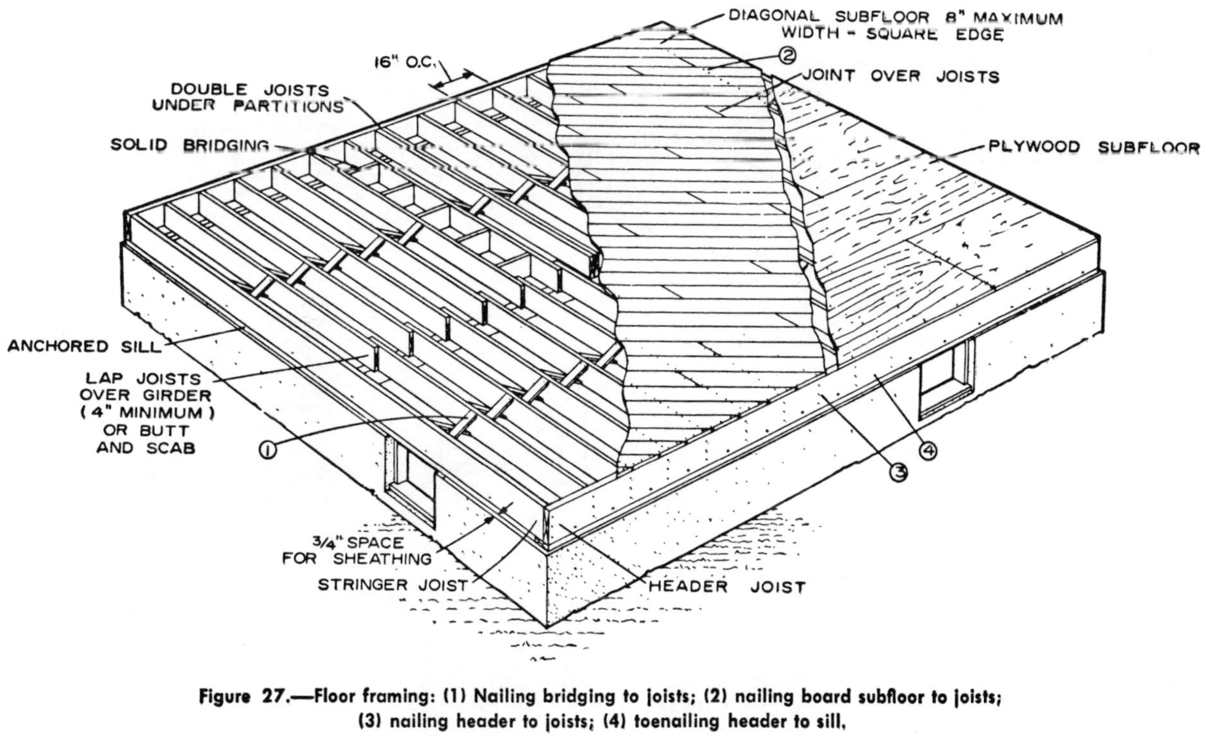

Figure 27.—Floor framing: (1) Nailing bridging to joists; (2) nailing board subfloor to joists; (3) nailing header to joists; (4) toenailing header to sill.

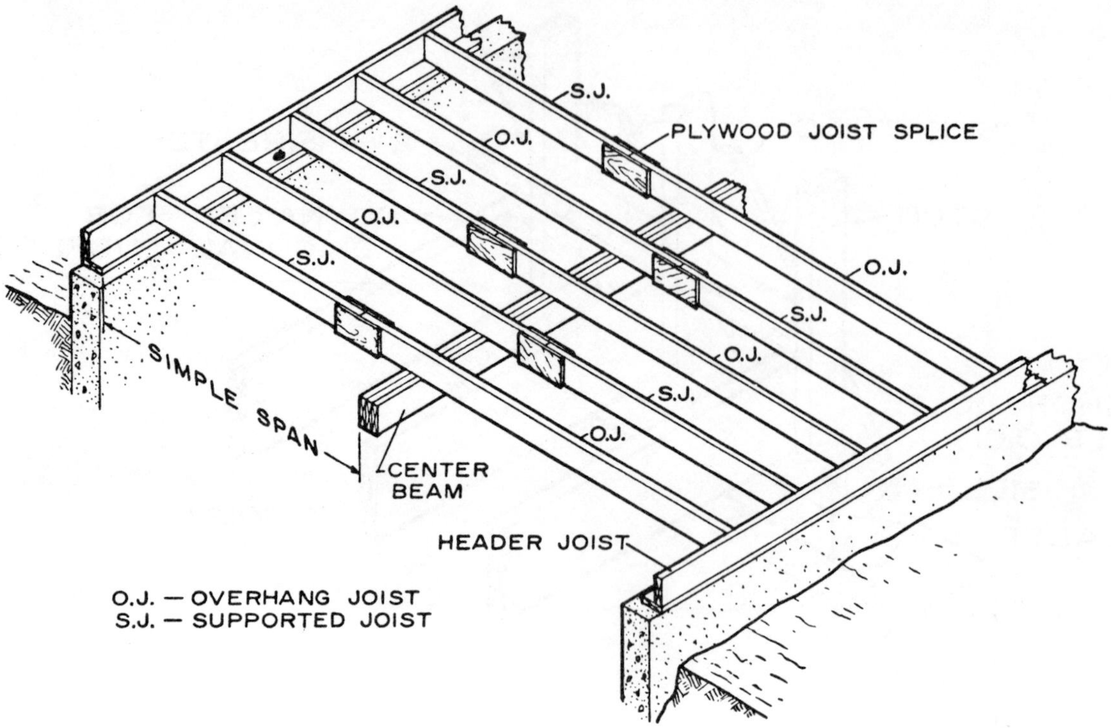

Figure 28.—"In-Line" joist system. Alternate extension of joists over the center support with plywood gusset joint allows the use of a smaller joist size.

Details At Floor Openings

When framing for large openings such as stairwells, fireplaces, and chimneys, the joists and headers around the opening should be doubled. A recommended method of framing and nailing is shown in figure 29.

Joist hangers and short sections of angle iron are often used to support headers and tail beams for large openings. For further details on stairwells, see Chapter 23—"Stairs."

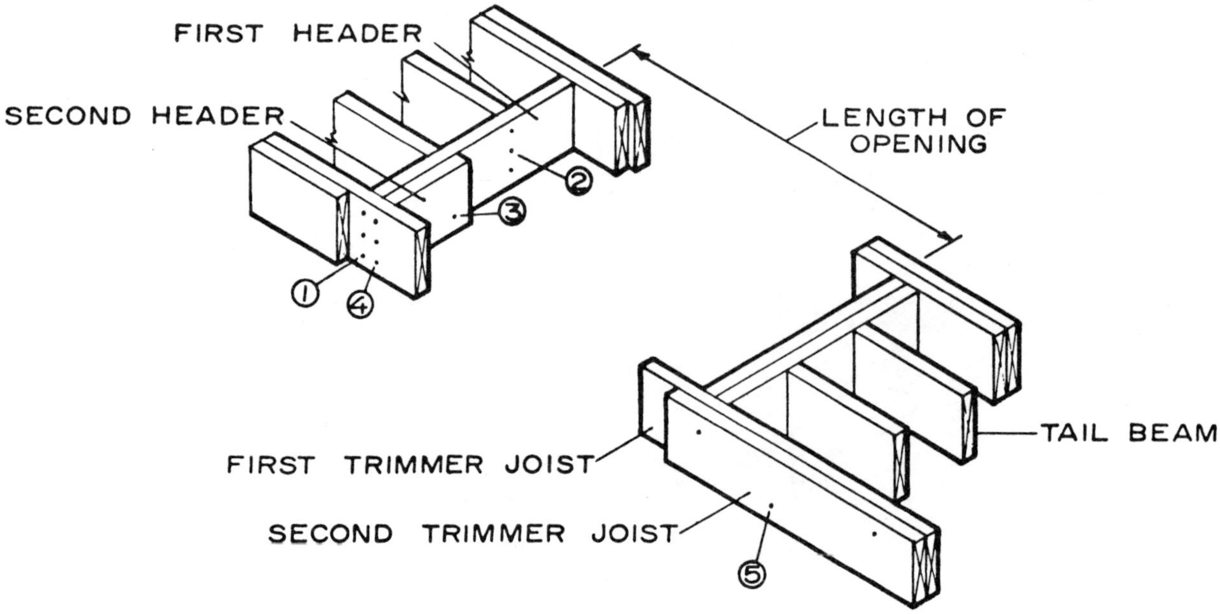

Figure 29.—Framing for floor openings: (1) Nailing trimmer to first header; (2) nailing header to tail beams; (3) nailing header together; (4) nailing trimmer to second header; (5) nailing trimmers together.

Bridging

Cross-bridging between wood joists has often been used in house construction, but research by several laboratories has questioned the benefits of bridging in relation to its cost, especially in normal house construction. Even with tight-fitting, well-installed bridgeing, there is no significant ability to transfer loads after subfloor and finish floor are installed. However, some building codes require the use of cross-bridging or *solid bridging* (table 2).

Solid bridging is often used between joists to provide a more rigid base for partitions located above joist spaces. Well-fitted solid bridging securely nailed to the joists will aid in supporting partitions above them (fig. 27). Load-bearing partitions should be supported by doubled joists.

Subfloor

Subflooring is used over the floor joists to form a working platform and base for finish flooring. It usually consists of (a) square-edge or tongued-and-grooved boards no wider than 8 inches and not less than 3/4 inch thick or (b) plywood 1/2 to 3/4 inch thick, depending on species, type of finish floor, and spacing of joists (fig. 27).

Boards

Subflooring may be applied either *diagonally* (most common) or at *right angles* to the joists. When subflooring is placed at right angles to the joists, the finish floor should be laid at right angles to the subflooring. Diagonal subflooring permits finish flooring to be laid either parallel or at right angles (most common) to the joists. End joints of the boards should always be made directly over the joists. Subfloor is nailed to each joist with two eightpenny nails for widths under 8 inches and three eightpenny nails for 8-inch widths.

The joist spacing should not exceed 16 inches on center when finish flooring is laid parallel to the joists, or where parquet finish flooring is used; nor exceed 24 inches on center when finish flooring at least 25/32 inch thick is at right angles to the joists.

Where balloon framing is used, blocking should be installed between ends of joists at the wall for nailing the ends of diagonal subfloor boards (fig. 26).

Plywood

Plywood can be obtained in a number of grades designed to meet a broad range of end-use requirements. All Interior-type grades are also available with fully waterproof adhesive identical with those used in Exterior plywood. This type is useful where a hazard of prolonged moisture exists, such as in underlayments or subfloors adjacent to plumbing fixtures and for roof sheathing which may be exposed for long periods during construction. Under normal conditions and for sheathing used on walls, Standard sheathing grades are satisfactory.

Plywood suitable for subfloor, such as Standard sheathing, Structural I and II, and C-C Exterior grades, has a panel identification index marking on each sheet. These markings indicate the allowable spacing of rafters and floor joists for the various thicknesses when the plywood is used as roof sheathing or subfloor. For example, an index mark of 32/16 indicates that the plywood panel is suitable for a maximum spacing of 32 inches for rafters and 16 inches for floor joists. Thus, no problem of strength differences between species is involved as the correct identification is shown for each panel.

Normally, when some type of underlayment is used over the plywood subfloor, the minimum thickness of the subfloor for species such as Douglas-fir and southern pine is 1/2 inch when joists are spaced 16 inches on center, and 5/8-inch thick for such plywood as western hemlock, western white pine, ponderosa pine, and similar species. These thicknesses of plywood might be used for 24-inch spacing of joists when a finish 25/32-inch strip flooring is installed at right angles to the joists. However, it is important to have a solid and safe platform for workmen during construction of the remainder of the house. For this reason, some builders prefer a slightly thicker plywood subfloor especially when joist spacing is greater than 16 inches on center.

Plywood can also serve as combined plywood subfloor and underlayment, eliminating separate underlayment because the plywood functions as both structural subfloor and a good substrate. This applies to thin resilient floorings, carpeting, and other nonstructural finish flooring. The plywood used in this manner must be tongued and grooved or blocked with 2-inch lumber along the unsupported edges. Following are recommendations for its use:

Grade: Underlayment, underlayment with exterior glue, C-C plugged

Spacing and thickness: (a) For species such as Douglas-fir (coast type), and southern pine—1/2 inch minimum thickness for 16-inch joist spacing, 5/8 inch for 20-inch joist spacing, and 3/4 inch for 24-inch joist spacing.

(b) For species such as western hemlock, western white pine, and ponderosa pine—5/8 inch minimum thickness for 16-inch joist spacing, 3/4 inch for 20-inch joist spacing, and 7/8 inch for 24-inch joist spacing.

Plywood should be installed with the grain direction of the outer plies at right angles to the joists and be staggered so that end joints in adjacent panels break over different joists. Plywood should be nailed to the joist at each bearing with eightpenny common or sevenpenny threaded nails for plywood 1/2 inch to 3/4 inch thick. Space nails 6 inches apart along all edges and 10 inches along intermediate members. When plywood serves as both subfloor and underlay-

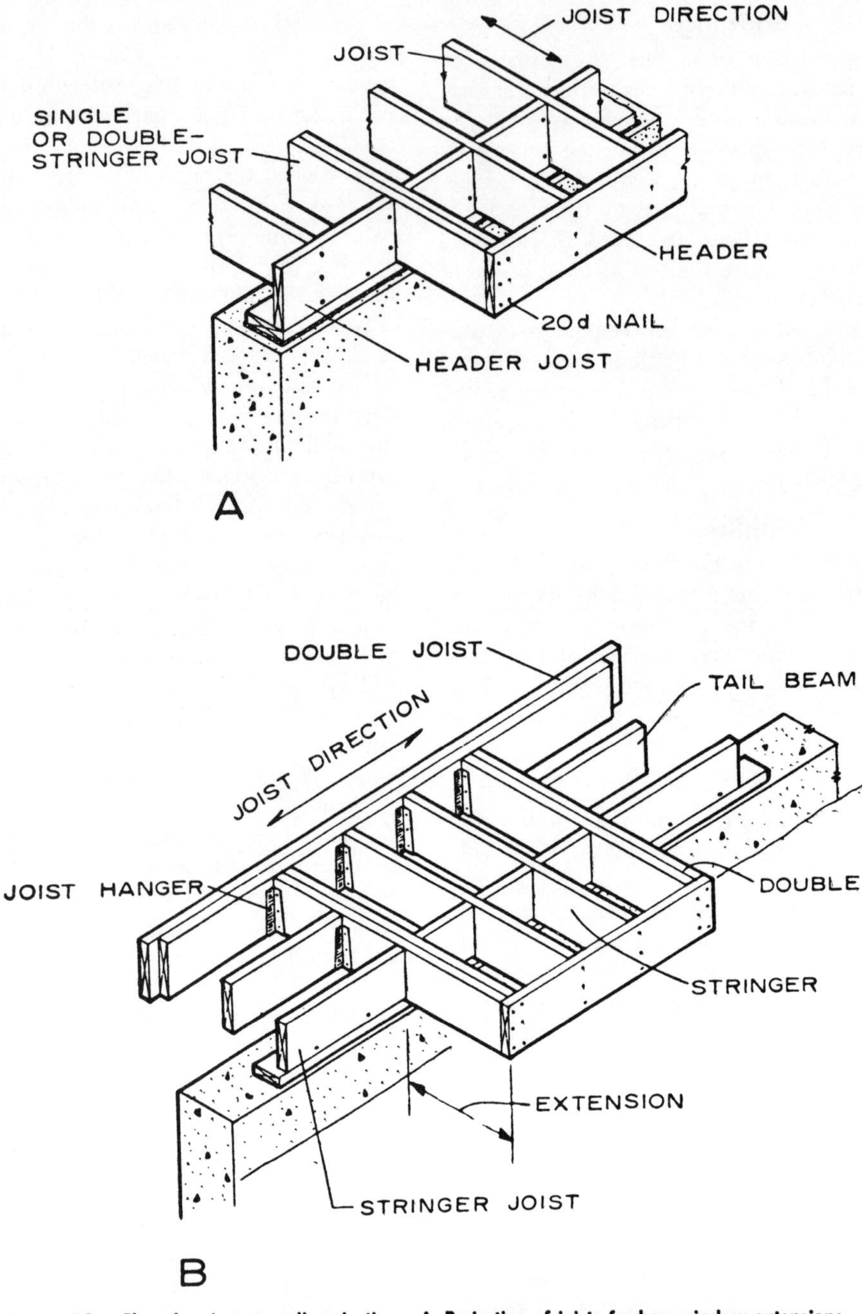

Figure 30.—Floor framing at wall projections: *A*, Projection of joists for bay window extensions; *B*, projection at right angles to joists.

ment, nails may be spaced 6 to 7 inches apart at all joists and blocking. Use eight- or ninepenny common nails or seven- or eightpenny threaded nails.

For the best performance, plywood should *not* be laid up with tight joints whether used on the interior or exterior. The following spacings are recommendations by the American Plywood Association on the basis of field experience:

Plywood location and use	*Spacing*	
	Edges (In.)	*Ends* (In.)
Underlayment or interior wall lining	1/32	1/32
Panel sidings and combination subfloor underlayment	1/16	1/16
Roof sheathing, subflooring, and wall sheathing (Under wet or humid conditions, spacing should be doubled.)	1/8	1/16

Floor Framing at Wall Projections

The framing for wall projections such as a *bay window* or first or second floor extensions beyond the lower wall should generally consist of projection of the floor joists (fig. 30). This extension normally should not exceed 24 inches unless designed specifically for greater projections, which may require special anchorage at the opposite ends of the joists. The joists forming each side of the bay should be doubled. Nailing, in general, should conform to that for stair openings. The subflooring is carried to and sawed flush with the outer framing member. Rafters are often carried by a header constructed in the main wall over the bay area, which supports the roofload. Thus the wall of the bay has less load to support.

Projections at right angles to the length of the floor joists should generally be limited to small areas and extensions of not more than 24 inches. In this construction, the stringer should be carried by doubled joists (fig. 30B). Joist hangers or a ledger will provide good connections for the ends of members.

CHAPTER 6

WALL FRAMING

The floor framing with its subfloor covering has now been completed and provides a convenient working platform for construction of the wall framing. The term "wall framing" includes primarily the vertical studs and horizontal members (soleplates, top plates, and window and door headers) of exterior and interior walls that support ceilings, upper floors, and the roof. The wall framing also serves as a nailing base for wall covering materials.

The wall framing members used in conventional construction are generally nominal 2- by 4-inch studs spaced 16 inches on center (*11*). Depending on thickness of covering material, 24-inch spacing might be considered. Top plates and soleplates are also nominal 2 by 4 inches in size. Headers over doors or windows in load-bearing walls consist of doubled 2- by 6-inch and deeper members, depending on span of the opening.

Requirements

The requirements for wall-framing lumber are good stiffness, good nail-holding ability, freedom from warp, and ease of working (*5*). Species used may include Douglas-fir, the hemlocks, southern pine, the spruces, pines, and white fir. As outlined under "Floor Framing," the grades vary by species, but it is common practice to use the third grade for studs and plates and the second grade for headers over doors and windows.

All framing lumber for walls should be reasonably dry. Material at about 15 percent moisture content is desirable, with the maximum allowable considered to be 19 percent. When the higher moisture content material is used (as studs, plates, and headers), it is advisable to allow the moisture content to reach in-service conditions before applying interior trim.

Ceiling height for the first floor is 8 feet under most conditions. It is common practice to rough-frame the wall (subfloor to top of upper plate) to a height of 8 feet 1½ inches. In platform construction, precut studs are often supplied to a length of 7 feet 8⅝ inches for plate thickness of 1⅝ inches. When dimension material is 1½ inches thick, precut studs would be 7 feet 9 inches long. This height allows the use of 8-foot-high dry-wall sheets, or six courses of rock lath, and still provides clearance for floor and ceiling finish or for plaster grounds at the floor line.

Second-floor ceiling heights should not be less than 7 feet 6 inches in the clear, except that portion under sloping ceilings. One-half of the floor area, however, should have at least a 7-foot 6-inch clearance.

As with floor construction, two general types of wall framing are commonly used—platform construction and balloon-frame construction. The platform method is more often used because of its simplicity. Balloon framing is generally used where stucco or masonry is the exterior covering material in two-story houses, as outlined in the chapter "Floor Framing."

Platform Construction

The wall framing in platform construction is erected above the subfloor which extends to all edges of the building (fig. 31). A combination of platform construction for the first floor sidewalls and full-length studs for end walls extending to end rafters of the gable ends is commonly used in single-story houses.

One common method of framing is the horizontal assembly (on the subfloor) or "tilt-up" of wall sections. When a sufficient work crew is available, full-length wall sections are erected. Otherwise, shorter length sections easily handled by a smaller crew can be used. This system involves laying out precut studs, window and door headers, cripple studs (short-length studs), and windowsills. Top and soleplates are then nailed to all vertical members and adjoining studs to headers and sills with sixteenpenny nails. Let-in corner bracing should be provided when required. The entire section is then erected, plumbed, and braced (fig. 31).

A variation of this system includes fastening the studs only at the top plate and, when the wall is erected, toenailing studs to the soleplates which have been previously nailed to the floor. Corner studs and headers are usually nailed together beforehand to form a single unit. Many contractors will also install sheathing before the wall is raised in place. Complete finished walls with windows and door units in place and most of the siding installed can also be fabricated in this manner.

When all exterior walls have been erected, plumbed, and braced, the remaining nailing is completed. Soleplates are nailed to the floor joists and headers or stringers (through the subfloor), corner braces (when used) are nailed to studs and plates, door and window headers are fastened to adjoining studs, and corner studs are nailed together. These and other recommended nailing practices are shown in table 2 and figure 31.

In hurricane areas or areas with high winds, it is often advisable to fasten wall and floor framing to the anchored foundation sill when sheathing does not provide this tie. Figure 32 illustrates one system of anchoring the studs to the floor framing with steel straps (7).

Several arrangements of studs at outside corners can be used in framing the walls of a house. Figure 31 shows one method commonly used. Blocking between two corner studs is used to provide a nailing edge for

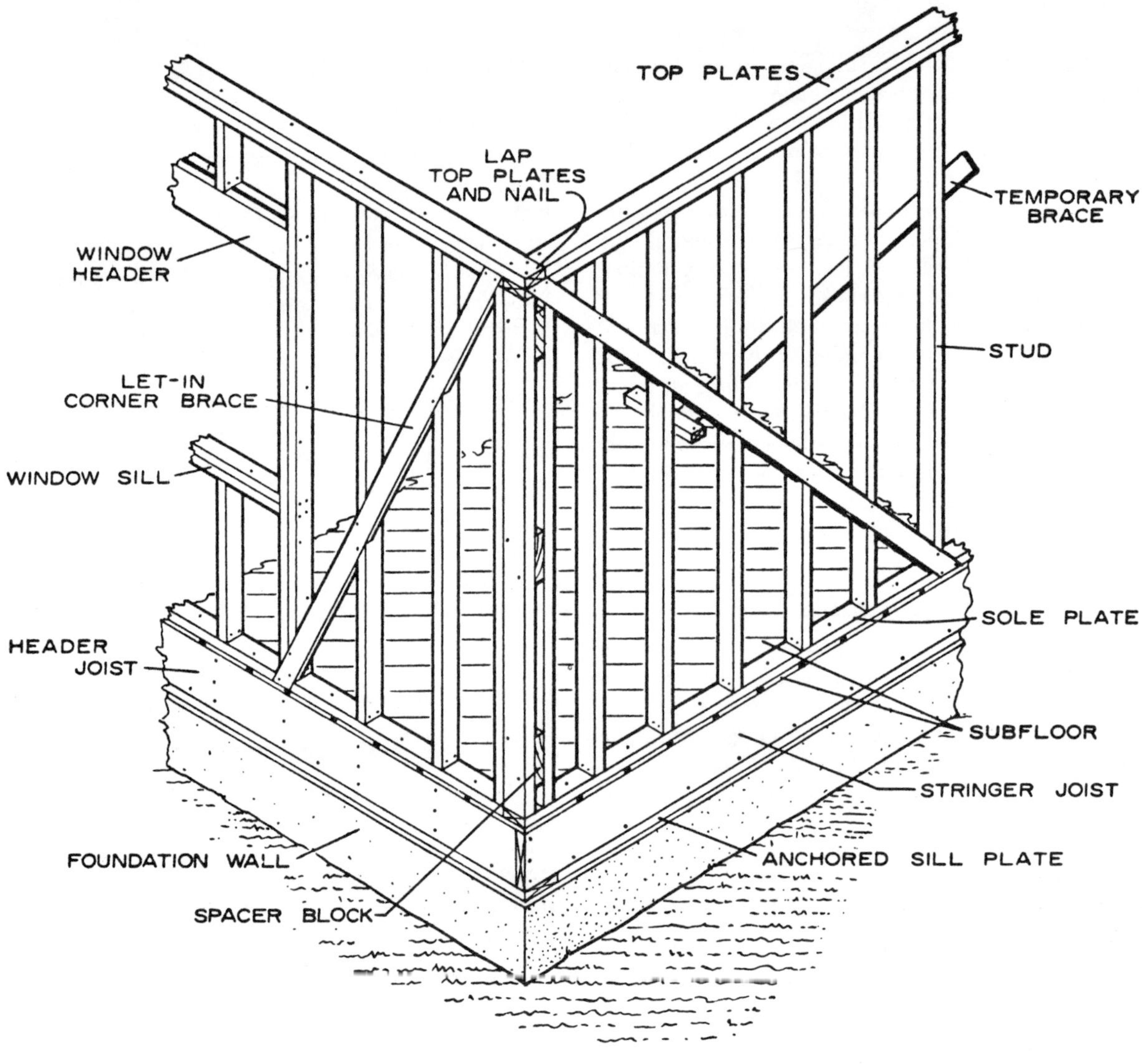

Figure 31.—Wall framing used with platform construction.

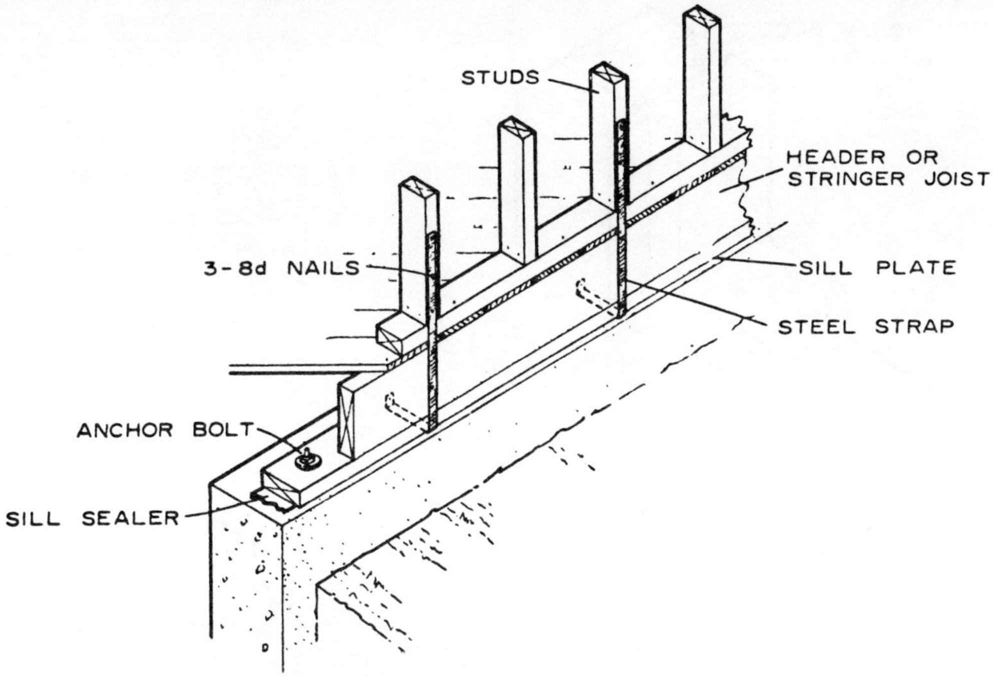

Figure 32.—Anchoring wall to floor framing.

interior finish (fig. 33,*A*). Figure 33,*B* and *C* show other methods of stud arrangement to provide the needed interior nailing surfaces as well as good corner support.

Interior walls should be well fastened to all exterior walls they intersect. This intersection should also provide nailing surfaces for the plaster base or dry-wall finish. This may be accomplished by doubling the outside studs at the interior wall line (fig. 34,*A*). Another method used when the interior wall joins the exterior wall between studs is shown in figure 34,*B*.

Short sections of 2- by 4-inch blocking are used between studs to support and provide backing for a 1- by 6 inch nailer. A 2- by 6-inch vertical member might also be used.

The same general arrangement of members is used at the intersection or crossing of interior walls. Nailing surfaces must be provided in some form or another at all interior corners.

After all walls are erected, a second top plate is added that laps the first at corners and wall intersections (fig. 31). This gives an additional tie to the framed walls. These top plates can also be partly fastened in place when the wall is in a horizontal position. Top plates are nailed together with sixteenpenny nails spaced 16 inches apart and with two nails at each wall interesection (table 2). Walls are normally plumbed and alined before the top plate is added. By using 1- by 6- or 1- by 8-inch temporary braces on the studs between intersecting partitions, a straight wall is assured. These braces are nailed to the studs at the top of the wall and to a 2- by 4-inch block fastened to the subfloor or joists. The temporary bracing is left in place until the ceiling and the roof framing are completed and sheathing is applied to the outside walls.

Balloon Construction

As described in the chapter on "Floor Framing," the main difference between platform and balloon framing is at the floor-lines. The balloon wall studs extend from the sill of the first floor to the top plate or end rafter of the second floor, whereas the platform-framed wall is complete for each floor.

In balloon-frame construction, both the wall studs and the floor joists rest on the anchored sill (fig. 35). The studs and joists are toenailed to the sill with eightpenny nails and nailed to each other with at least three tenpenny nails.

The ends of the second-floor joists bear on a 1- by 4-inch ribbon that has been let into the studs. In addition, the joists are nailed with four tenpenny nails to the studs at these connections (fig. 35). The end joists parallel to the exterior on both the first and second floors are also nailed to each stud.

Other nailing details should conform in general to those described for platform construction and in table 2.

In most areas, building codes require that *firestops* be used in balloon framing to prevent the spread of fire through the open wall passages. These firestops are ordinarily of 2- by 4-inch blocking placed between the studs (fig. 35) or as required by local regulations.

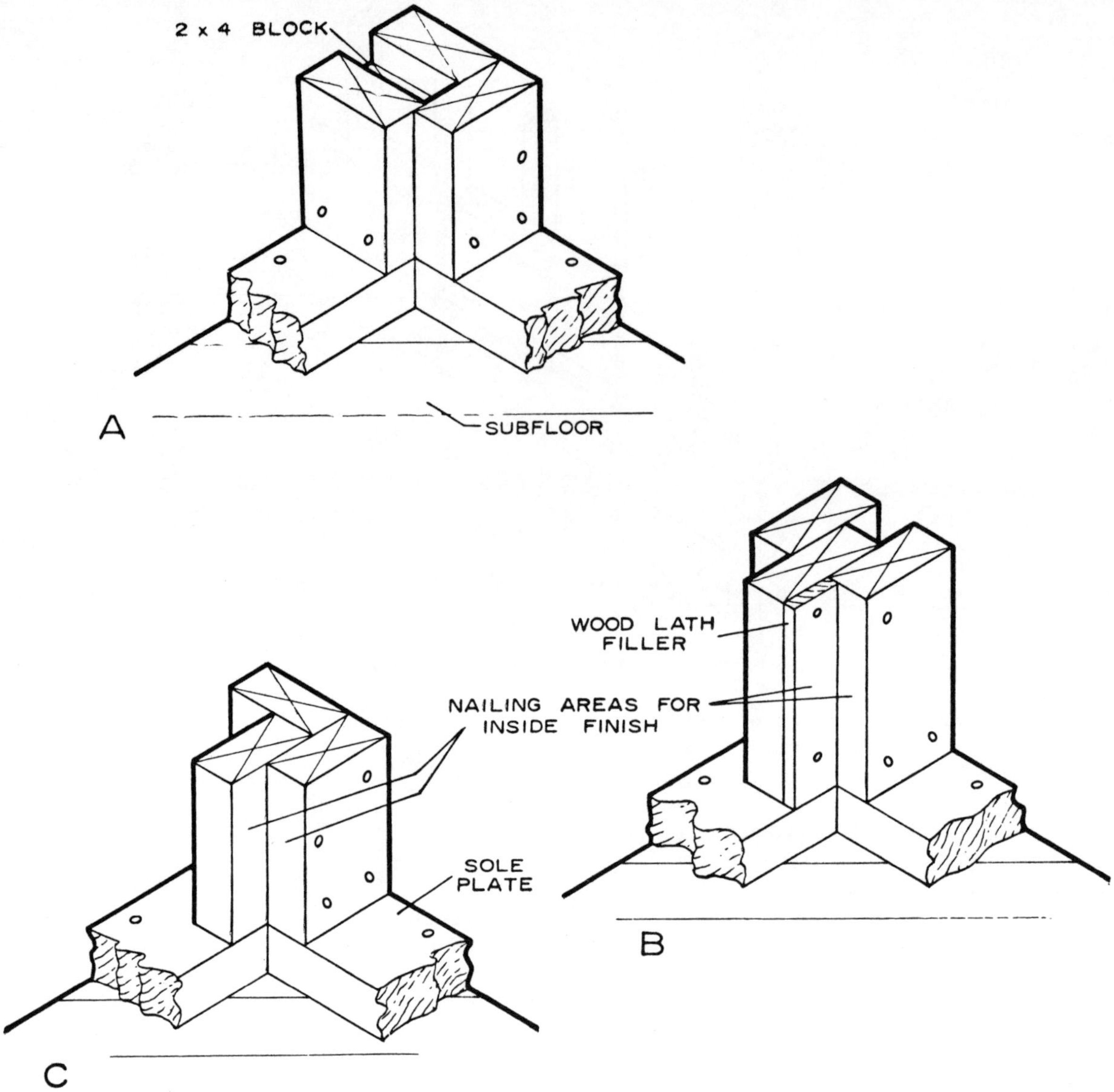

Figure 33.—Examples of corner stud assembly: *A*, Standard outside corner; *B*, special corner with lath filler; *C*, special corner without lath filler.

Window and Door Framing

The members used to span over window and door openings are called *headers* or *lintels* (fig. 36). As the span of the opening increases, it is necessary to increase the depth of these members to support the ceiling and roofloads. A header is made up of two 2-inch members, usually spaced with ⅜-inch lath or wood strips, all of which are nailed together. They are supported at the ends by the inner studs of the double-stud joint at exterior walls and interior bearing walls. Two headers of species normally used for floor joists are usually appropriate for these openings in normal light-frame construction. The following sizes might be used as a guide for headers:

Maximum span (*Ft.*)	*Header size* (*In.*)
3½	2 by 6
5	2 by 8
6½	2 by 10
8	2 by 12

For other than normal light-frame construction, independent design may be necessary. Wider openings often require trussed headers, which may also need special design.

Location of the studs, headers, and sills around window openings should conform to the rough open-

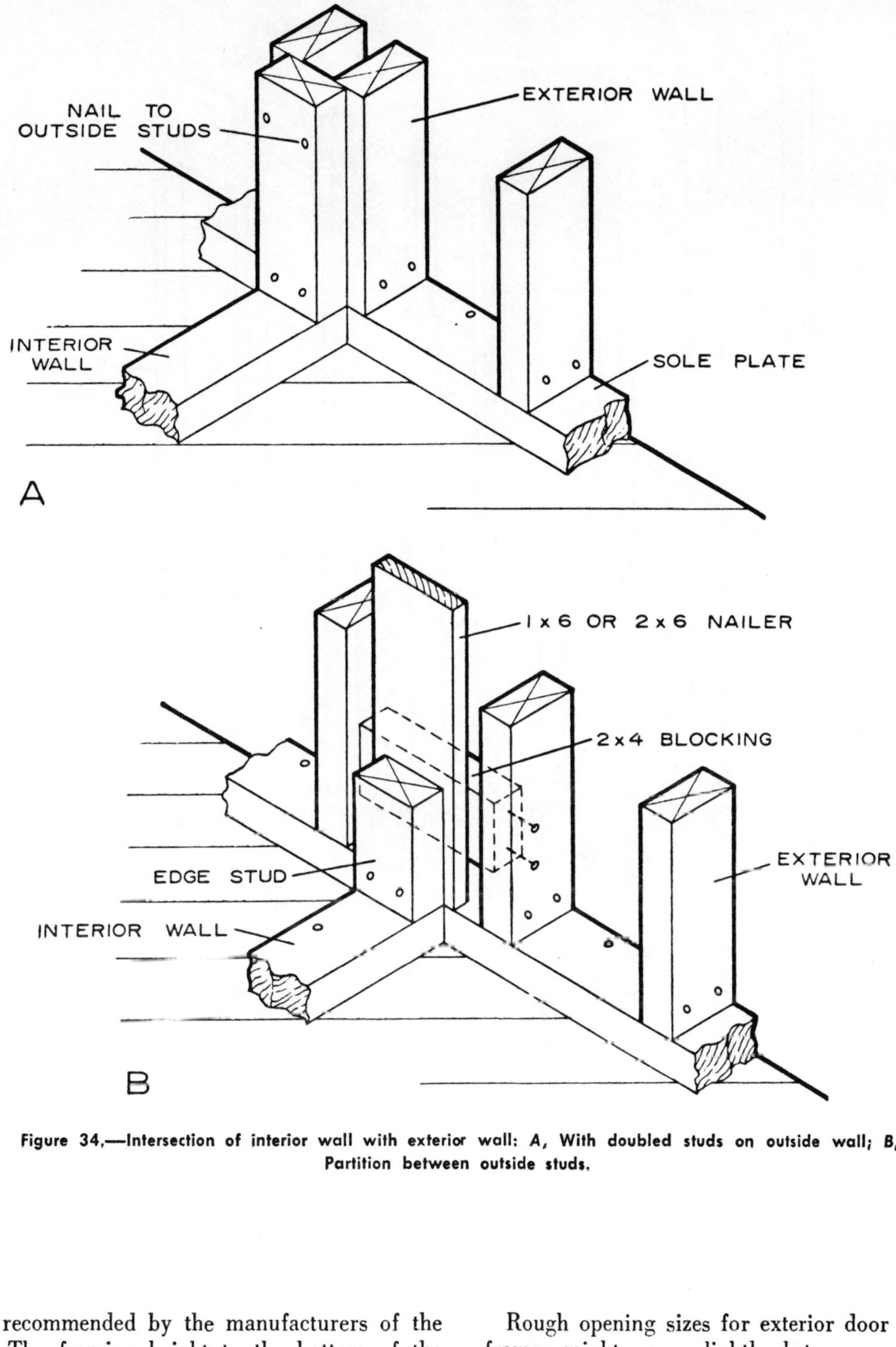

Figure 34.—Intersection of interior wall with exterior wall: *A*, With doubled studs on outside wall; *B*, Partition between outside studs.

ing sizes recommended by the manufacturers of the millwork. The framing height to the bottom of the window and door headers should be based on the door heights, normally 6 feet 8 inches for the main floor. Thus to allow for the thickness and clearance of the head jambs of window and door frames and the finish floor, the bottoms of the headers are usually located 6 feet 10 inches to 6 feet 11 inches above the subfloor, depending on the type of finish floor used.

Rough opening sizes for exterior door and window frames might vary slightly between manufacturers, but the following allowances should be made for the stiles and rails, thickness of jambs, and thickness and slope of the sill:

Double-Hung Window (Single Unit)

Rough opening width = glass width plus 6 inches

Rough opening height = total glass height plus 10 inches

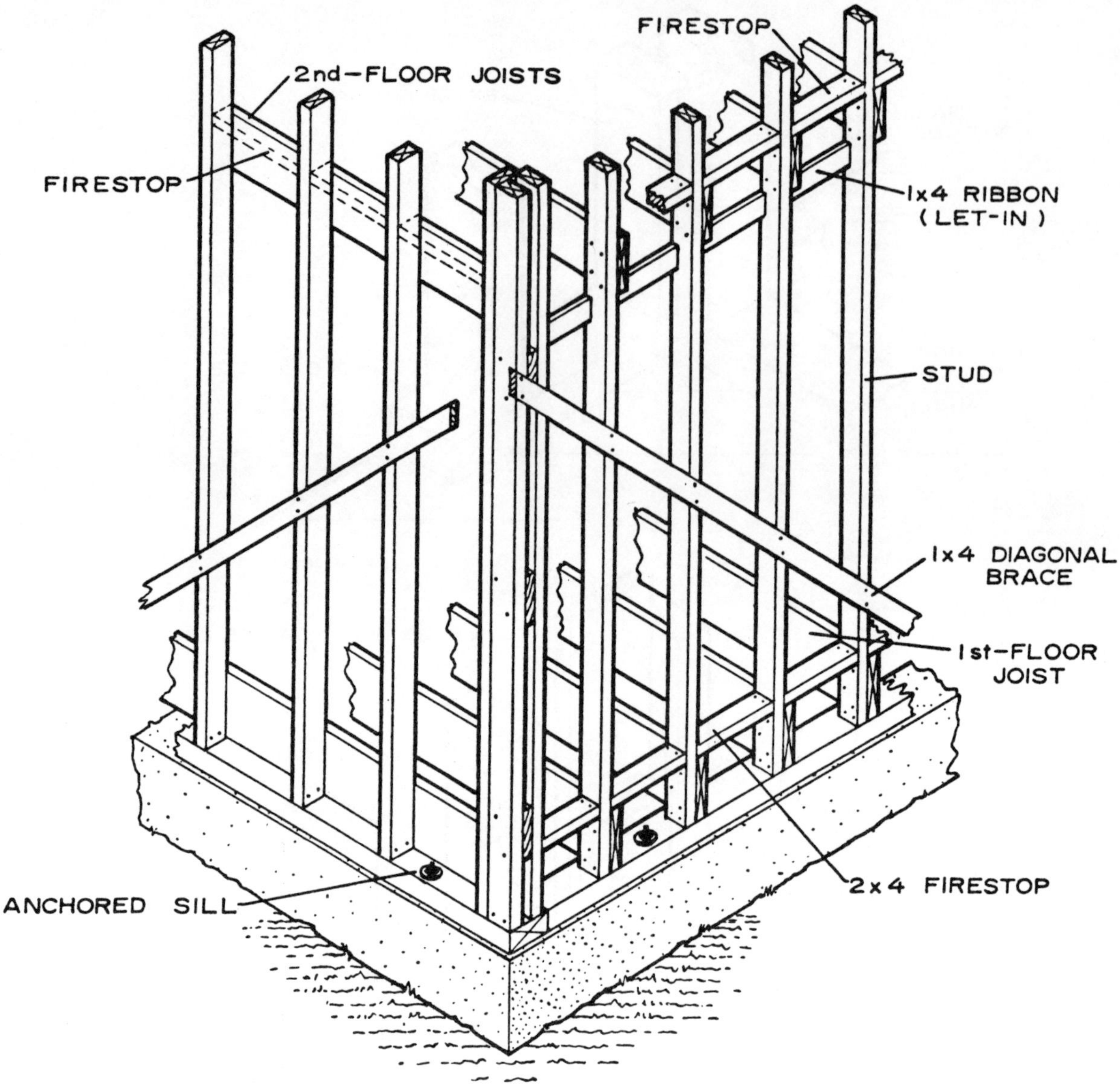

Figure 35.—Wall framing used in balloon construction.

For example, the following tabulation illustrates several glass and rough opening sizes for double-hung windows:

Window glass size (each sash) *Width* (*In.*)	*Height* (*In.*)	*Rough frame opening* *Width* (*In.*)	*Height* (*In.*)
24 by	16	30 by	42
28 by	20	34 by	50
32 by	24	38 by	58
36 by	24	42 by	58

Casement Window (One Pair—Two Sash)

Rough opening width = total glass width plus 11¼ inches

Rough opening height = total glass height plus 6⅜ inches

Doors

Rough opening width = door width plus 2½ inches

Rough opening height = door height plus 3 inches

End-wall Framing

The framing for the end walls in platform and balloon construction varies somewhat. Figure 37 shows a commonly used method of wall and ceiling framing for platform construction in 1½- or 2-story houses with finished rooms above the first floor. The edge floor joist is toenailed to the top wall plate with eightpenny nails spaced 16 inches on center. The subfloor,

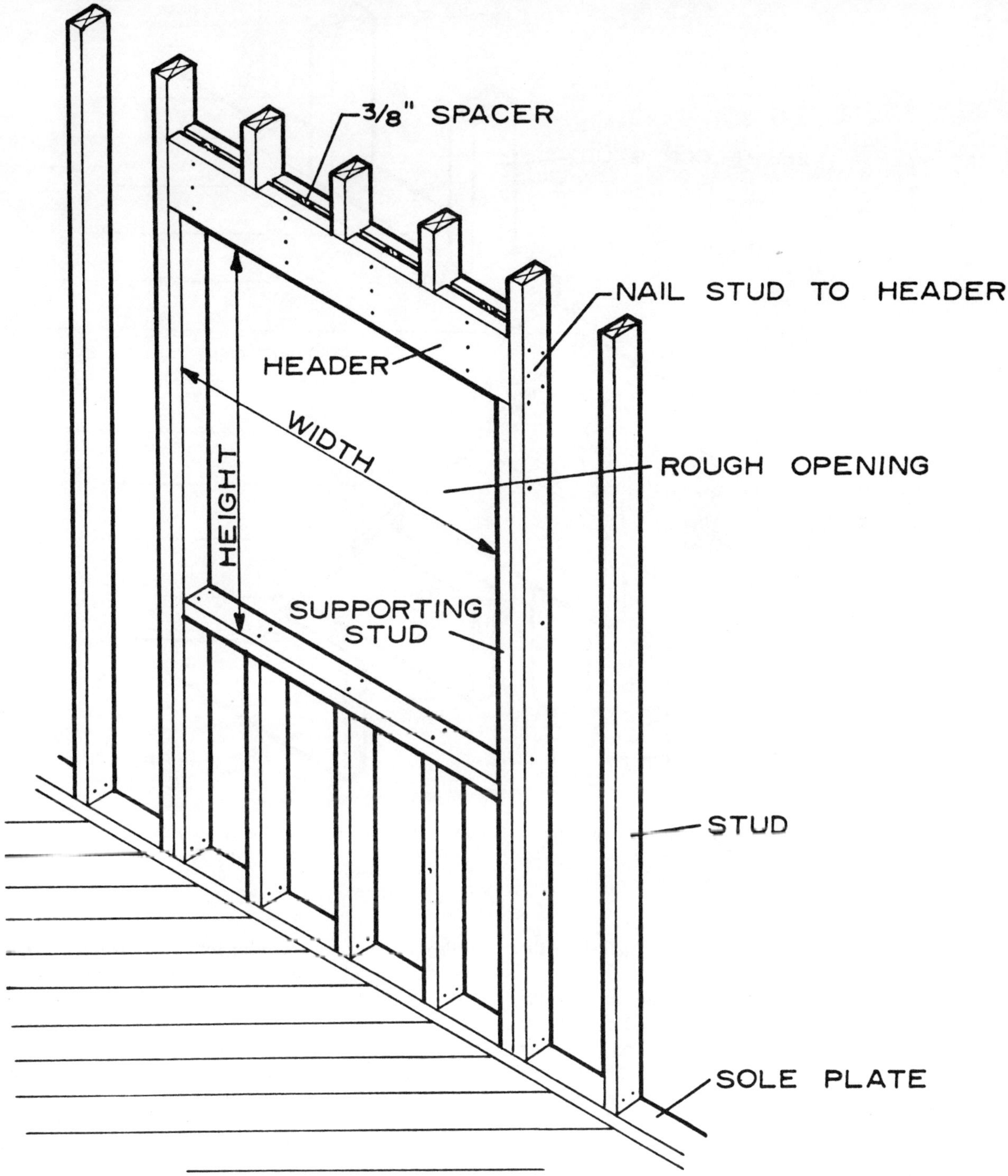

Figure 36.—Headers for windows and door openings.

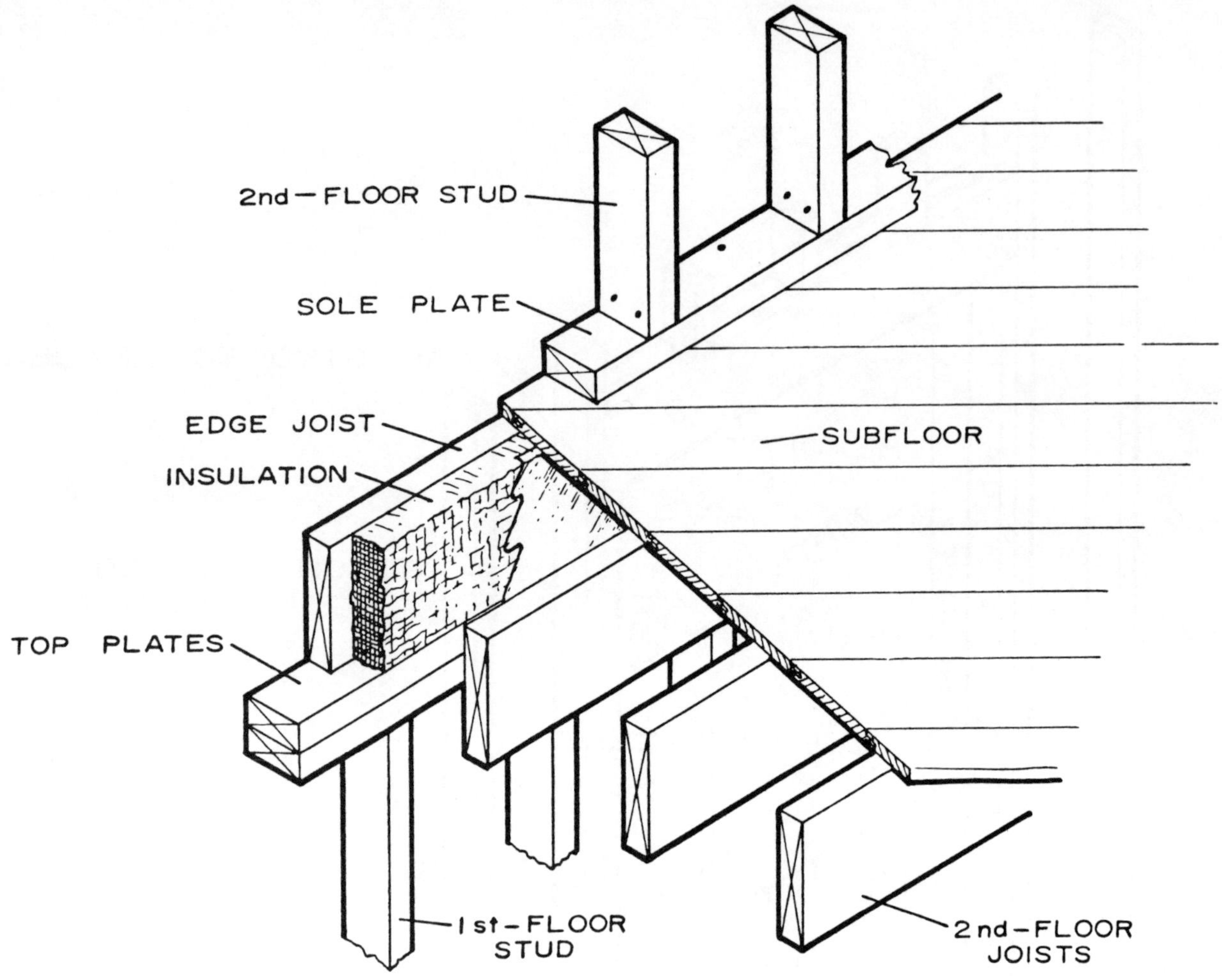

Figure 37.—End-wall framing for platform construction (junction of first-floor ceiling with upper-story floor framing).

soleplate, and wall framing are then installed in the same manner used for the first floor.

In balloon framing, the studs continue through the first and second floors (fig. 38). The edge joist can be nailed to each stud with two or three tenpenny nails. As for the first floor, 2- by 4-inch firestops are cut between each stud. Subfloor is applied in a normal manner. Details of the sidewall supporting the ends of the joists are shown in figure 35.

Interior Walls

The interior walls in a house with conventional joist and rafter roof construction are normally located to serve as bearing walls for the ceiling joists as well as room dividers. Walls located parallel to the direction of the joists are commonly nonload bearing. Studs are nominal 2 by 4 inches in size for load-bearing walls but can be 2 by 3 inches in size for nonload-bearing walls. However, most contractors use 2 by 4's throughout. Spacing of the studs is usually controlled by the thickness of the covering material. For example, 24-inch stud spacing will require ½-inch gypsum board for dry wall interior covering.

The interior walls are assembled and erected in the same manner as exterior walls, with a single bottom (sole) plate and double top plates. The upper top plate is used to tie intersecting and crossing walls to each other. A single framing stud can be used at each side of a door opening in nonload-bearing partitions. They must be doubled for load-bearing walls, however, as shown in figure 36. When trussed rafters (roof trusses) are used, no load-bearing interior partitions are required. Thus, location of the walls and size and spacing of the studs are determined by the room size desired and type of interior covering selected. The bottom chords of the trusses are used to fasten and

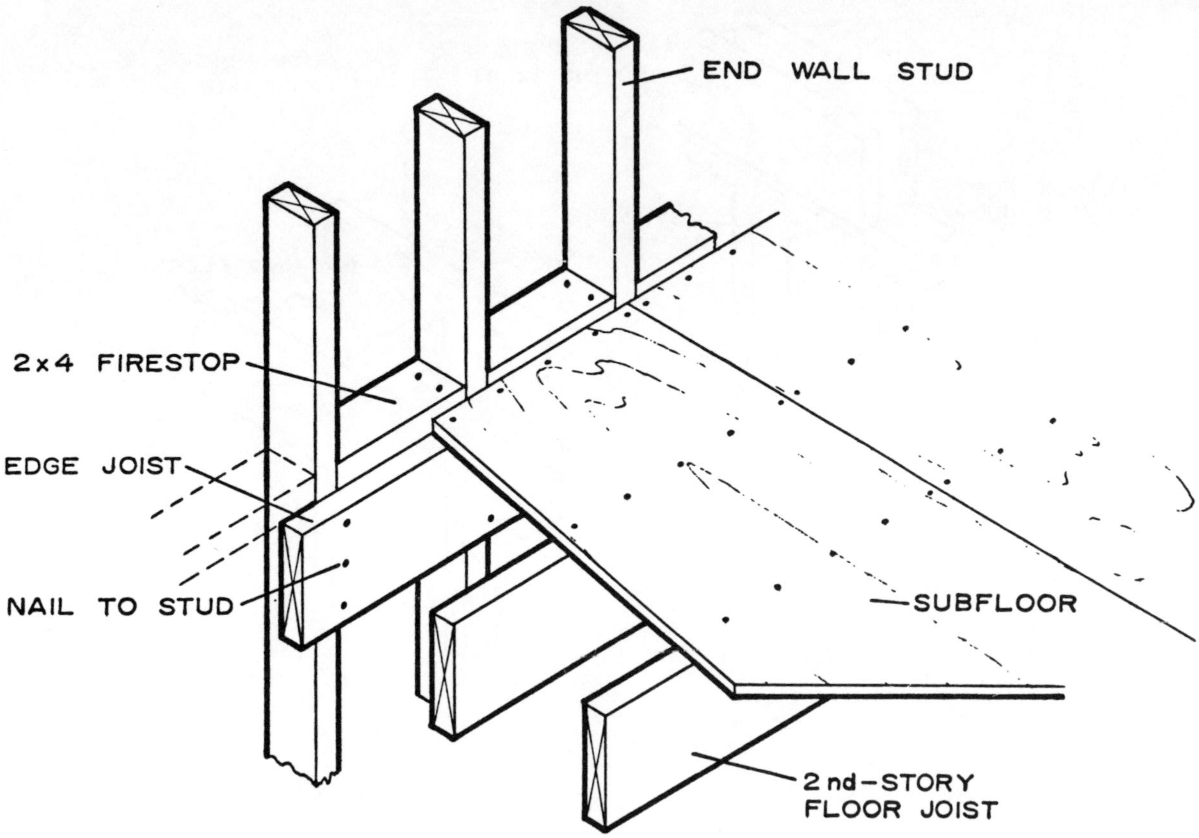

Figure 38.—End-wall framing for balloon construction (junction of first-floor ceiling and upper-story floor framing).

anchor crossing partitions. When partition walls are parallel to and located between trusses, they are fastened to 2- by 4-inch blocks which are nailed between the lower chords.

Lath Nailers

During the framing of walls and ceilings, it is necessary to provide for both vertical and horizontal fastening of plaster-base lath or dry wall at all inside corners. Figures 33 and 34, which illustrate corner and intersecting wall construction, also show methods of providing lath nailers at these areas.

Horizontal lath nailers at the junction of wall and ceiling framing may be provided in several ways. Figure 39,*A* shows doubled ceiling joists above the wall, spaced so that a nailing surface is provided by each joist. In figure 39,*B* the parallel wall is located between two ceiling joists. A 1- by 6-inch lath nailer is placed and nailed to the top plates with backing blocks spaced on 3- to 4-foot centers. A 2- by 6-inch member might also be used here in place of the 1 by 6.

When the partition wall is at a right angle to the ceiling joists, one method of providing lath nailers is to let in 2- by 6-inch blocks between the joists (fig. 39,*C*). They are nailed directly to the top plate and toenailed to the ceiling joists.

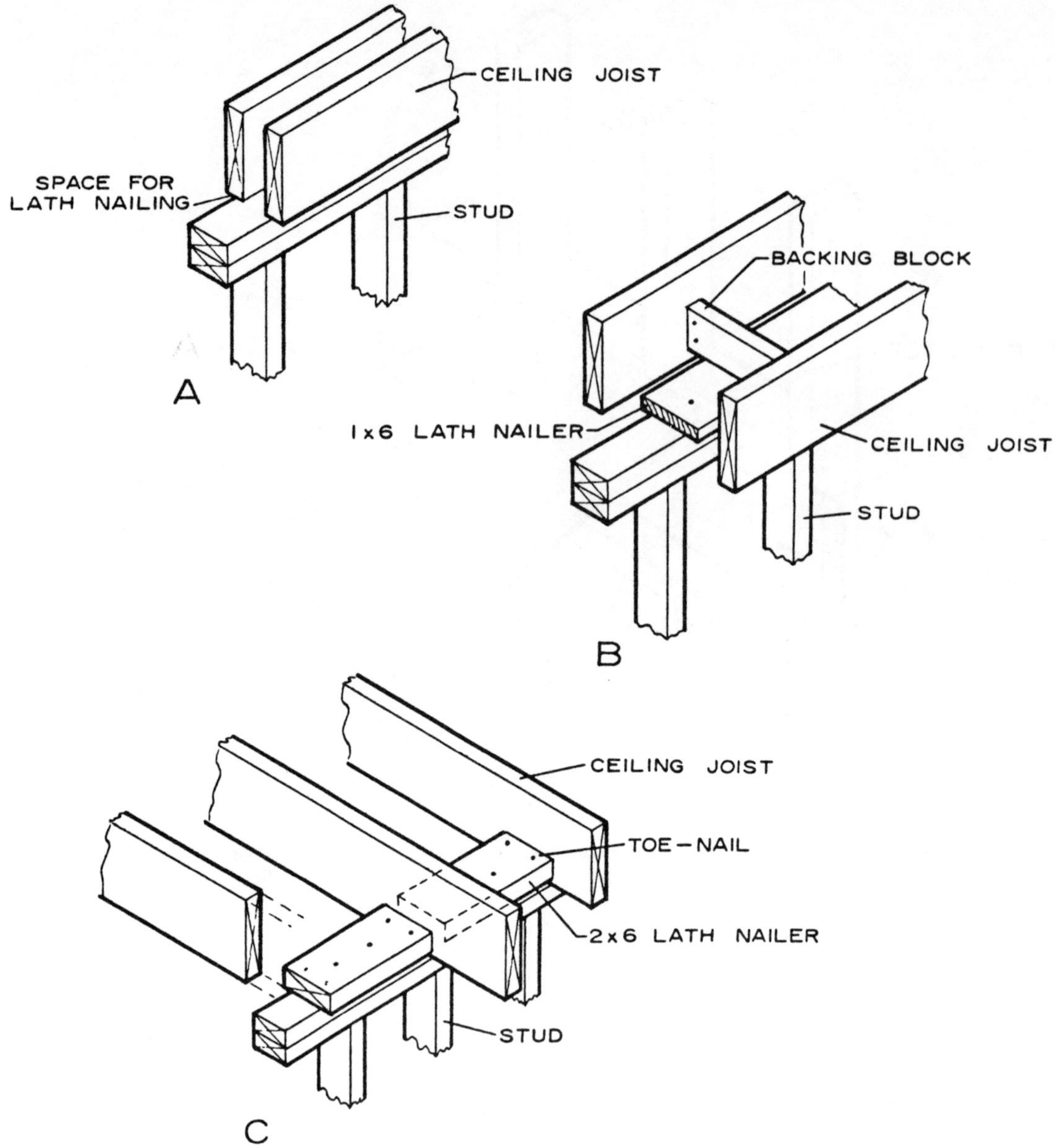

Figure 39.—Horizontal lath catchers at ceiling: *A*, Using ceiling joists over stud wall; *B*, lath nailer between ceiling joists; C, stud wall at right angle to joist.

CHAPTER 7

CEILING AND ROOF FRAMING

Ceiling Joists

After exterior and interior walls are plumbed, braced, and top plates added, ceiling joists can be positioned and nailed in place. They are normally placed across the width of the house, as are the rafters. The partitions of the house are usually located so that ceiling joists of even lengths (10, 12, 14, and 16 ft. or longer) can be used without waste to span from exterior walls to load-bearing interior walls. The sizes of the joists depend on the span, wood species, spacing between joists, and the load on the second floor or attic. The correct sizes for various conditions can be found in joist tables or designated by local building requirements (7). When preassembled trussed rafters (roof trusses) are used, the lower chord acts as the ceiling joist. The truss also eliminates the need for load-bearing partitions.

Second grades of the various species are commonly used for ceiling joists and rafters (5). This has been more fully described in Chapter 5, "Floor Framing." It is also desirable, particularly in two-story houses and when material is available, to limit the moisture content of the second-floor joists to no more than 15 percent. This applies as well to other lumber used throughout the house. Maximum moisture content for dimension material should be 19 percent.

Ceiling joists are used to support ceiling finishes. They often act as floor joists for second and attic floors and as ties between exterior walls and interior partitions. Since ceiling joists also serve as tension members to resist the thrust of the rafters of pitched roofs, they must be securely nailed to the plate at outer and inner walls. They are also nailed together, directly or with wood or metal cleats, where they cross or join at the load-bearing partition (fig. 40,*A*) and to the rafter at the exterior walls (fig. 40,*B*). Toenail at each wall.

In areas of severe windstorms, the use of metal strapping or other systems of anchoring ceiling and roof framing to the wall is good practice. When ceiling joists are perpendicular to rafters, collar beams and cross ties should be used to resist thrust. Recommended sizes and spacing of nails for the framing are listed in table 2. The in-line joist system as shown in Figure 28 and described in the section on joist installation can also be adapted to ceiling or second floor joists.

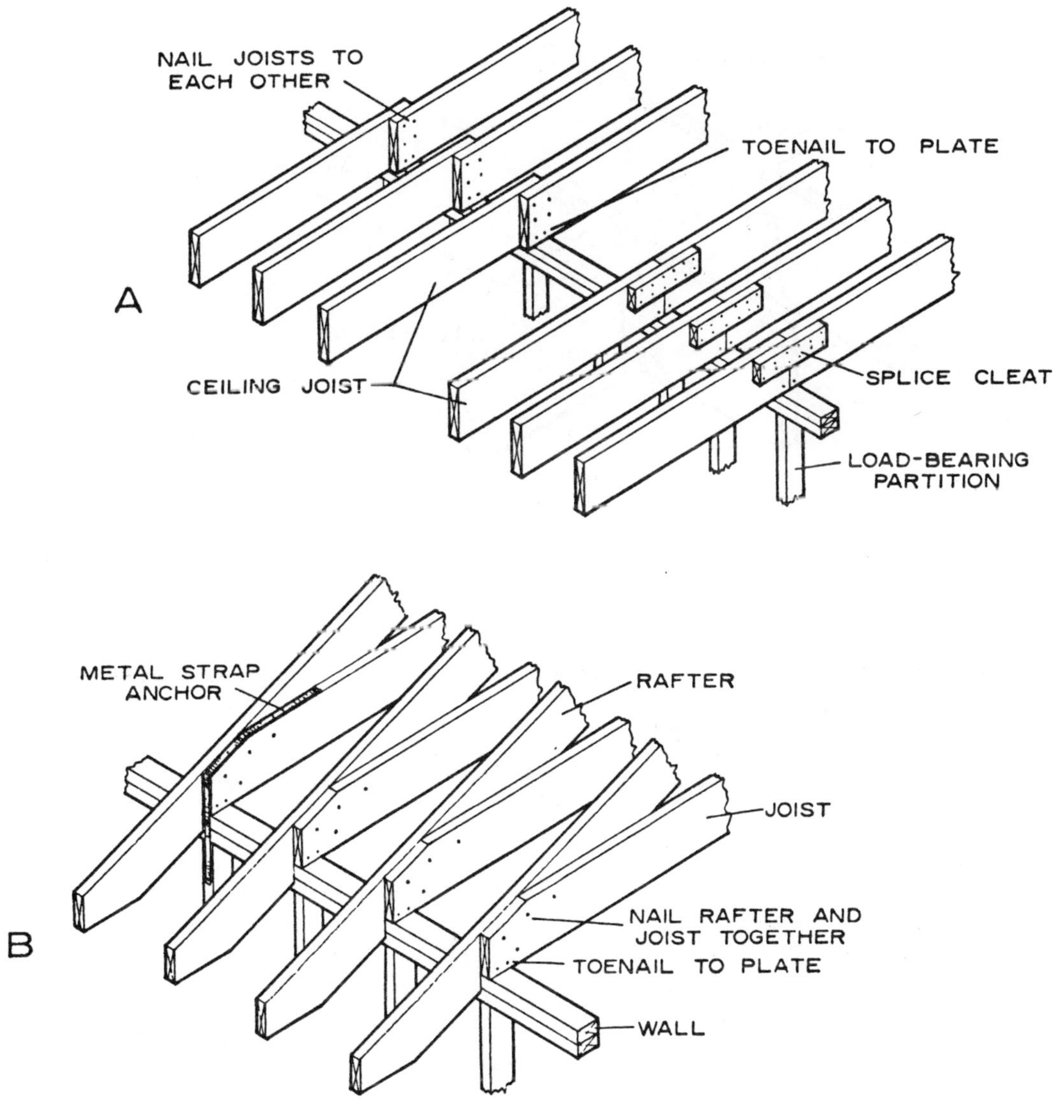

Figure 40.—Ceiling joist connections: *A*, At center partition with joists lapped or butted; *B*, at outside wall.

Flush Ceiling Framing

In many house designs, the living room and the dining or family room form an open "L." A wide, continuous ceiling area between the two rooms is often desirable. This can be created with a flush beam, which replaces the load-bearing partitions used in the remainder of the house. A nail-laminated beam, designed to carry the ceiling load, supports the ends of the joists. Joists are toenailed into the beam and supported by metal joist hangers (fig. 41,*A*) or wood hangers (fig. 41,*B*). To resist the thrust of the rafters for longer spans, it is often desirable to provide added resistance by using metal strapping. Strapping should be nailed to each opposite joist with three or four eightpenny nails.

Post and Beam Framing

In contemporary houses, *exposed beams* are often a part of the interior design and may also replace interior and exterior load-bearing walls. With post and beam construction, exterior walls can become fully glazed panels between posts, requiring no other support. Areas below interior beams within the house can remain open or can be closed in with wardrobes, cabinets, or light curtain walls.

This type of construction, while not adaptable to many styles of architecture, is simple and straightforward. However, design of the house should take into account the need for shear or racking resistance of the exterior walls. This is usually accomplished by solid masonry walls or fully sheathed frame walls

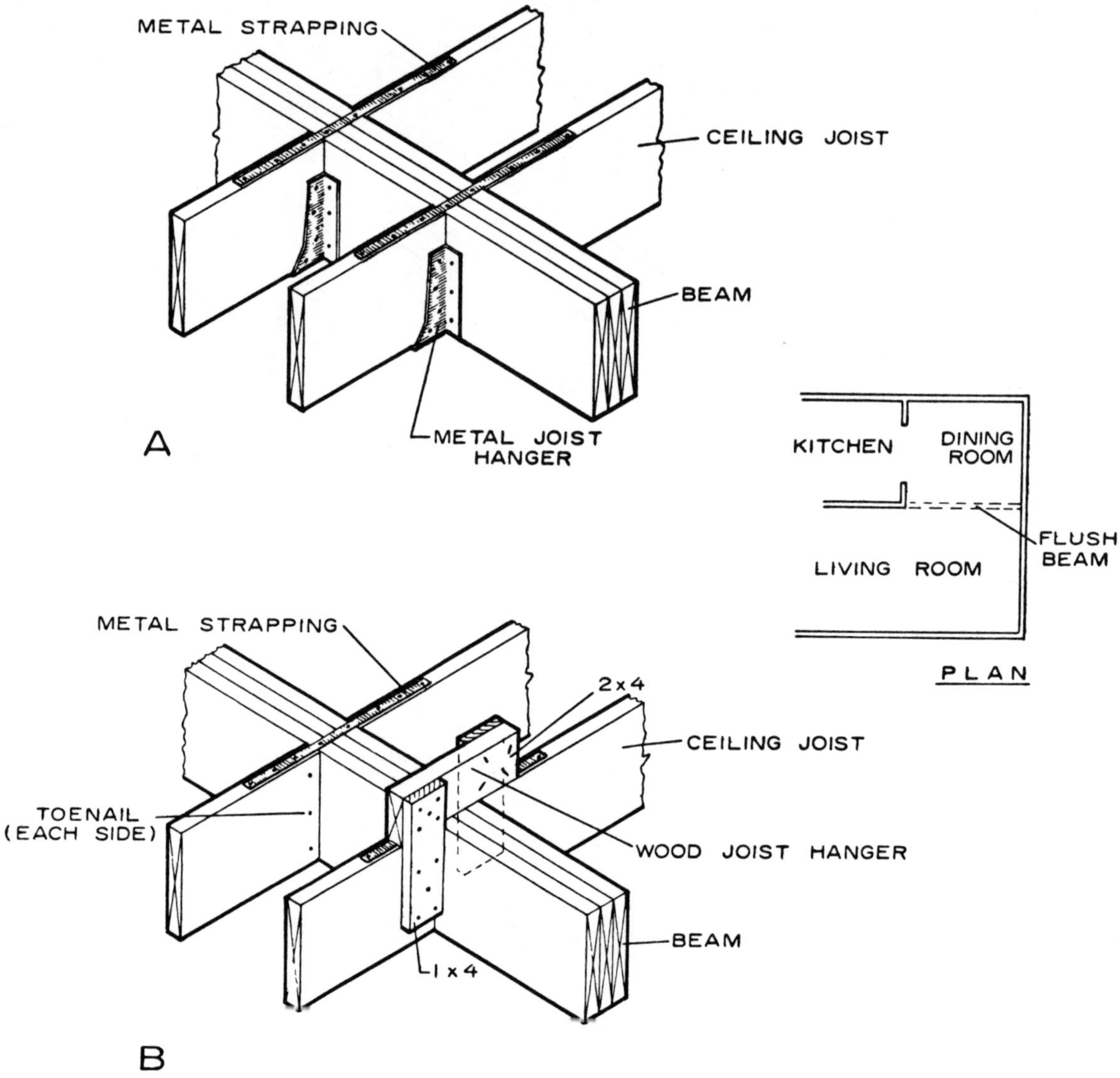

Figure 41.—Flush ceiling framing: *A*, Metal joist hanger; *B*, wood hanger.

between open glazed areas.

Roofs of such houses are often either flat or low-pitched, and may have a conventional rafter-joist combination or consist of thick wood decking spanning between beams. The need for a well-insulated roof often dictates the type of construction that might be used.

The connection of the supporting posts at the floor plate and beam is important to provide uplift resistance. Figure 42 shows connections at the soleplate and at the beam for solid or spaced members. The solid post and beam are fastened together with metal angles nailed to the top plate and to the soleplate as well as the roof beam (fig. 42,*A*). The spaced beam and post are fastened together with a ⅜-inch or thicker plywood cleat extending between and nailed to the spaced members (fig. 42,*B*). A wall header member between beams can be fastened with joist hangers.

Continuous headers are often used with spaced posts in the construction of framed walls or porches requiring large glazed openings. The beams should be well fastened and reinforced at the corners with lag screws or metal straps. Figure 43,*A* illustrates one connection method using metal strapping.

In low-pitch or flat roof construction for a post and beam system, wood or fiberboard decking is often used. Wood decking, depending on thickness, is frequently used for beam spacings up to 10 or more feet. However, for the longer spans, special application instructions are required (*1*). Depending on the type,

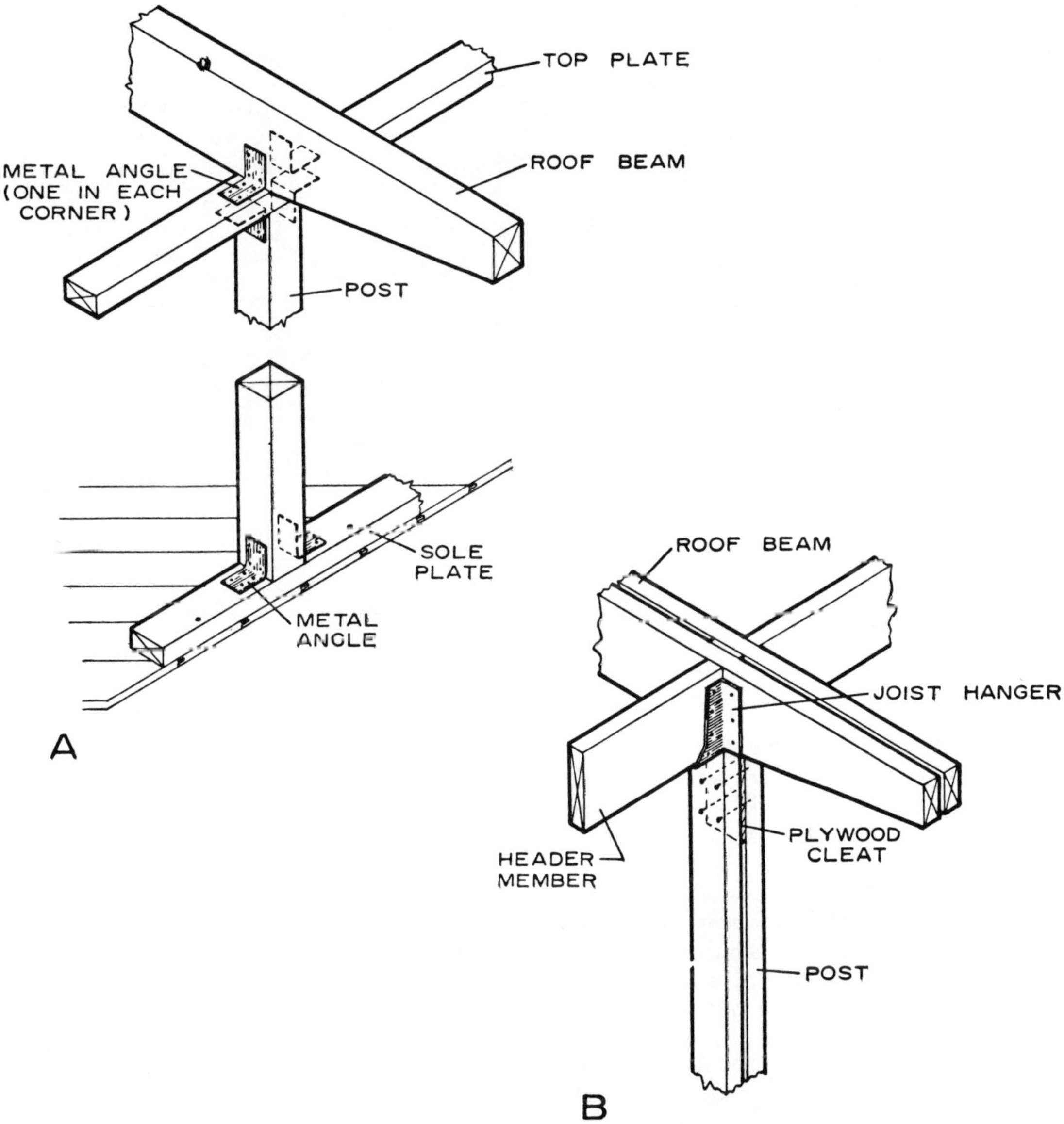

Figure 42.—Post and beam connections; *A*, Solid post and beam; *B*, spaced post and beam.

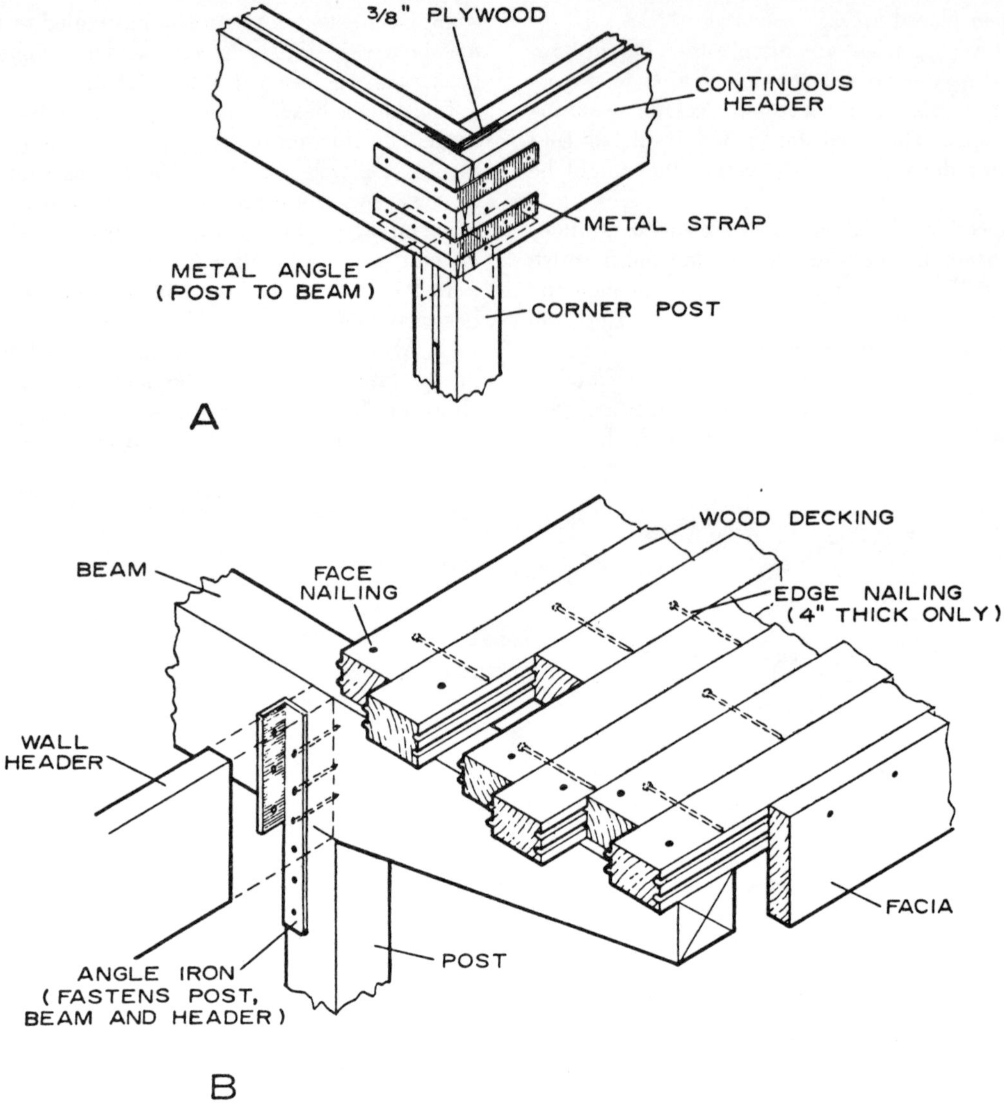

Figure 43.—Post and beam details: *A*, Corner connection with continuous header; *B*, with roof decking.

2- to 3-inch thick fiberboard decking normally is limited to a beam or purlin spacing of 4 feet.

Tongued-and-grooved solid wood decking, 3 by 6 and 4 by 6 inches in size, should be toe-nailed and face-nailed directly to the beams and edge-nailed to each other with long nails used in predrilled holes (fig. 43,*B*). Thinner decking is usually only face-nailed to the beams. Decking is usually square end-trimmed to provide a good fit. If additional insulation is required for the roof, fiberboard or an expanded foamed plastic in sheet form is fastened to the decking before the built-up or similar type of roof is installed. The moisture content of the decking should be near its in-service condition to prevent joints opening later as the wood dries.

Roof Slopes

The architectural style of a house often determines the type of roof and roof slope which are best suited. A contemporary design may have a flat or slightly pitched roof, a rambler or ranch type an intermediate slope, and a Cape Cod cottage a steep slope. Generally, however, the two basic types may be called *flat* or *pitched*, defined as (a) flat or slightly pitched roofs in which roof and ceiling supports are furnished by one type of member, and (b) pitched roofs where both ceiling joists and rafters or trusses are required.

The slope of the roof is generally expressed as the number of inches of vertical rise in 12 inches of horizontal run. The rise is given first, for example, 4 in 12.

A further consideration in choosing a roof slope is

the type of roofing to be used. However, modern methods and roofing materials provide a great deal of leeway in this. For example, a built-up roof is usually specified for flat or very low-pitched roofs, but with different types of asphalt or coal-tar pitch and aggregate surfacing materials, slopes of up to 2 in 12 are sometimes used. Also, in sloped roofs where wood or asphalt shingles might be selected, doubling the underlay and decreasing the exposure distance of the shingles will allow slopes of 4 in 12 and less.

Second grades of the various wood species are normally used for rafters. Most species of softwood framing lumber are acceptable for roof framing, subject to maximum allowable spans for the particular species, grade, and use. Because all species are not equal in strength properties, larger sizes, as determined from the design, must be used for weaker species for a given span (*11*).

All framing lumber should be well seasoned. Lumber 2 inches thick and less should have a moisture content not over 19 percent, but when obtainable, lumber at about 15 percent is more desirable because less shrinkage will occur when moisture equilibrium is reached.

Flat Roofs

Flat or low-pitched roofs, sometimes known as *shed roofs*, can take a number of forms, two of which are shown in figure 44. Roof joists for flat roofs are commonly laid level or with a slight pitch, with roof sheathing and roofing on top and with the underside utilized to support the ceiling. Sometimes a slight roof slope may be provided for roof drainage by tapering the joist or adding a cant strip to the top.

The house design usually includes an overhang of the roof beyond the wall. Insulation is sometimes used in a manner to provide for an airways just under the roof sheathing to minimize condensation problems in winter. Flat or low-pitched roofs of this type require larger sized members than steeper pitched roofs because they carry both roof and ceiling loads.

The use of solid wood decking often eliminates the need for joists. Roof decking used between beams serves as: (a) Supporting members, (b) interior finish, and (c) roof sheathing. It also provides a moderate amount of insulation. In cold climates, rigid insulating materials are used over the decking to further reduce heat loss.

When overhang is involved on all sides of the flat roof, lookout rafters are ordinarily used (fig. 45). Lookout rafters are nailed to a doubled header and toenailed to the wallplate. The distance from the doubled header to the wall line is usually twice the overhang. Rafter ends may be finished with a nailing header which serves for fastening soffit and facia boards. Care should be taken to provide some type of ventilation at such areas.

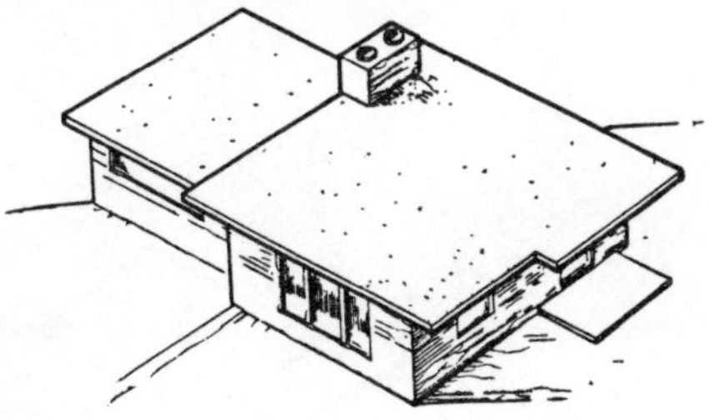

A

B

Figure 44.—Roofs using single roof construction: *A*, Flat roof; *B*, low-pitched roof.

Pitched Roofs

Gable Roof

Perhaps the simplest form of the pitched roof, where both rafters and ceiling joists are required because of the attic space formed, is the gable roof (fig. 46,*A*). All rafters are cut to the same length and pattern and erection is relatively simple, each pair being fastened at the top to a *ridge board*. The ridge board is usually a 1- by 8-inch member for 2- by 6-inch rafters and provides support and a nailing area for the rafter ends.

A variation of the gable roof, used for Cape Cod or similar styles, includes the use of shed and gable *dormers* (fig. 46,*B*). Basically, this is a one-story house because the majority of the rafters rest on the first-floor plate. Space and light are provided on the second floor by the shed and gable dormers for bedrooms and bath. Roof slopes for this style may vary from 9 in 12 to 12 in 12 to provide the needed headroom.

A third style in roof designs is the *hip roof* (fig. 46,*C*). Center rafters are tied to the ridge board, while hip rafters supply the support for the shorter jack rafters. Cornice lines are carried around the perimeter of the building.

While these roof types are the most common, others may include such forms as the *mansard* and the *A-frame* (where wall and roof members are the same members).

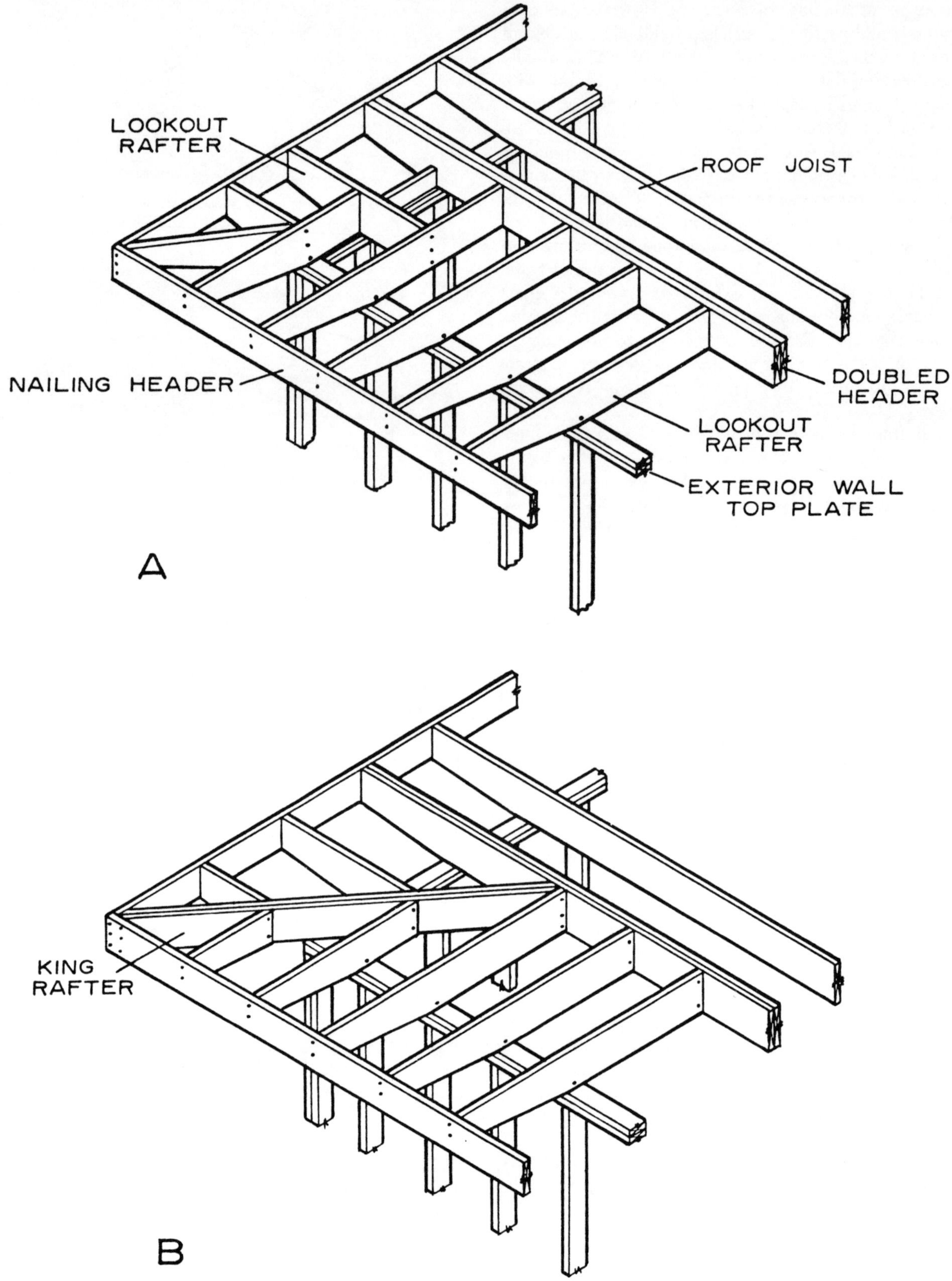

Figure 45.—Typical construction of flat or low-pitched roof with side and end overhang of: *A*, Less than 3 feet; *B*, more than 3 feet.

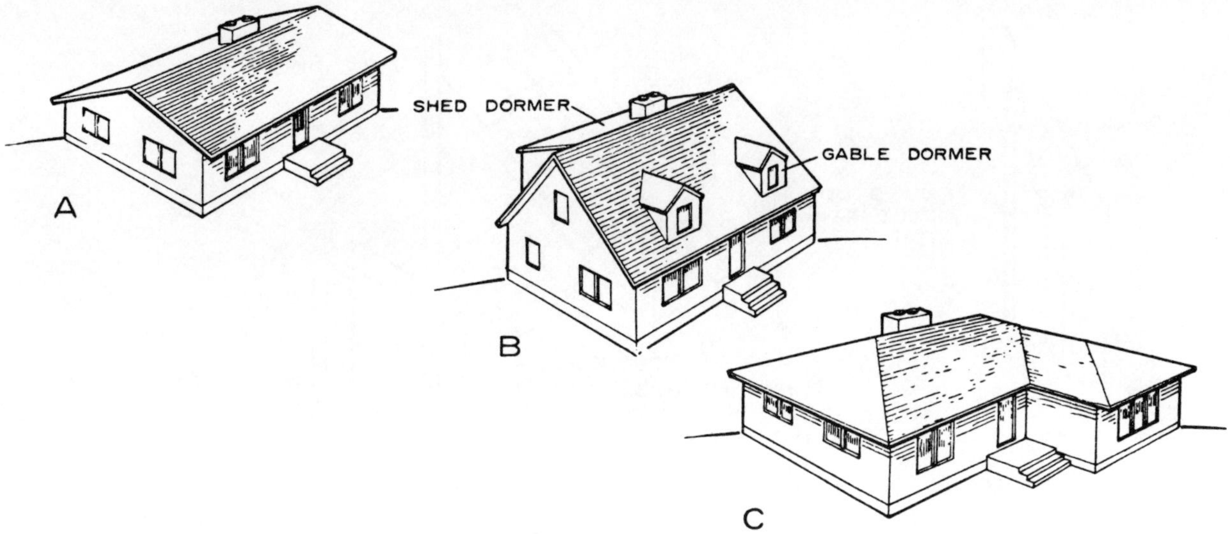

Figure 46.—Types of pitched roofs: *A*, Gable; *B*, gable with dormers; *C*, hip.

In normal pitched-roof construction, the ceiling joists are nailed in place after the interior and the exterior wall framing are complete. Rafters should not be erected until ceiling joists are fastened in place, as the thrust of the rafters will otherwise tend to push out the exterior walls.

Rafters are usually precut to length with proper angle cut at the ridge and eave, and with notches provided for the top plates (fig. 47,*A*). Rafters are erected in pairs. Studs for gable end walls are cut to fit and nailed to the end rafter and the topplate of the end wall soleplate (fig. 47,*B*). With a gable (rake) overhang, a fly rafter is used beyond the end rafter and is fastened with blocking and by the sheathing. Additional construction details applicable to roof framing are given in Chapter 10, "Exterior Trim and Millwork."

Hip Roof

Hip roofs are framed the same as a gable roof at the center section of a rectangular house. The ends are framed with hip rafters which extend from each outside corner of the wall to the ridge board at a 45°

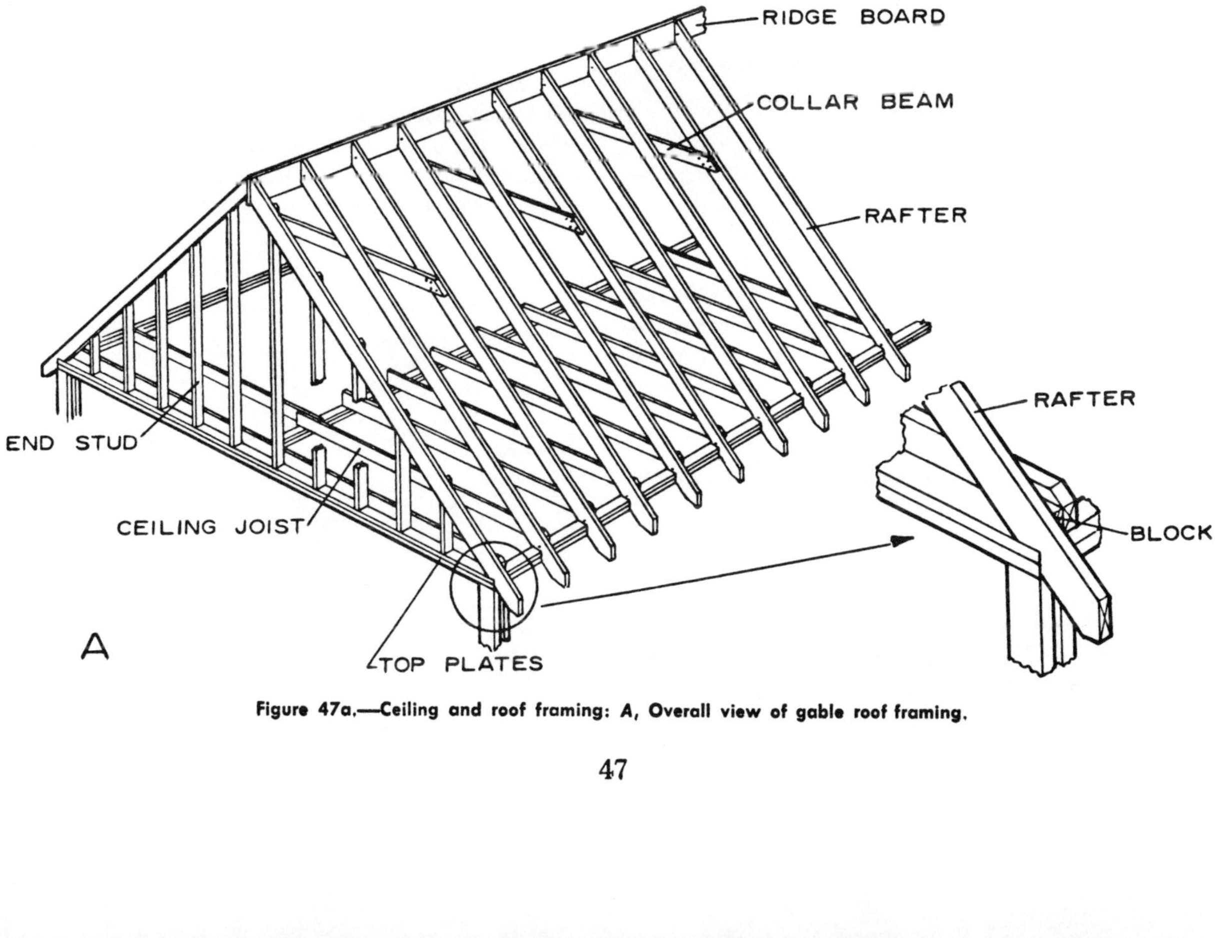

Figure 47a.—Ceiling and roof framing: *A*, Overall view of gable roof framing.

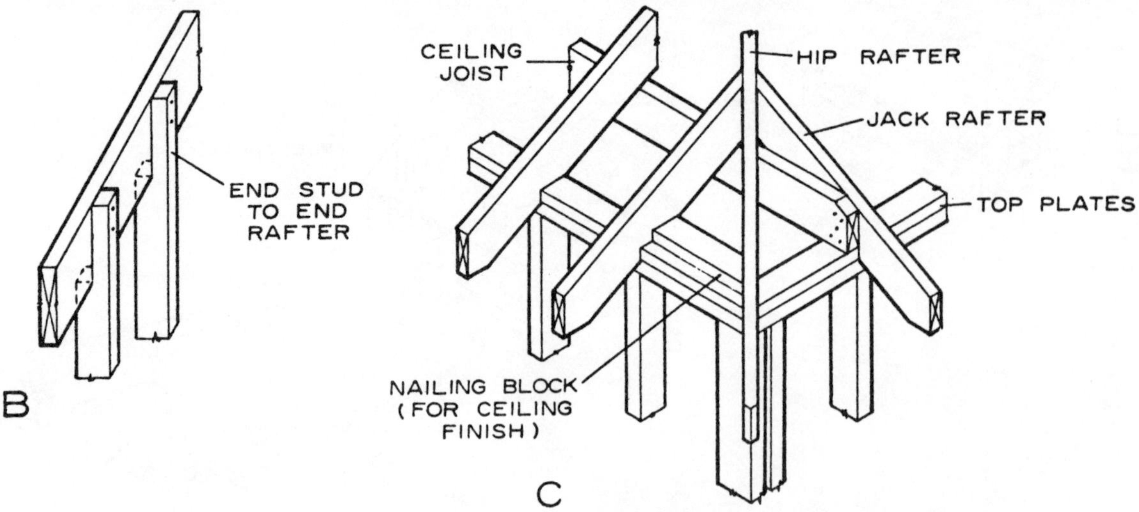

Figure 47b.—Ceiling and roof framing: *B*, Connection of gable end studs to end rafter; *C*, detail of corner of hip roof.

angle. Jack rafters extend from the top plates to the hip rafters (fig. 47,*C*).

When roofs spans are long and slopes are flat, it is common practice to use collar beams between opposing rafters. Steeper slopes and shorter spans may also require collar beams but only on every third rafter. Collar beams may be 1- by 6-inch material. In 1½-story houses, 2- by 4-inch members or larger are used at each pair of rafters which also serve as ceiling joists for the finished rooms.

Good practices to be followed in the nailing of rafters, ceiling joists, and end studs are shown in figure 47 and table 2.

Valleys

The *valley* is the internal angle formed by the junction of two sloping sides of a roof. The key member of valley construction is the *valley rafter*. In the intersection of two equal-size roof sections, the valley rafter is doubled (fig. 48) to carry the roofload, and is 2 inches deeper than the common rafter to provide full contact with jack rafters. Jack rafters are nailed to the ridge and toenailed to the valley rafter with three tenpenny nails.

Dormers

In construction of small gable dormers, the rafters at each side are doubled and the side studs and the short valley rafter rest on these members (fig. 49). Side studs may also be carried past the rafter and bear on a soleplate nailed to the floor framing and subfloor. This same type of framing may be used for the sidewalls of shed dormers. The valley rafter is also tied to the roof framing at the roof by a header. Methods of fastening at top plates conform to those previously described. Where future expansion is contemplated or additional rooms may be built in an attic, consideration should be given to framing and enclosing such dormers when the house is built.

Overhangs

In two-story houses, the design often involves a projection or overhang of the second floor for the purpose of architectural effect, to accommodate brick veneer on the first floor, or for other reasons. This overhang may vary from 2 to 15 inches or more. The overhang should ordinarily extend on that side of the house where joist extensions can support the wall framing (fig. 50). This extension should be provided with insulation and a vapor barrier.

When the overhang parallels the second-floor joists, a doubled joist should be located back from the wall at a distance about twice the overhang. These details

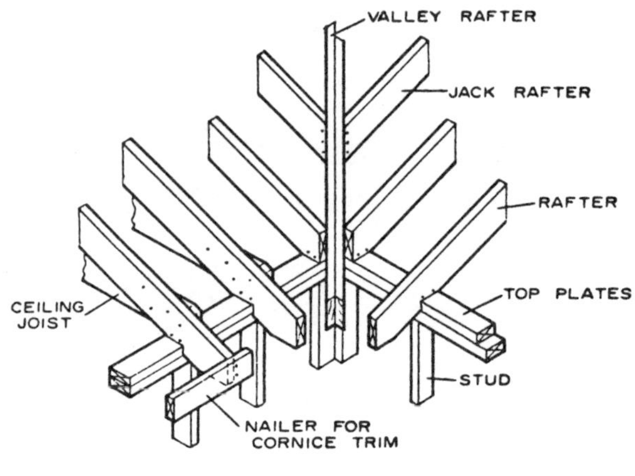

Figure 48.—Framing at a valley.

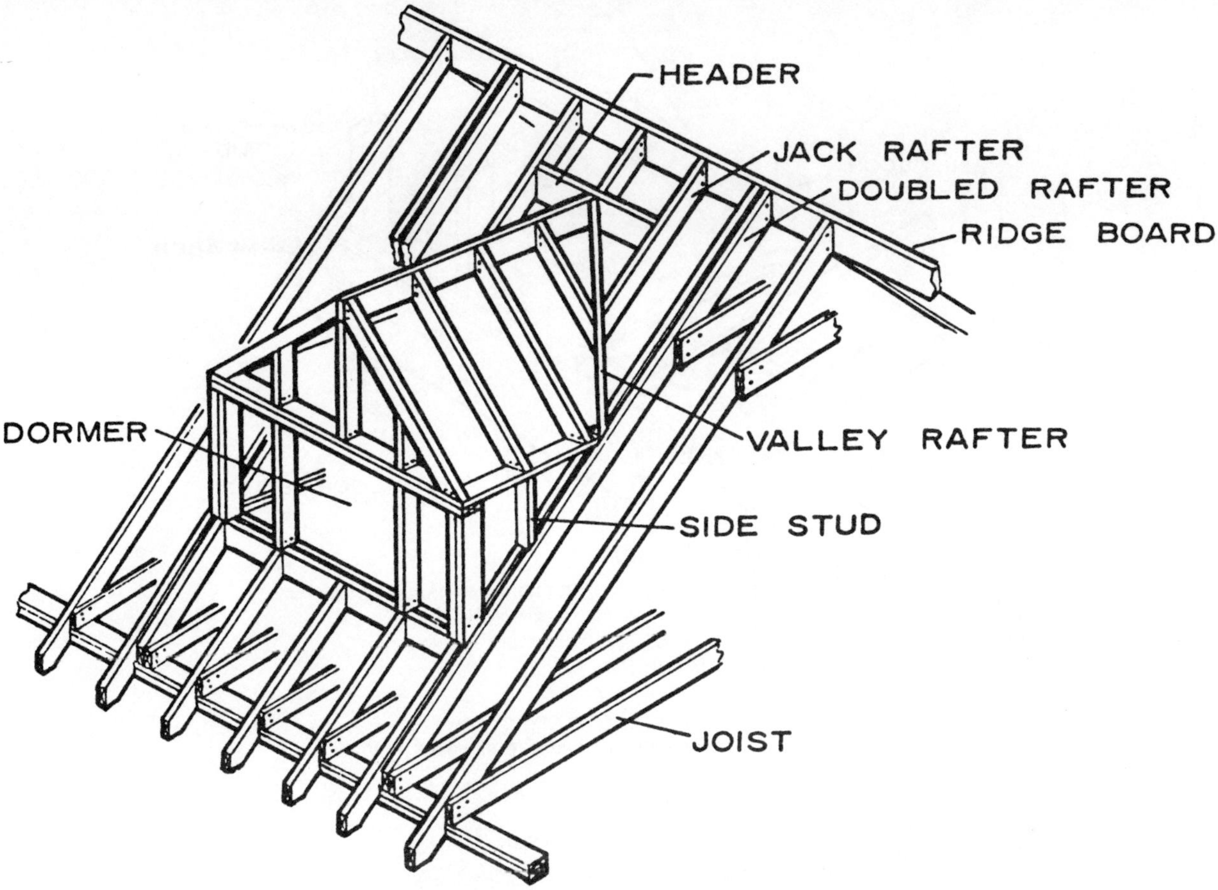

Figure 49.—Typical dormer framing.

are similar to those shown in figure 30 under "Floor Framing."

Ridge Beam Roof Details

In low-slope roof designs, the style of architecture often dictates the use of a *ridge beam*. These solid, glue-laminated, or nail-laminated beams span the open area and are usually supported by an exterior wall at one end and an interior partition wall or a post at the other. The beam must be designed to support the roof load for the span selected. Wood decking can serve both as supporting and sheathing. Spaced rafters placed over the ridge beam or hung on metal joist hangers serve as alternate framing methods. When a ridge beam and wood decking are used (fig. 51,*A*), good anchoring methods are needed at the ridge and outer wall. Long ringshank nails and supplemental metal strapping or angle iron can be used at both bearing areas.

A combination of large spaced rafters (purlin rafters) which serve as beams for longitudinal wood or structural fiberboard decking is another system which might be used with a ridge beam. Rafters can be supported by metal hangers at the ridge beam (fig. 51*B*) and extend beyond the outer walls to form an overhang. Fastenings should be supplemented by strapping or metal angles.

Lightweight Wood Roof Trusses

The *simple truss* or *trussed rafter* is an assembly of members forming a rigid framework of triangular shapes capable of supporting loads over long spans without intermediate support. It has been greatly refined during its development over the years, and the *gusset* and other preassembled types of wood trusses are being used extensively in the housing field. They save material, can be erected quickly, and the house can be enclosed in a short time.

Trusses are usually designed to span from one exterior wall to the other with lengths from 20 to 32 feet or more. Because no interior bearing walls are required. the entire house becomes one large workroom. This allows increased flexibility for interior planning, as partitions can be placed without regard to structural requirements.

Wood trusses most commonly used for houses include the *W-type* truss, the *King-post*, and the *scissors*

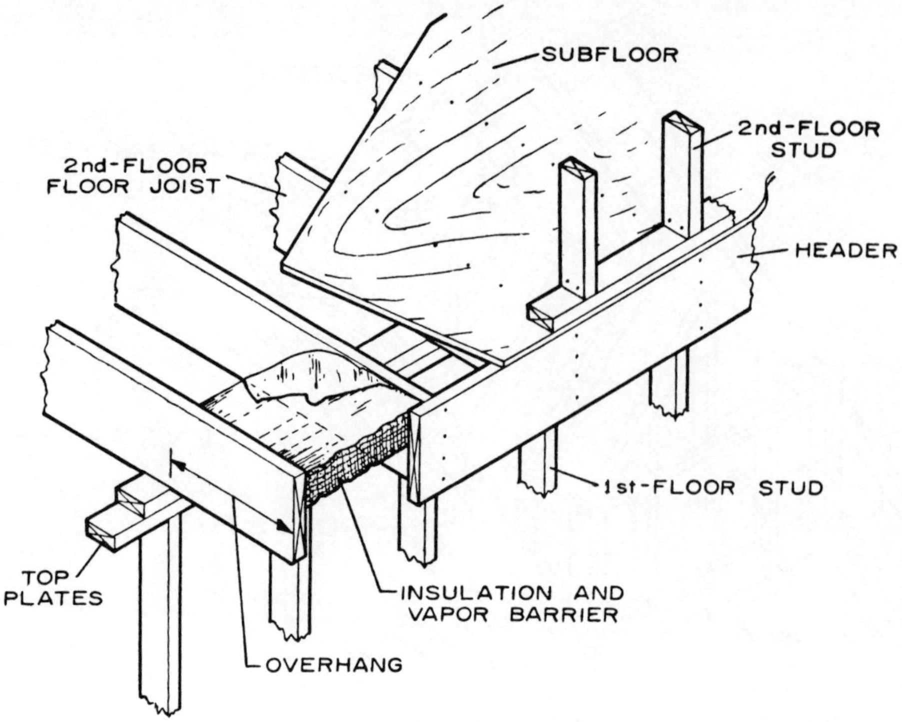

Figure 50.—Construction of overhang at second floor.

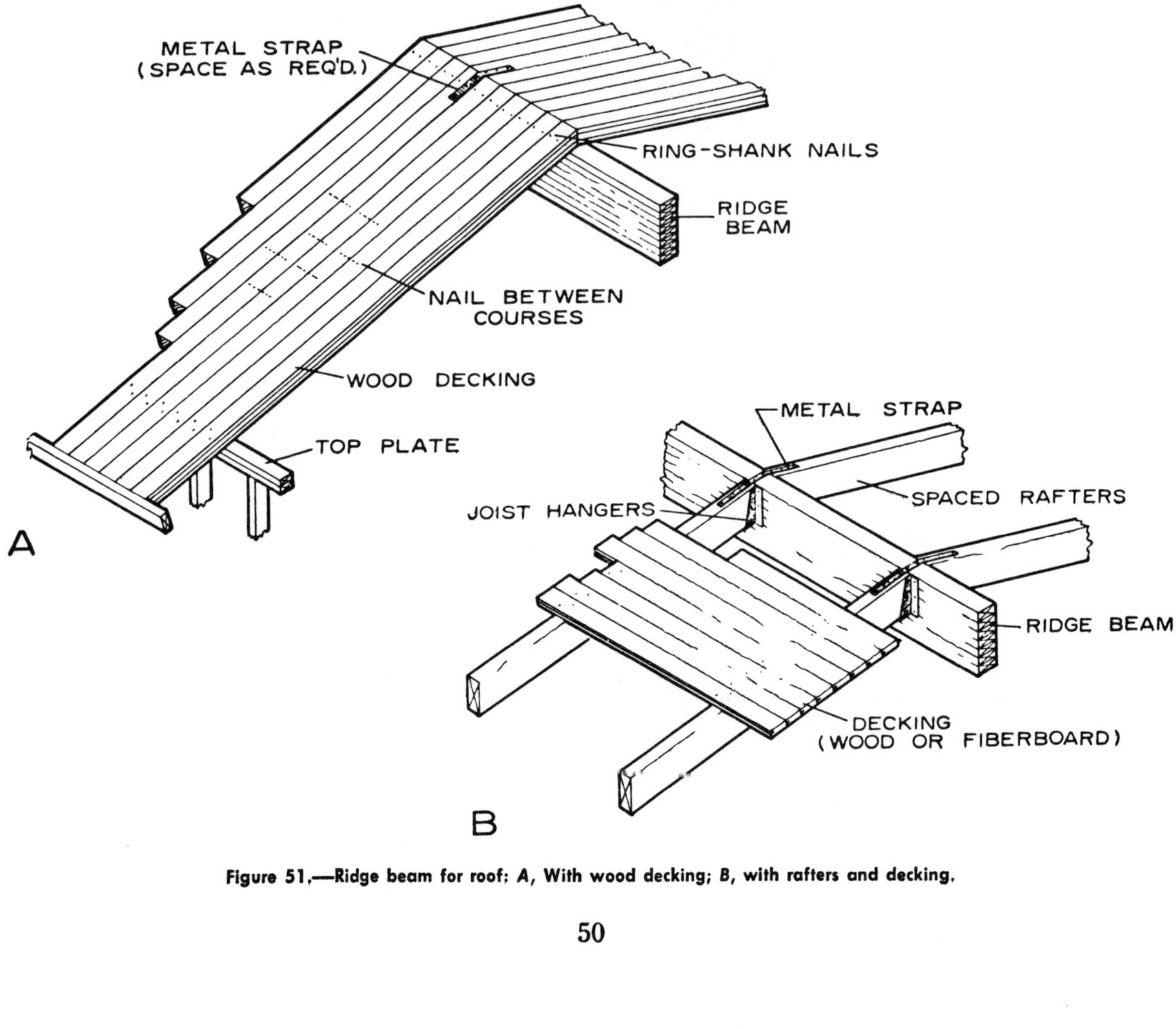

Figure 51.—Ridge beam for roof: *A*, With wood decking; *B*, with rafters and decking.

(fig. 52). These and similar trusses are most adaptable to houses with rectangular plans so that the constant width requires only one type of truss. However, trusses can also be used for L plans and for hip roofs as special hip trusses can be provided for each end and valley area.

Trusses are commonly designed for 2-foot spacing, which requires somewhat thicker interior and exterior sheathing or finish material than is needed for conventional joist and rafter construction using 16-inch spacing. Truss designs, lumber grades, and construction details are available from several sources including the American Plywood Association (*2*).

W-Type Truss

The W-type truss (fig. 52,*A*) is perhaps the most popular and extensively used of the light wood trusses. Its design includes the use of three more members than the King-post truss, but distances between connections are less. This usually allows the use of lower grade lumber and somewhat greater spans for the same member size.

King-post truss.—The King-post is the simplest form of truss used for houses, as it is composed only of upper and lower chords and a center vertical post (fig. 52,*B*). Allowable spans are somewhat less than for the W-truss when the same size members are used, because of the unsupported length of the upper chord. For short and medium spans, it is probably more economical than other types because it has fewer pieces and can be fabricated faster. For example, under the same conditions, a plywood gusset King-post truss with 4 in 12 pitch and 2-foot spacing is limited to about a 26-foot span for 2- by 4-inch members, while the W-type truss with the same size members and spacing could be used for a 32-foot span. Furthermore, the grades of lumber used for the two types might also vary.

Local prices and design load requirements (for snow, wind, etc.) as well as the span should likely govern the type of truss to be used.

Scissors Truss

The scissors truss (fig. 52,*C*) is a special type used for houses in which a sloping living room ceiling is desired. Somewhat more complicated than the W-type truss, it provides good roof construction for a "cathedral" ceiling with a saving in materials over conventional framing methods.

Design and Fabrication

The design of a truss not only includes snow and windload considerations but the weight of the roof itself. Design also takes into account the slope of the roof. Generally, the flatter the slope, the greater the stresses. This results not only in the need for larger members but also in stronger connections. Consequently, all conditions must be considered before the type of truss is selected and designed.

A great majority of the trusses used are fabricated with gussets of plywood (nailed, glued, or bolted in place) or with metal gusset plates. Others are assembled with split-ring connectors. Designs for standard W-type and King-post trusses with plywood gussets are usually available through a local lumber dealer (*2*). Information on metal plate connectors for wood trusses is also available. Many lumber dealers are able to provide the builder or homeowner with completed trusses ready for erection.

To illustrate the design and construction of a typical wood W-truss more clearly, the following example is given:

The span for the nail-glued gusset truss (fig. 53) is 26 feet, the slope 4 in 12, and the spacing 24 inches. Total roof load is 40 pounds per square foot, which is usually sufficient for moderate to heavy snow belt areas. Examination of tables and charts (*2*) shows that the upper and lower chords can be 2 by 4 inches in size, the upper chord requiring a slightly higher grade of material. It is often desirable to use dimension material with a moisture content of about 15 percent with a maximum of 19 percent.

Plywood gussets can be made from ⅜- or ½-inch

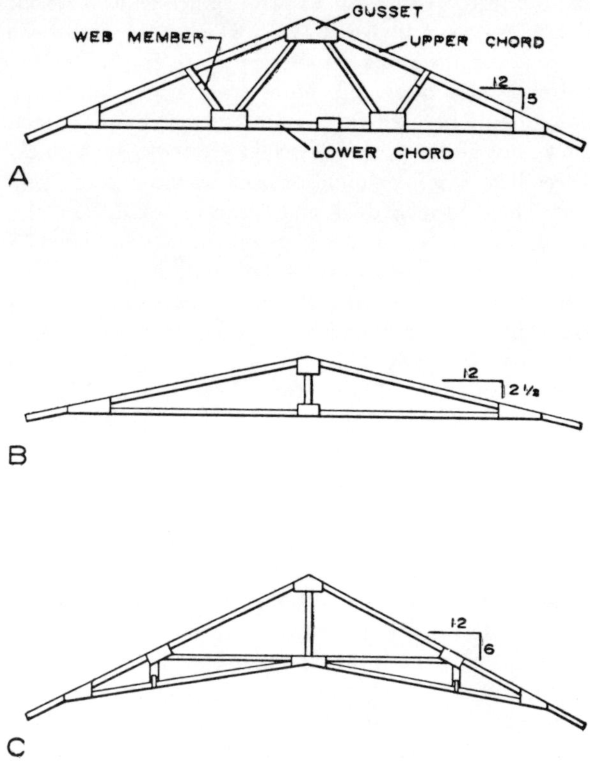

Figure 52.—Light wood trusses: *A*, W-type; *B*, King-post; *C*, scissors.

standard plywood with exterior glueline or exterior sheathing grade plywood (*18*). The cutout size of the gussets and the general nailing pattern for nail-gluing are shown in figure 53. More specifically, fourpenny nails should be used for plywood gussets up to ⅜ inch thick and sixpenny for plywood ½ to ⅞ inch thick. Three-inch spacing should be used when plywood is no more than ⅜ inch thick and 4 inches for thicker plywood. When wood truss members are nominal 4 inches wide, use two rows of nails with a ¾-inch edge distance. Use three rows of nails when truss members are 6 inches wide. Gussets are used on both sides of the truss.

For normal conditions and where relative humidities in the attic area are inclined to be high, such as might occur in the southern and southeastern States, a *resorcinol glue* should be used for the gussets. In dry and arid areas where conditions are more favorable, a casein or similar glue might be considered.

Glue should be spread on the clean surfaces of the gusset and truss members. Either nails or staples might be used to supply pressure until the glue has set, although only nails are recommended for plywood ½ inch and thicker. Use the nail spacing previously outlined. Closer or intermediate spacing may be used to insure "squeezeout" at all visible edges. Gluing should be done under closely controlled temperature conditions. This is especially true if using the resorcinol adhesives. Follow the assembly temperatures recommended by the manufacturer.

Handling

In handling and storage of completed trusses, avoid placing unusual stresses on them. They were designed to carry roofloads in a vertical position, and it is important that they be lifted and stored in an upright position. If they must be handled in a flat position, enough men or supports should be used along their length to minimize bending deflections. Never support them only at the center or only at each end when in a flat position.

Erection

One of the important details in erecting trusses is the method of *anchoring*. Because of their single member thickness and the presence of plywood guessets at the wallplates, it is usually desirable to use some type of metal connector to supplement the toenailings. Plate anchors are available commercially or can be formed from sheet metal. Resistance to uplift stresses as well as thrust must be considered. Many dealers supply trusses with a 2- by 4-inch soffit return at the end of each upper chord to provide nailing areas for the soffit.

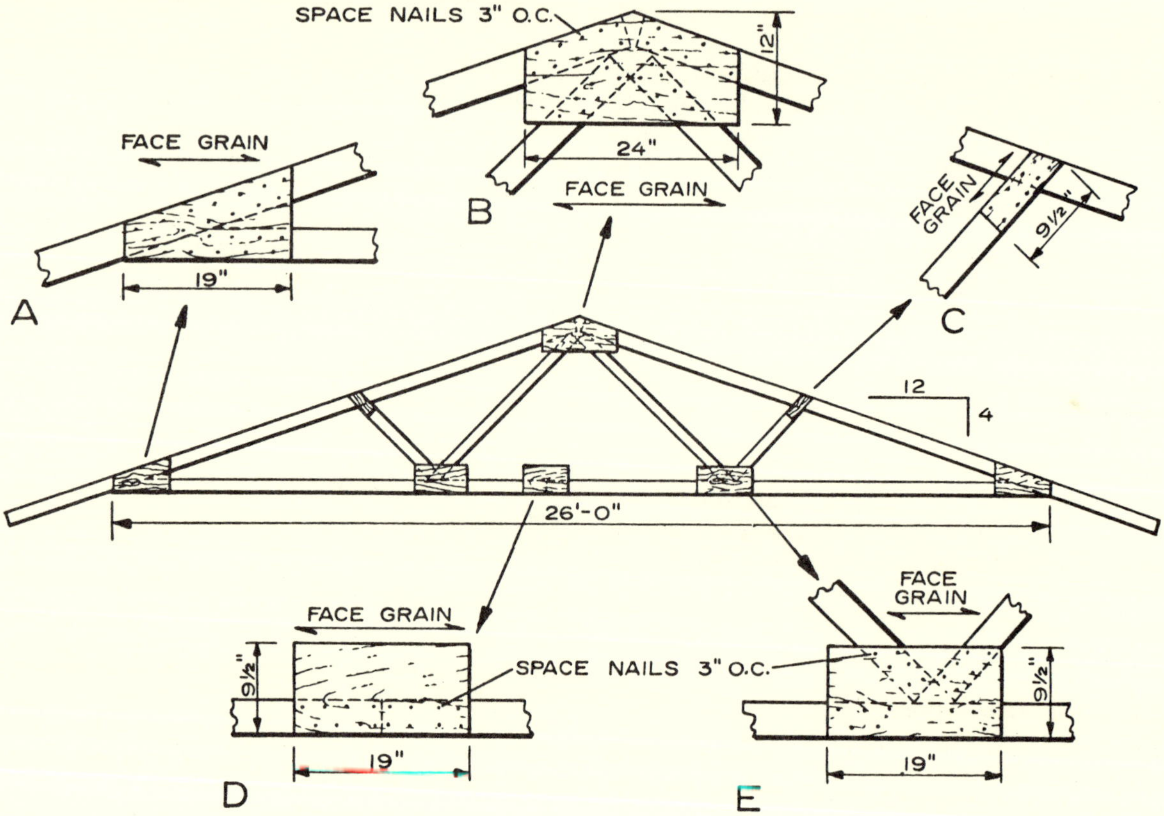

Figure 53.—Construction of a 26-foot W truss: *A*, Bevel-heel gusset; *B*, peak gusset; *C*, upper chord intermediate gusset; *D*, splice of lower chord; *E*, lower chord intermediate gusset.

CHAPTER 8

WALL SHEATHING

Wall sheathing (*7*) is the outside covering used over the wall framework of studs, plates, and window and door headers. It forms a flat base upon which the exterior finish can be applied. Certain types of sheathing and methods of application can provide great rigidity to the house, eliminating the need for corner bracing. Sheathing serves to minimize air infiltration and, in certain forms, provides some insulation.

Some sheet materials serve both as sheathing and siding. Sheathing is sometimes eliminated from houses in the mild climates of the South and West. It is a versatile material and manufacturers produce it in many forms. Perhaps the most common types used in construction are: boards, plywood, structural insulating board, and gypsum sheathing.

Types of Sheathing

Wood Sheathing

Wood sheathing is usually of nominal 1-inch boards in a shiplap, a tongued-and-grooved, or a square-edge pattern. Resawn $1\frac{1}{16}$-inch boards are also allowed under certain conditions. The requirements for wood sheathing are easy working, easy nailing, and moderate shrinkage. Widths commonly used are 6, 8, and 10 inches. It may be applied horizontally or diagonally (fig. 54,*A*). Sheathing is sometimes carried only to the subfloor (fig. 54,*B*), but when diagonal sheathing or sheet materials are placed as shown in figure 54,*C*, greater strength and rigidity result. It is desirable to limit the wood moisture content to 15 percent to minimize openings between matched boards when shrinkage occurs.

Some manufacturers produce random-length side and end-matched boards for sheathing. Most softwood species, such as the spruces, Douglas-fir, southern pine, hemlock, the soft pines, and others, are suitable for sheathing. Grades vary between species, but sheathing is commonly used in the third grade (*5*).

Refer to the chapter on "Floor Framing."

Plywood Sheathing

Plywood is used extensively for sheathing of walls, applied vertically, normally in 4- by 8-foot and longer sheets (fig. 55). This method of sheathing eliminates the need for diagonal corner bracing; but, as with all sheathing materials, it should be well nailed (table 2).

Standard sheathing grade (*18*) is commonly used for sheathing. For more severe exposures, this same plywood is furnished with an exterior glueline. While the minimum plywood thickness for 16-inch stud spacing is $\frac{5}{16}$ inch, it is often desirable to use $\frac{3}{8}$ inch and thicker, especially when the exterior finish must be nailed to the sheathing. The selection of plywood thickness is also influenced somewhat by standard jamb widths in window and exterior door frames. This may occasionally require sheathing of $\frac{1}{2}$-inch or greater thicknesses. Some modification of jambs is required and readily accomplished when other plywood thicknesses are used.

Structural Insulating Board Sheathing

The three common types of insulating board (structural fiberboards) used for sheathing include *regular density*, *intermediate density*, and *nail-base*. Insulating board sheathings are coated or impregnated with asphalt or given other treatment to provide a water-resistant product. Occasional wetting and drying that occur during construction will not damage the sheathing materially.

Regular-density sheathing is manufactured in $\frac{1}{2}$- and $\frac{25}{32}$-inch thicknesses and in 2- by 8-, 4- by 8-, and 4- by 9-foot sizes. Intermediate-density and nail-base sheathing are denser products than regular-density. They are regularly manufactured only in $\frac{1}{2}$-inch thickness and in 4- by 8- and 4- by 9-foot sizes. While 2- by 8-foot sheets with matched edges are used horizontally, 4- by 8-foot and longer sheets are usually installed with the long dimension vertical.

Corner bracing is required on horizontally applied sheets and usually on applications of $\frac{1}{2}$-inch regular-density sheathing applied vertically. Additional corner bracing is usually not required for regular-density insulating board sheathing $\frac{25}{32}$ inch thick or for intermediate-density and nail-base sheathing when properly applied (fig. 55) with long edges vertical. Naturally fastenings must be adequate around the perimeter and at intermediate studs, and adequately fastened (nails, staples, or other fastening system). Nail-base sheathing also permits shingles to be applied directly to it as siding if fastened with special annular-grooved nails. Galvanized or other corrosion-resistant fasteners are recommended for installation of insulating-board sheathing.

Gypsum Sheathing

Gypsum sheathing is $\frac{1}{2}$ inch thick, 2 by 8 feet in size, and is applied horizontally for stud spacing of 24 inches or less (fig. 56). It is composed of treated gypsum filler faced on two sides with water-resistant paper, often having one edge grooved, and the other with a matched *V* edge. This makes application easier, adds a small amount of tie between sheets, and provides some resistance to air and moisture penetration.

Corner Bracing

The purpose of corner bracing is to provide rigidity to the structure and to resist the racking forces of wind. Corner bracing should be used at all external corners of houses where the type of sheathing used does not provide the bracing required (figs. 31 and 54,*A*). Types of sheathing that provide adequate bracing are: (a) Wood sheathing, when applied diagonally; (b) plywood, when applied vertically in sheets 4 feet wide by 8 or more feet high and where attached with nails or staples spaced not more than 6 inches apart on all edges and not more than 12 inches at intermediate supports; and (c) structural insulating board sheathing 4 feet wide and 8 feet or longer (25/32-inch-thick regular grade and 1/2-inch-thick intermediate-density or nail-base grade) applied with long edges vertical with nails or staples spaced 3 inches along all edges and 6 inches at intermediate studs.

Another method of providing the required rigidity and strength for wall framing consists of a 1/2-inch plywood panel at each side of each outside corner and 1/2-inch regular-density fiberboard at intermediate areas. The plywood must be in 4-foot-wide sheets and applied vertically with full perimeter and intermediate stud nailing.

Where corner bracing is required, use 1- by 4-inch or wider members let into the outside face of the studs, and set at an angle of 45° from the bottom of the soleplate to the top of the wallplate or corner stud. Where window openings near the corner interfere with 45° braces, the angle should be increased but the full-length brace should cover at least three stud spaces. Tests conducted at the Forest Products Laboratory showed a full-length brace to be much more effective than a K-brace, even though the angle was greater than that of a 45° K-brace.

Installation of Sheathing

Wood Sheathing

The minimum thickness of wood sheathing is generally 3/4 inch. However, for particular uses, depending on exterior coverings, resawn boards of 11/16-inch thickness may be used as sheathing. Widths commonly used are 6, 8, and 10 inches. The 6- and 8-inch widths will have less shrinkage than greater widths, so that smaller openings will occur between boards.

The boards should be nailed at each stud crossing with two nails for the 6- and 8-inch widths and three nails for the 10- and 12-inch widths. When diagonal sheathing is used, one more nail can be used at each stud; for example, three nails for 8-inch sheathing. Joints should be placed over the center of studs (fig. 54,*A*) unless end-matched (tongued-and-grooved) boards are used. End-matched tongued-and-grooved boards are applied continuously, either horizontally or diagonally, allowing end joints to fall where they may, even if between studs (fig. 54,*A*). However, when end-matched boards are used, no two adjoining boards should have end joints over the same stud space and each board should bear on at least two studs.

Two arrangements of floor framing and soleplate location may be used which affect wall sheathing application. The first method has the soleplate set in from the outside wall line so that the sheathing is flush with the floor framing (fig. 54,*B*). This does not provide a positive tie between wall and floor framing and in high wind areas should be supplemented with metal strapping (fig. 32) placed over the sheathing. The second method has the sill plate located the thickness of the sheathing in from the edge of the foundation wall (fig. 54,C). When vertically applied plywood or diagonal wood sheathing is used, a good connection between the wall and floor framing is obtained. This method is usually preferred where good wall-to-floor-to-foundation connections are desirable.

Wood sheathing (fig. 54,*A*) is commonly applied horizontally because it is easy to apply and there is less lumber waste than with diagonal sheathing. Horizontal sheathing, however, requires diagonal corner bracing for wall framework.

Diagonal sheathing (fig. 54,*A*) should be applied at a 45° angle. This method of sheathing adds greatly to the rigidity of the wall and eliminates the need for corner bracing. There is more lumber waste than with horizontal sheathing because of angle cuts, and application is somewhat more difficult. End joints should be made over studs. This method is often specified in hurricane areas along the Atlantic Coast and in Florida.

Structural Insulating Board, Plywood, and Other Sheathing in 4-foot and Longer Sheets

Vertical application of structural insulating board (fig. 55) in 4- by 8-foot sheets is usually recommended by the manufacturer because perimeter nailing is possible. Depending on local building regulations, spacing nails 3 inches on edges and 6 inches at intermediate framing members usually eliminates the need for corner bracing when 25/32-inch structural insulating board sheathing or 1/2-inch medium-density structural insulating board sheathing is used. Use 1 3/4-inch galvanized roofing nails for the 25/32-inch sheathing and 1 1/2-inch nails for the 1/2-inch sheathing (table 2). The manufacturers usually recommend 1/8-inch spacing between sheets. Joints are centered on framing members.

Plywood used for sheathing should be 4 by 8 feet or longer and applied vertically with perimeter nailing to eliminate the need for corner bracing (fig. 55). Sixpenny nails are used for plywood 3/8 inch or less in thickness. Use eightpenny nails for plywood 1/2 inch and more in thickness. Spacing should be a minimum of 6 inches at all edges and 12 inches at intermediate framing members (table 2).

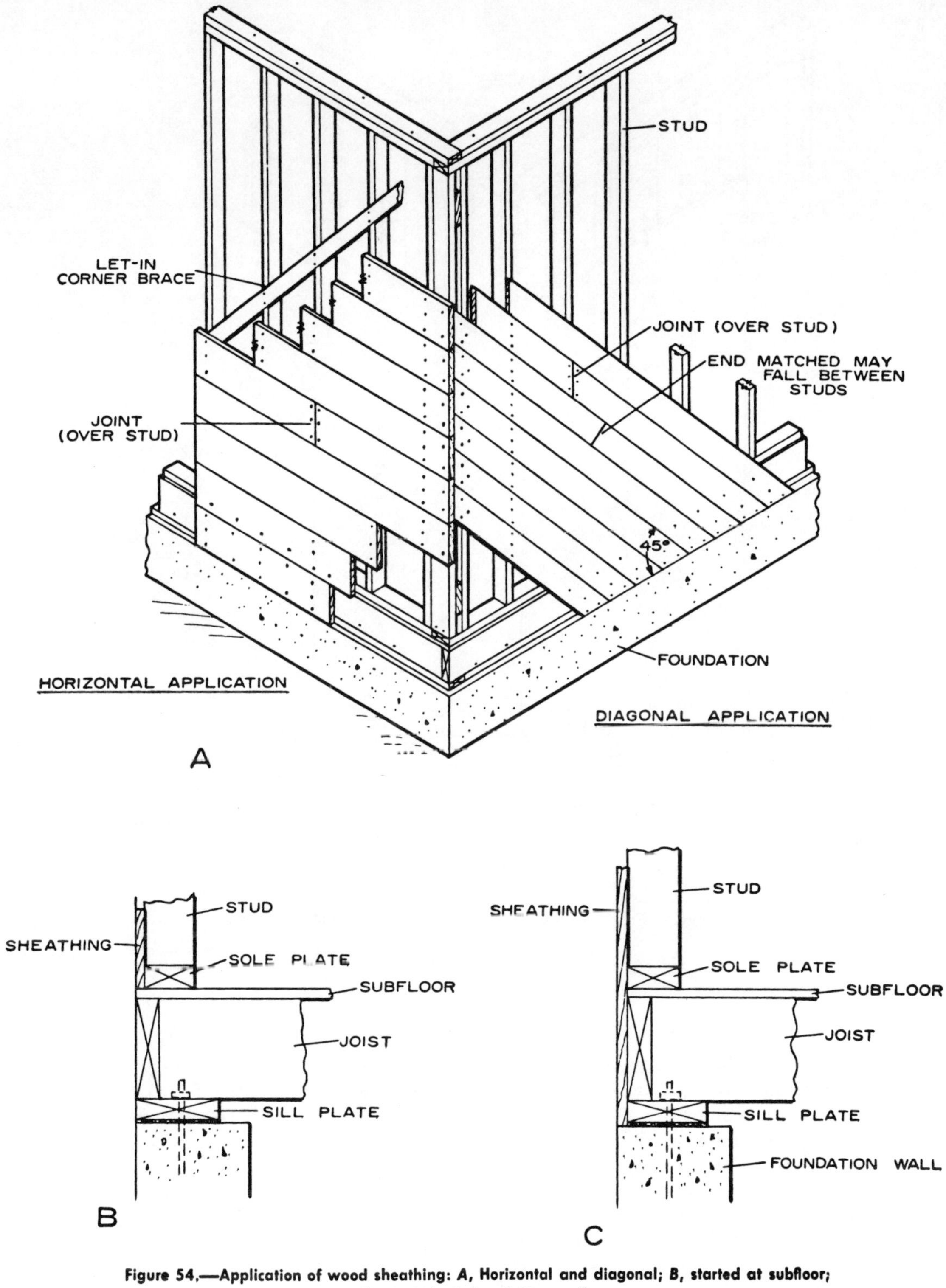

Figure 54.—Application of wood sheathing: *A*, Horizontal and diagonal; *B*, started at subfloor; *C*, started at foundation wall.

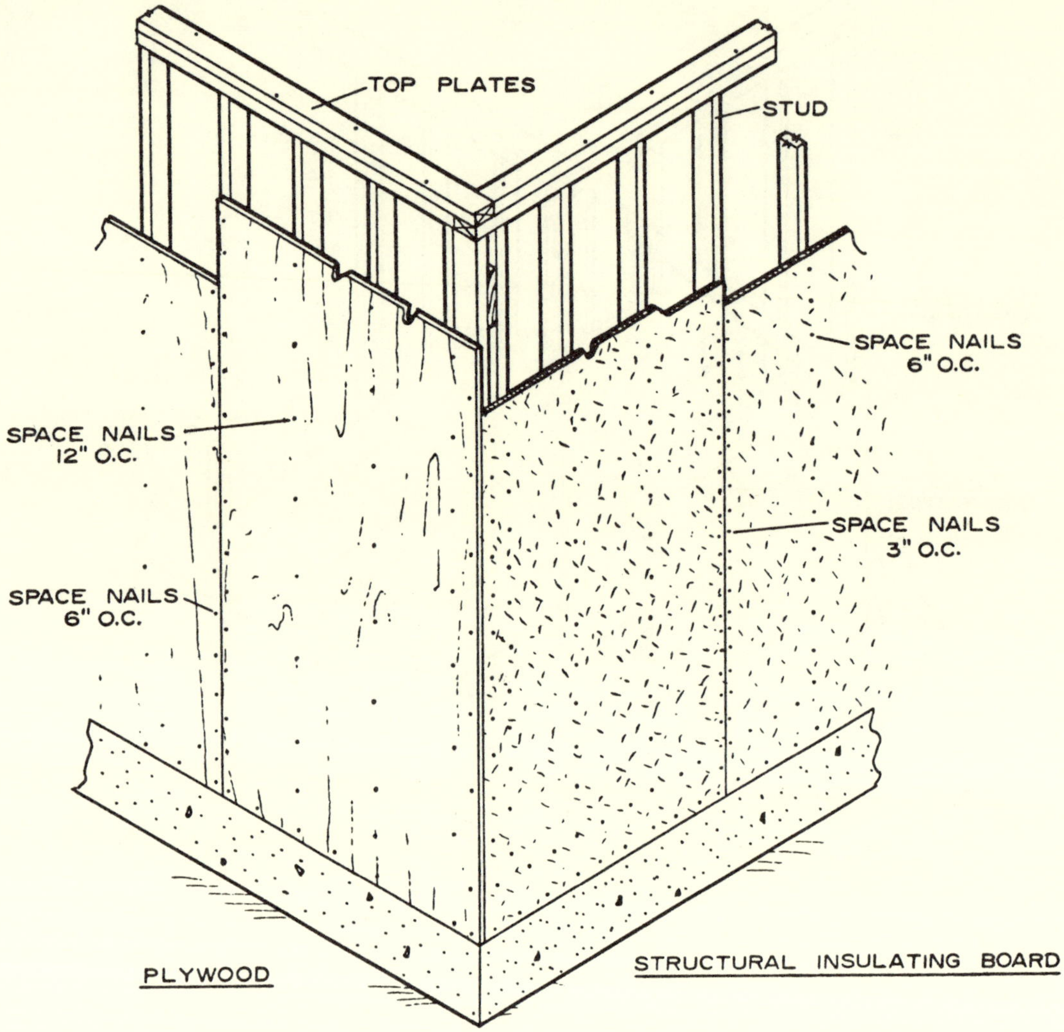

Figure 55.—Vertical application of plywood or structural insulating board sheathing.

Plywood may also be applied horizontally, but not being as efficient from the standpoint of rigidity and strength, it normally requires diagonal bracing. However, blocking between studs to provide for horizontal edge nailing will improve the rigidity and usually eliminate the need for bracing. When shingles or similar exterior finishes are employed, it is necessary to use threaded nails for fastening when plywood is only $5/16$ or $3/8$-inch thick. Allow $1/8$-inch edge spacing and $1/16$-inch end spacing between plywood sheets when installing.

Particleboard, hardboard, and other sheet materials may also be used as a sheathing. However, their use is somewhat restricted because cost is usually substantially higher than the sheet materials previously mentioned.

Insulating Board and Gypsum Sheathing in 2- by 8-foot Sheets

Gypsum and insulating board sheathing in 2- by 8-foot sheets applied horizontally require corner bracing (fig. 56). Vertical joints should be staggered. The $25/32$-inch board should be nailed to each crossing stud with $1\frac{3}{4}$-inch galvanized roofing nails spaced about $4\frac{1}{2}$ inches apart (six nails in the 2-foot height).

The $1/2$-inch gypsum and insulating board sheathing should be nailed to the framing members with $1\frac{1}{2}$-inch galvanized roofing nails spaced about $3\frac{1}{2}$ inches apart (seven nails in the 2-foot height).

When wood bevel or similar sidings are used over plywood sheathing less than $5/8$ inch thick, and over insulating board and gypsum board, nails must usually be located so as to contact the stud. When wood

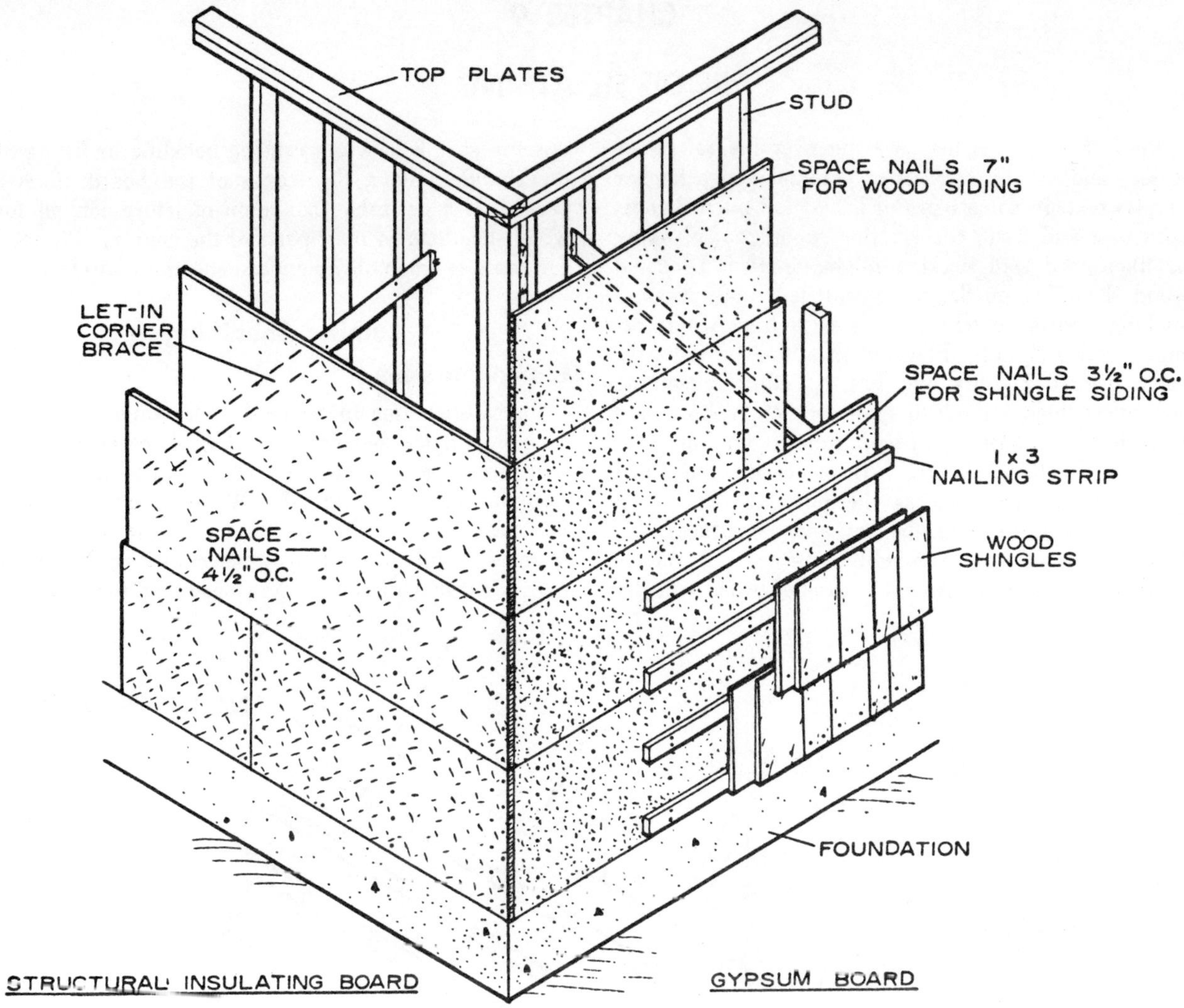

Figure 56.—Horizontal application of 2-by 8-foot structural insulating board or gypsum sheathing.

shingles and similar finishes are used over gypsum and regular density insulating board sheathing, the walls are stripped with 1- by 3-inch horizontal strips spaced to conform to the shingle exposure. The wood strips are nailed to each stud crossing with two eightpenny or tenpenny threaded nails, depending on the sheathing thickness (fig. 56). Nail-base sheathing board usually does *not* require stripping when threaded nails are use.

Sheathing Paper

Sheathing paper should be water-resistant but not vapor-resistant. It is often called "breathing" paper as it allows the movement of water vapor but resists entry of direct moisture. Materials such as 15-pound asphalt felt, rosin, and similar papers are considered satisfactory. Sheathing paper should have a "perm" value of 6.0 or more (*4*),(*17*). It also serves to resist air infiltration.

Sheathing paper should be used behind a stucco or masonry veneer finish and over wood sheathing. It should be installed horizontally starting at the bottom of the wall. Succeeding layers should lap about 4 inches. Ordinarily, it is not used over plywood, fiberboard, or other sheet materials that are water-resistant. However, 8-inch or wider strips of sheathing paper should be used around window and door openings to minimize air infiltration.

CHAPTER 9

ROOF SHEATHING

Roof sheathing is the covering over the rafters or trusses and usually consists of nominal 1-inch lumber or plywood. In some types of flat or low-pitched roofs with post and beam construction, wood roof planking or fiberboard roof decking might be used. Diagonal wood sheathing on flat or low-pitched roofs provides racking resistance where areas with high winds demand added rigidity. Plywood sheathing provides the same desired rigidity and bracing effect. Sheathing should be thick enough to span between supports and provide a solid base for fastening the roofing material.

Roof sheathing boards are generally the third grades of species such as the pines, redwood, the hemlocks, western larch, the firs, and the spruces (5). It is important that thoroughly seasoned material be used with asphalt shingles. Unseasoned wood will dry out and shrink in width causing buckling or lifting of the shingles, along the length of the board. Twelve percent is a desirable maximum moisture content for wood sheathing in most parts of the country. Plywood for roofs is commonly standard sheathing grade.

Lumber Sheathing

Closed Sheathing

Board sheathing to be used under such roofing as asphalt shingles, metal-sheet roofing, or other materials that require continuous support should be laid closed (without spacing) (fig. 57). Wood shingles can also be used over such sheathing. Boards should be matched, shiplapped, or square-edged with joints made over the center of rafters. Not more than two adjacent

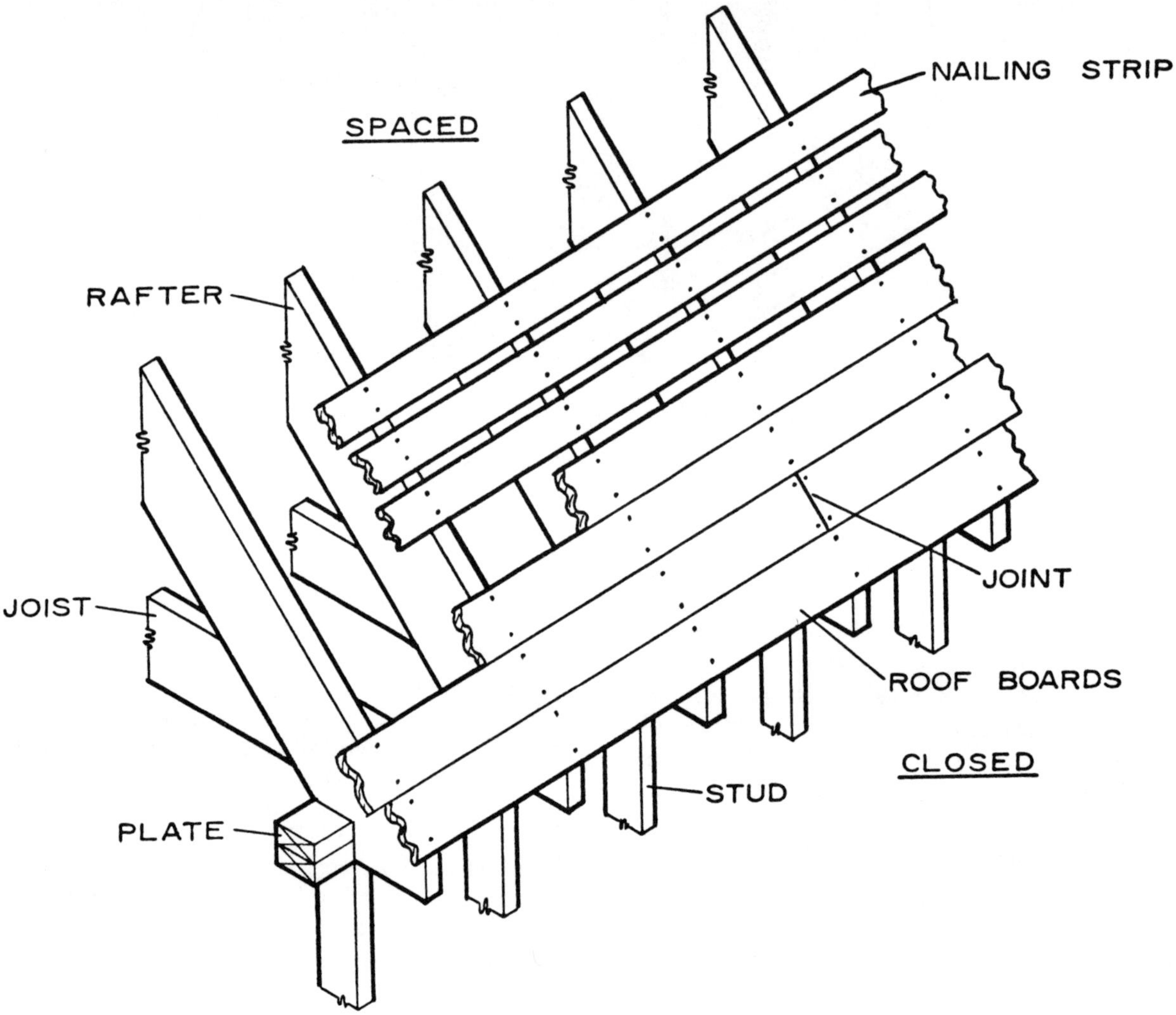

Figure 57.—Installation of board roof sheathing, showing both closed and spaced types.

boards should have joints over the same support. It is preferable to use boards no wider than 6 or 8 inches to minimize problems which can be caused by shrinkage. Boards should have a minimum thickness of ¾ inch for rafter spacing of 16 to 24 inches, and be nailed with two eightpenny common or sevenpenny threaded nails for each board at each bearing. End-matched tongued-and-grooved boards can also be used and joints made between rafters. However, in no case should the joints of adjoining boards be made over the same rafter space. Each board should be supported by at least two rafters.

Use of long sheathing boards at roof ends is desirable to obtain good framing anchorage, especially in gable roofs where there is a substantial rake overhang.

Spaced Sheathing

When wood shingles or shakes are used in damp climates, it is common to have spaced roof boards (fig. 57). Wood nailing strips in nominal 1- by 3- or 1- by 4-inch size are spaced the same distance on centers as the shingles are to be laid to the weather. For example, if shingles are laid 5 inches to the weather and nominal (surfaced) 1- by 4-inch strips are used, there would be spaces of 1⅜ to 1½ inches between each board to provide the needed ventilation spaces.

Plywood Roof Sheathing

When plywood roof sheathing is used, it should be laid with the face grain perpendicular to the rafters (fig. 58). Standard sheathing grade plywood is commonly specified but, where damp conditions occur, it is desirable to use a standard sheathing grade with exterior glueline. End joints are made over the center of the rafters and should be staggered by at least one rafter 16 or 24 inches, or more.

For wood shingles or shakes and for asphalt shingles, 5/16-inch-thick plywood is considered to be a minimum thickness for 16-inch spacing of rafters. When edges are blocked to provide perimeter nailing, ⅜-inch-thick plywood can be used for 24-inch rafter spacing. A system which reduces costs by eliminating the blocking is acceptable in most areas for ⅜-inch plywood when rafters are spaced 24 inches on center. This is with the use of plyclips or similar H clips between rafters instead of blocking.

To provide better penetration for nails used for the shingles, better racking resistance, and a smoother roof appearance, it is often desirable to increase the minimum thicknesses to ⅜ and ½ inch. U.S. Product Standard PS 1–66 (*18*) provides that standard grades be marked for allowable spacing of rafters. For slate and similar heavy roofing materials, ½-inch plywood is considered minimum for 16-inch rafter spacing.

Plywood should be nailed at each bearing, 6 inches on center along all edges and 12 inches on center along intermediate members. A sixpenny common nail or fivepenny threaded nail should be used for 5/16- and ⅜-inch plywood, and eightpenny common or sevenpenny threaded nail for greater thicknesses. Unless plywood has an exterior glueline, raw edges should not be exposed to the weather at the gable end or at the cornice, but should be protected by the trim. Allow a ⅛-inch edge spacing and 1/16-inch end spacing between sheets when installing.

Plank Roof Decking

Plank roof decking, consisting of 2-inch and thicker tongued-and-grooved wood planking, is commonly used in flat or low-pitched roofs in post and beam construction. Common sizes are nominal 2- by 6-, 3- by 6-, and 4- by 6-inch V-grooved members, the thicker planking being suitable for spans up to 10 or 12 feet. Maximum span for 2-inch planking is 8 feet when continuous over two supports, and 6 feet over single spans in grades and species commonly used for this purpose. Special load requirements may reduce these allowable spans. Roof decking can serve both as an interior ceiling finish and as a base for roofing. Heat loss is greatly reduced by adding fiberboard or other rigid insulation over the wood decking.

The decking is blind-nailed through the tongue and also face-nailed at each support. In 4- by 6-inch size, it is predrilled for edge nailing (fig. 43,*B*). For thinner decking, a vapor barrier is ordinarily installed between the top of the plank and the roof insulation when planking does not provide sufficient insulation.

Fiberboard Roof Decking

Fiberboard roof decking is used the same way as wood decking, except that supports are spaced much closer together. Planking is usually supplied in 2- by 8-foot sheets with tongued-and-grooved edges. Thicknesses of the plank and spacing of supports ordinarily comply with the following tabulation:

Minimum thickness (In.)	*Maximum joist spacing (In.)*
1½	24
2	32
3	48

Manufacturers of some types of roof decking recommend the use of 1⅞-inch thickness for 48-inch spacing of supports.

Nails used to fasten the fiberboard to the wood members are corrosion-resistant and spaced not more than 5 inches on center. They should be long enough to penetrate the joist or beam at least 1½ inches. A built-up roof is normally used for flat and low-pitched roofs having wood or fiberboard decking.

Extension of Roof Sheathing at Gable Ends

Method of installing board or plywood roof sheathing at the gable ends of the roof is shown in figure 59. Where the gable ends of the house have little or no

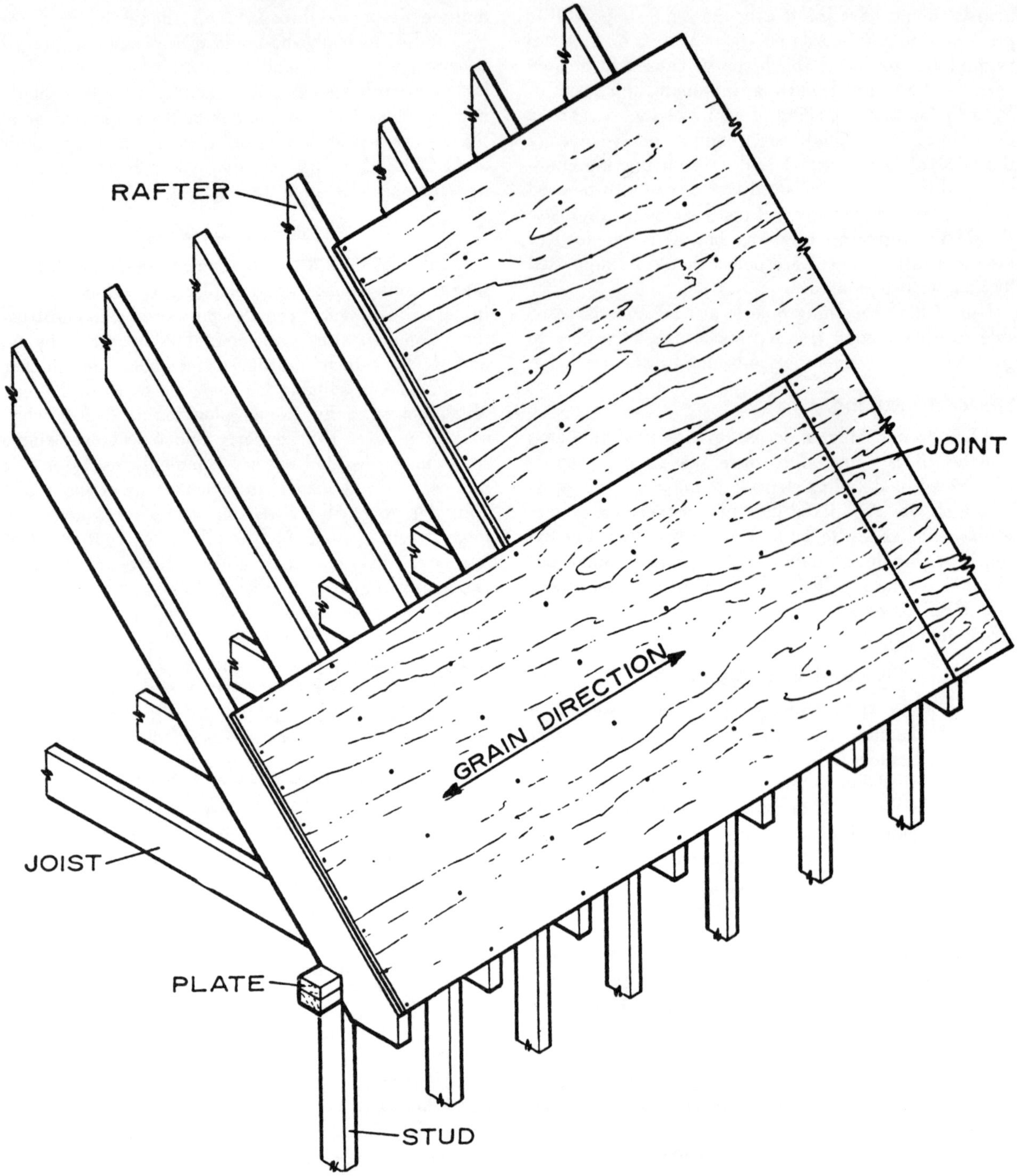

Figure 58.—Application of plywood roof sheathing.

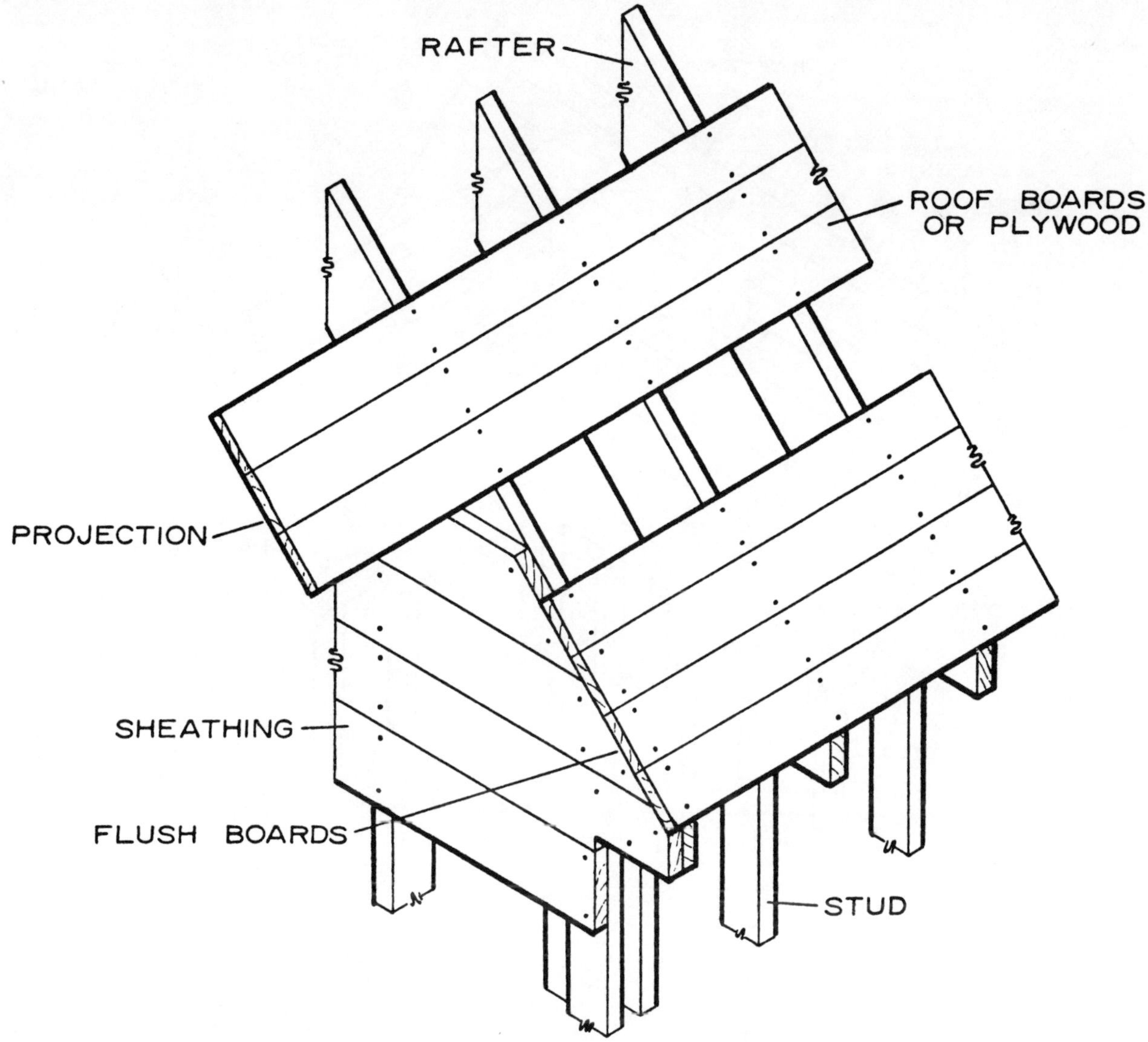

Figure 59.—Board roof sheathing at ends of gable.

extension (rake projection), roof sheathing is placed flush with the outside of the wall sheathing. (See Chapter 10, "Exterior Trim and Millwork.")

Roof sheathing that extends beyond end walls for a projected roof at the gables should span not less than three rafter spaces to insure anchorage to the rafters and to prevent sagging (fig. 59). When the projection is greater than 16 to 20 inches, special ladder framing is used to support the sheathing, as described in Chapter 10, "Exterior Trim and Millwork."

Plywood extension beyond the end wall is usually governed by the rafter spacing to minimize waste. Thus, a 16-inch rake projection is commonly used when rafters are spaced 16 inches on center. Butt joints of the plywood sheets should be alternated so they do not occur on the same rafter.

Sheathing at Chimney Openings

Where chimney openings occur within the roof area, the roof sheathing and subfloor should have a clearance of $\frac{3}{4}$ inch from the finished masonry on all sides (fig. 60, sec. *A–A*). Rafters and headers around the opening should have a clearance of 2 inches from the masonry for fire protection.

Sheathing at Valleys and Hips

Wood or plywood sheathing at the valleys and hips should be installed to provide a tight joint and should be securely nailed to hip and valley rafters (fig. 60). This will provide a solid and smooth base for metal flashing.

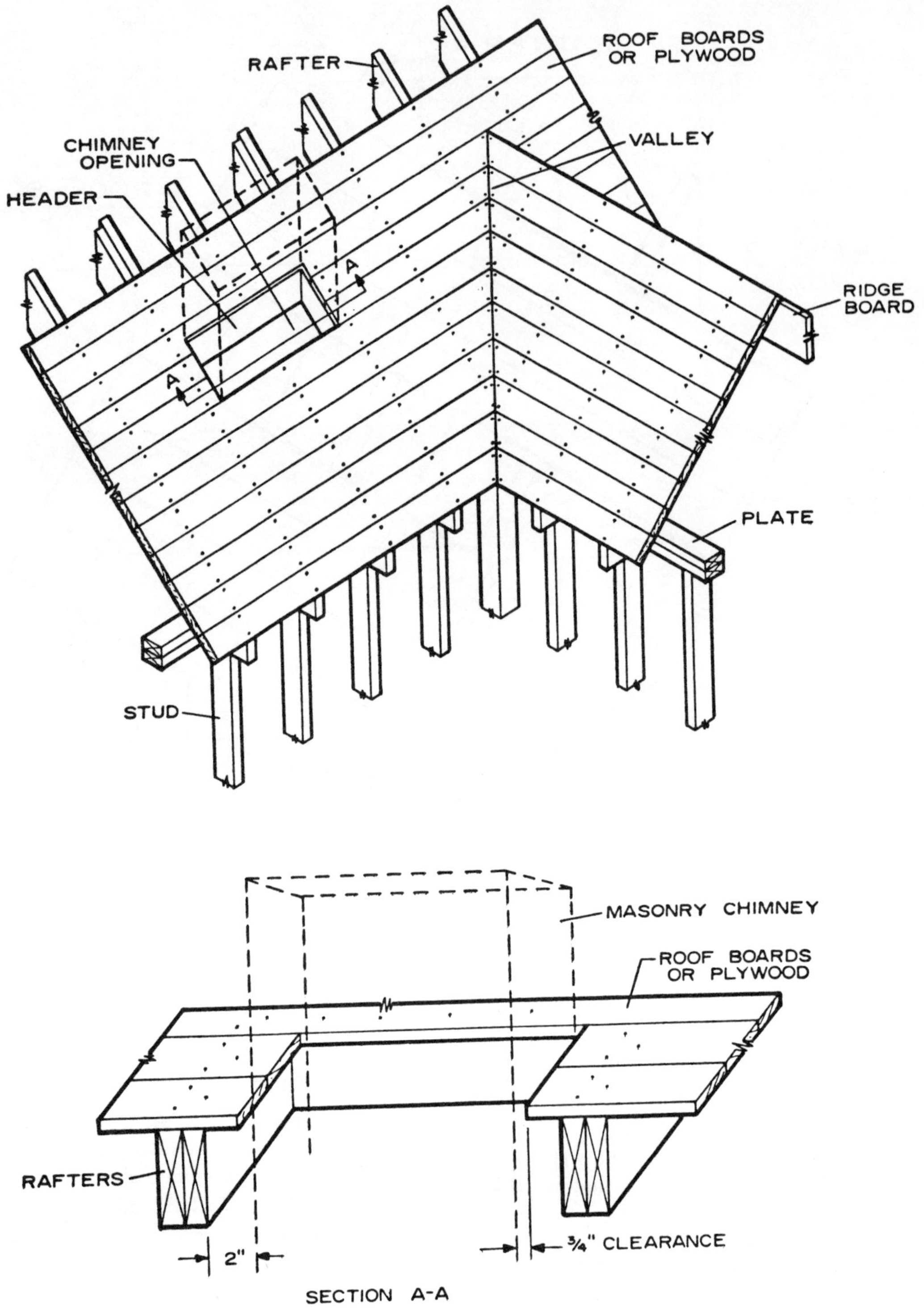

Figure 60.—Board roof sheathing detail at valley and chimney openings. *Section A-A* shows clearance from masonry.

CHAPTER 10

EXTERIOR TRIM AND MILLWORK

Exterior trim is usually considered as being that part of the exterior finish other than the wall covering. It includes such items as window and door trim, cornice moldings, facia boards and soffits, rake or gable-end trim, and porch trim and moldings. Contemporary house designs with simple cornices and moldings will contain little of this material, while traditionally designed houses will have considerably more. Much of the exterior trim, in the form of finish lumber and moldings, is cut and fitted on the job. Other materials or assemblies such as shutters, louvres, railings, and posts are shop-fabricated and arrive on the job ready to be fastened in place.

Material Used for Trim

The properties (5) desired in materials used for trim are good painting and weathering characteristics, easy working qualities, and maximum freedom from warp. Decay resistance is also desirable where materials may absorb moisture in such areas as the caps and the bases of porch columns, rails, and shutters. Heartwood of the cedars, cypress, and redwood has high decay resistance. Less durable species may be treated to make them decay resistant.

Many manufacturers predip at the factory such materials as siding, window sash, window and door frames, and trim using a water-repellent preservative. On-the-job dipping of end joints or miters cut at the building site is recommended when resistance to water entry and increased protection are desired.

Fastenings used for trim, whether nails or screws, should preferably be rust-resistant, i.e., galvanized, stainless steel, aluminum, or cadmium-plated. When a natural finish is used, nails should be stainless steel or aluminum to prevent staining and discoloration. Cement-coated nails are not rust-resistant.

Siding and trim are normally fastened in place with a standard siding nail, which has a small flat head. However, finish or casing nails might also be used for some purposes. If not rust-resistant, they should be set below the surface and puttied after the prime coat of paint has been applied. Most of the trim along the shingle line, such as at gable ends and cornices, is installed before the roof shingles are applied.

Material used for exterior trim should be of the better grade (5). Moisture content should be approximately 12 percent, except in the dry Southwestern States, where it should average about 9 percent.

Cornice Construction

The *cornice* of a building is the projection of the roof at the eave line that forms a connection between the roof and sidewalls. In gable roofs it is formed on each side of the house, and in hip roofs it is continuous around the perimeter. In flat or low-pitched roof designs, it is usually formed by the extension of the ceiling joists which also serve as rafters.

The three general cornice types might be considered to be the *box*, the *close* (no projection), and the *open*. The box cornice is perhaps the most commonly used in house design and not only presents a finished appearance, but also aids in protecting the sidewalls from rain. The close cornice with little overhang does not provide as much protection. The open cornice may be used in conjunction with exposed laminated or solid beams with wood roof decking and wide overhangs in contemporary or rustic designs or to provide protection to side walls at a reasonable cost in low-cost houses.

Narrow Box Cornice

The narrow box cornice is one in which the projection of the rafter serves as a nailing surface for the soffit board as well as the facia trim (fig. 61). Depending on the roof slope and the size of the rafters, this extension may vary between 6 and 12 or more inches. The soffit provides a desirable area for inlet ventilators. (See Chapter 16, "Ventilation.")

A frieze board or a simple molding is often used to terminate the siding at the top of the wall. Some builders slope the soffit slightly outward, leaving a $\frac{1}{4}$-inch open space behind the facia for drainage of water that might enter because of snow and ice dams on the overhang. However, good attic ventilation and proper cornice ventilators, in addition to good insulation, will minimize ice dams under normal conditions.

Wide Box Cornice (With Lookouts)

A wide box cornice normally requires additional members for fastening the soffit. This is often supplied by lookout members which can be toenailed to the wall and face-nailed to the ends of the rafter extensions (fig. 62). Soffit material is often lumber, plywood, paper-overlaid plywood, hardboard, medium-density fiberboard, or other sheet materials. Thicknesses should be based on the distance between supports, but $\frac{3}{8}$-inch plywood and $\frac{1}{2}$-inch fiberboard are often used for 16-inch rafter spacing. A nailing header at the ends of the joists will provide a nailing area for soffit and facia trim. The nailing header is sometimes eliminated in moderate cornice extensions when a rabbeted facia is used. Inlet ventilators, often narrow

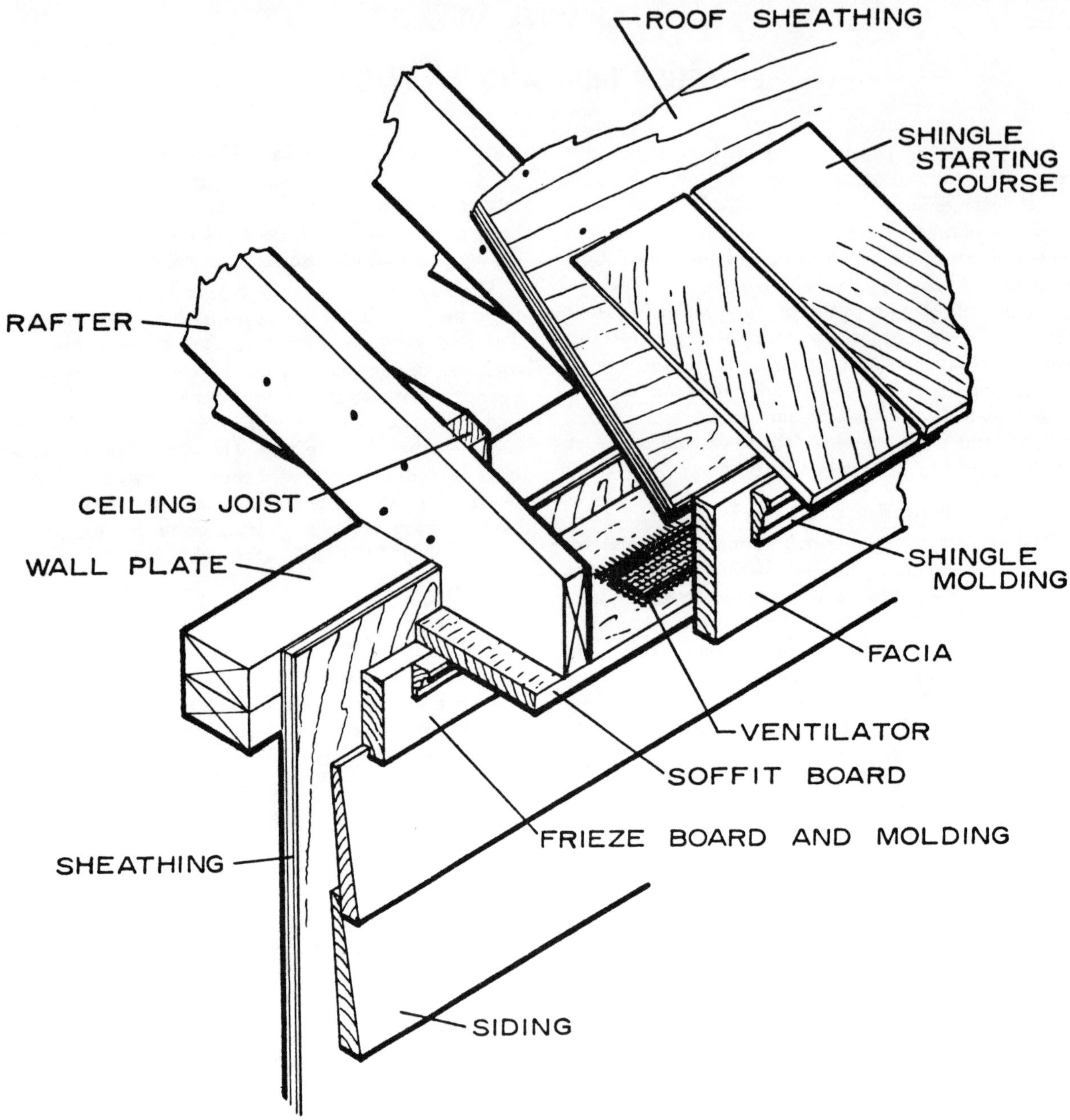

Figure 61.—Narrow box cornice.

continuous slots, can be installed in the soffit areas. This type cornice is often used in a hip-roofed house.

The projection of the cornice beyond the wall should not be so great as to prevent the use of a narrow frieze or a frieze molding above the top casing of the windows. A combination of a steeper slope and wide projection will bring the soffit in this type of cornice too low, and a box cornice, without the lookouts, should be used.

Boxed Cornice Without Lookouts

A wide boxed cornice without lookouts provides a sloped soffit and is sometimes used for houses with wide overhangs (fig. 63). The soffit material is nailed directly to the underside of the rafter extensions. In gable houses, this sloping soffit extends around the roof extension at each gable end. Except for elimination of the lookout members, this type of cornice is much the same as the wide box cornice previously described. Inlet ventilators, singly or a continuous screened slot, are installed in the soffit area.

Open Cornice

The open cornice is much the same structurally as the wide box cornice without lookouts (fig. 63), except that the soffit is eliminated. It might be used on post

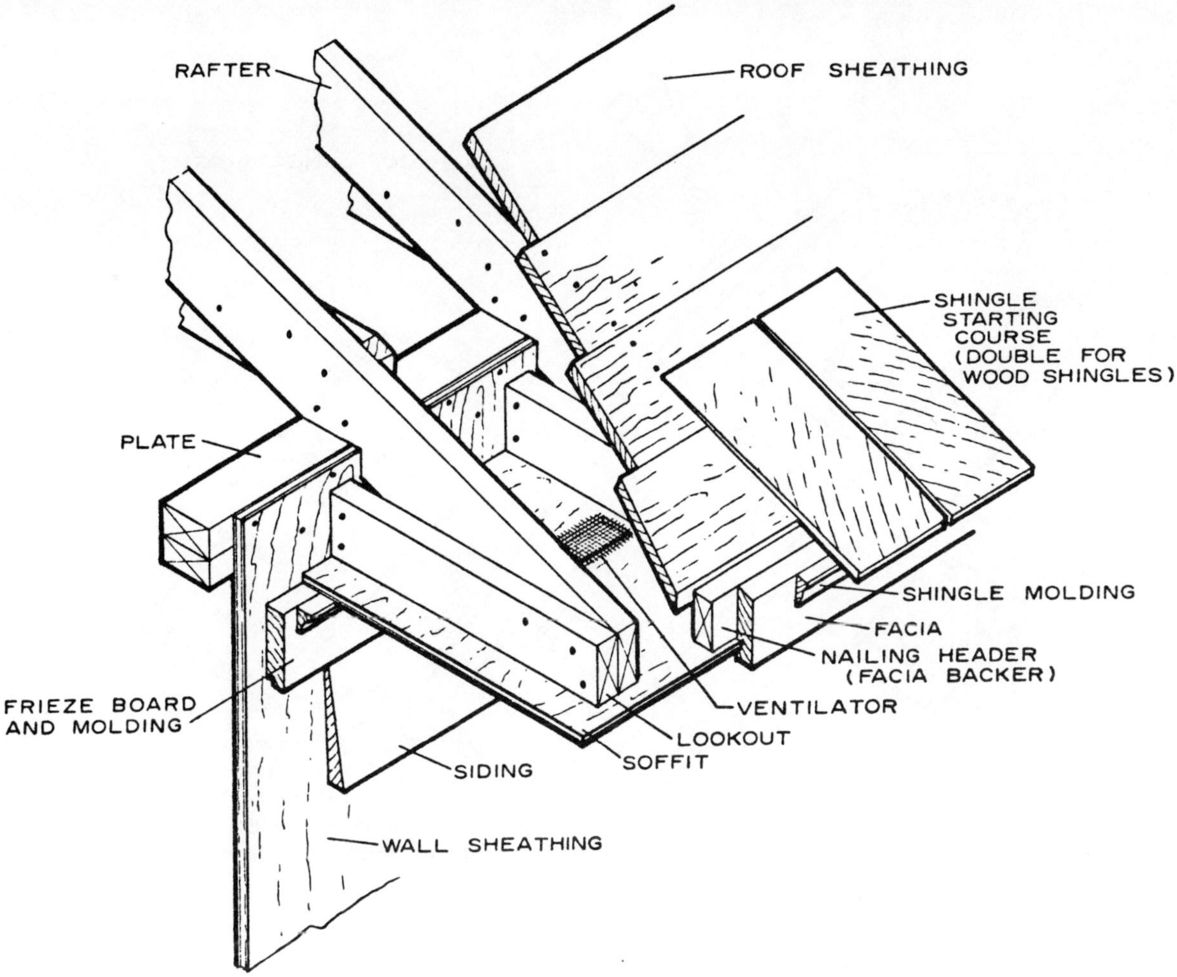

Figure 62.—Wide box cornice (with horizontal lookouts)

and beam construction in which spaced rafters extend beyond the wall line. In widely spaced rafters, the roof sheathing may consist of wood decking, the underside of which would be visible. When rafters are more closely spaced, paper-overlaid plywood or V-grooved boards might be used for roof sheathing at the overhanging section. This type of cornice might also be used for conventionally framed low-cost houses, utility buildings, or cottages, with or without a facia board.

Close Cornice

A close cornice is one in which there is no rafter projection beyond the wall (fig. 64). Sheathing is often carried to the ends of the rafters and ceiling joists. The roof is terminated only by a frieze board and shingle molding. While this cornice is simple to build, it is not too pleasing in appearance and does not provide much weather protection to the sidewalls or a convenient area for inlet ventilators. Appearance can be improved somewhat by the use of a formed wood gutter.

Rake or Gable-end Finish

The *rake section* is the extension of a gable roof beyond the end wall of the house. This detail might be classed as being (a) a close rake with little projection or (b) a boxed or open extension of the gable roof, varying from 6 inches to 2 feet or more. Sufficient projection of the roof at the gable is desirable to provide some protection to the sidewalls. This usually results in longer paint life.

When the rake extension is only 6 to 8 inches, the facia and soffit can be nailed to a series of short lookout blocks (fig. 65,*A*). In addition, the facia is further secured by nailing through the projecting roof sheath-

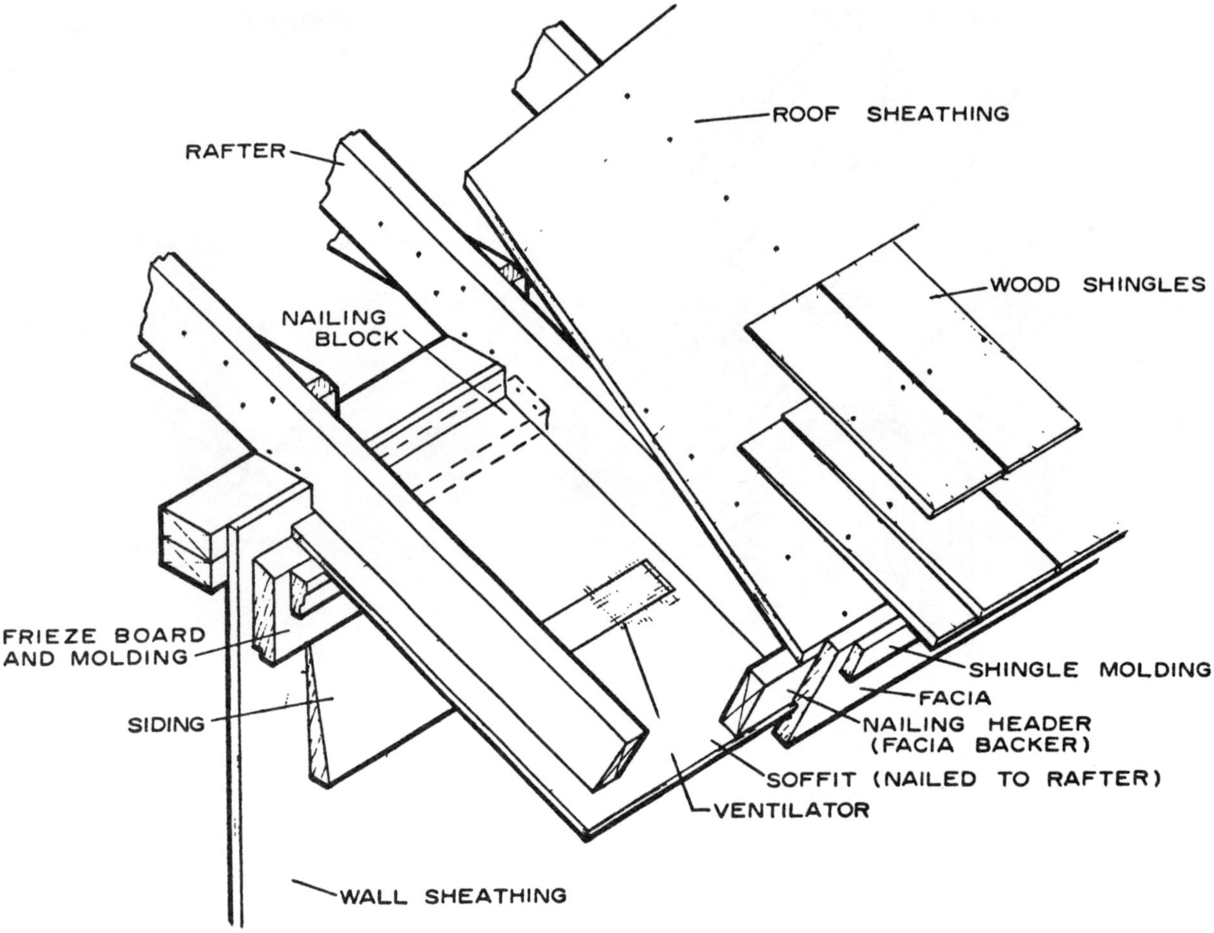

Figure 63.—Wide box cornice (without lookouts)

ing. A frieze board and appropriate moldings complete the construction.

In a moderate overhang of up to 20 inches, both the extending sheathing and a *fly rafter* aid in supporting the rake section (fig. 65,*B*). The fly rafter extends from the ridge board to the nailing header which connects the ends of the rafters. The roof sheathing boards or the plywood should extend from inner rafters to the end of the gable projection to provide rigidity and strength.

The roof sheathing is nailed to the fly rafter and to the lookout blocks which aid in supporting the rake section and also serve as a nailing area for the soffit. Additional nailing blocks against the sheathing are sometimes required for thinner soffit materials.

Wide gable extensions (2 feet or more) require rigid framing to resist roof loads and prevent deflection of the rake section. This is usually accomplished by a series of *purlins* or lookout members nailed to a fly rafter at the outside edge and supported by the end wall and a doubled interior rafter (fig. 66,*A* and *B*). This framing is often called a "ladder" and may be constructed in place or on the ground or other convenient area and hoisted in place.

When ladder framing is preassembled, it is usually made up with a header rafter on the inside and a fly rafter on the outside. Each is nailed to the ends of the lookouts which bear on the gable end wall. When the header is the same size as the rafter, be sure to provide a notch for the wall plates the same as for the regular rafters. In moderate width overhangs, nailing the header and fly rafter to the lookouts with supplemental toenailing is usually sufficiently strong to eliminate the need for the metal hangers shown in figure 66,*B*. The header rafters can be face-nailed directly to the end rafters with twelvepenny nails spaced 16 to 20 inches apart.

Other details of soffit, facia, frieze board, and moldings can be similar to those used for a wide gable overhang. Lookouts should be spaced 16 to 24 inches apart, depending on the thickness of the soffit material.

A close rake has no extension beyond the end wall

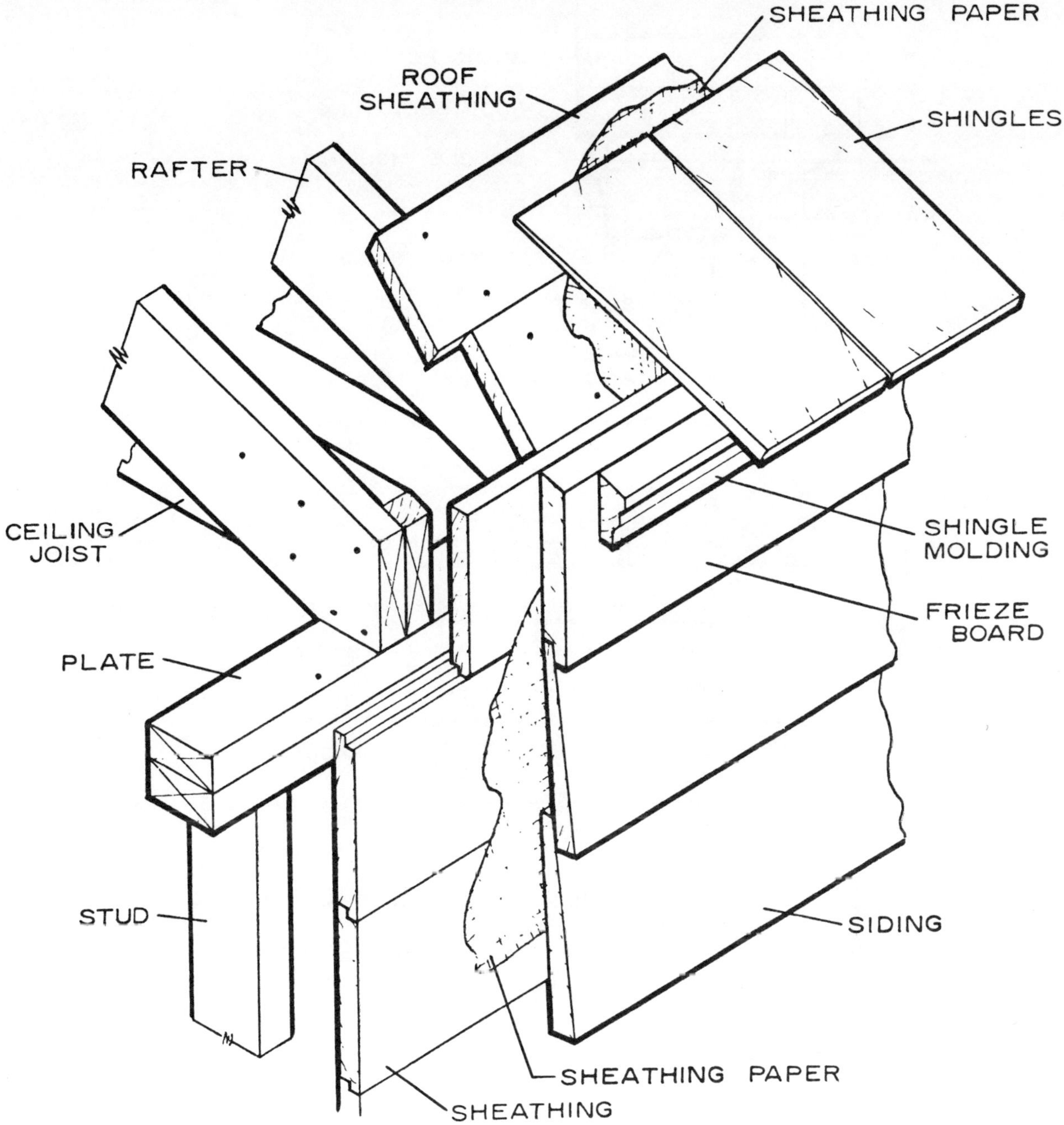

Figure 64.—Close cornice.

other than the frieze board and moldings. Some additional protection and overhang can be provided by using a 2- by 3- or 2- by 4-inch facia block over the sheathing (fig. 66,*C*). This member acts as a frieze board, as the siding can be butted against it. The facia, often 1 by 6 inches, serves as a trim member. Metal roof edging is often used along the rake section as flashing.

Cornice Return

The *cornice return* is the end finish of the cornice on a gable roof. In hip roofs and flat roofs, the cornice is usually continuous around the entire house. In a gable house, however, it must be terminated or joined with the gable ends. The type of detail selected depends to a great extent on the type of cornice and the projection of the gable roof beyond the end wall.

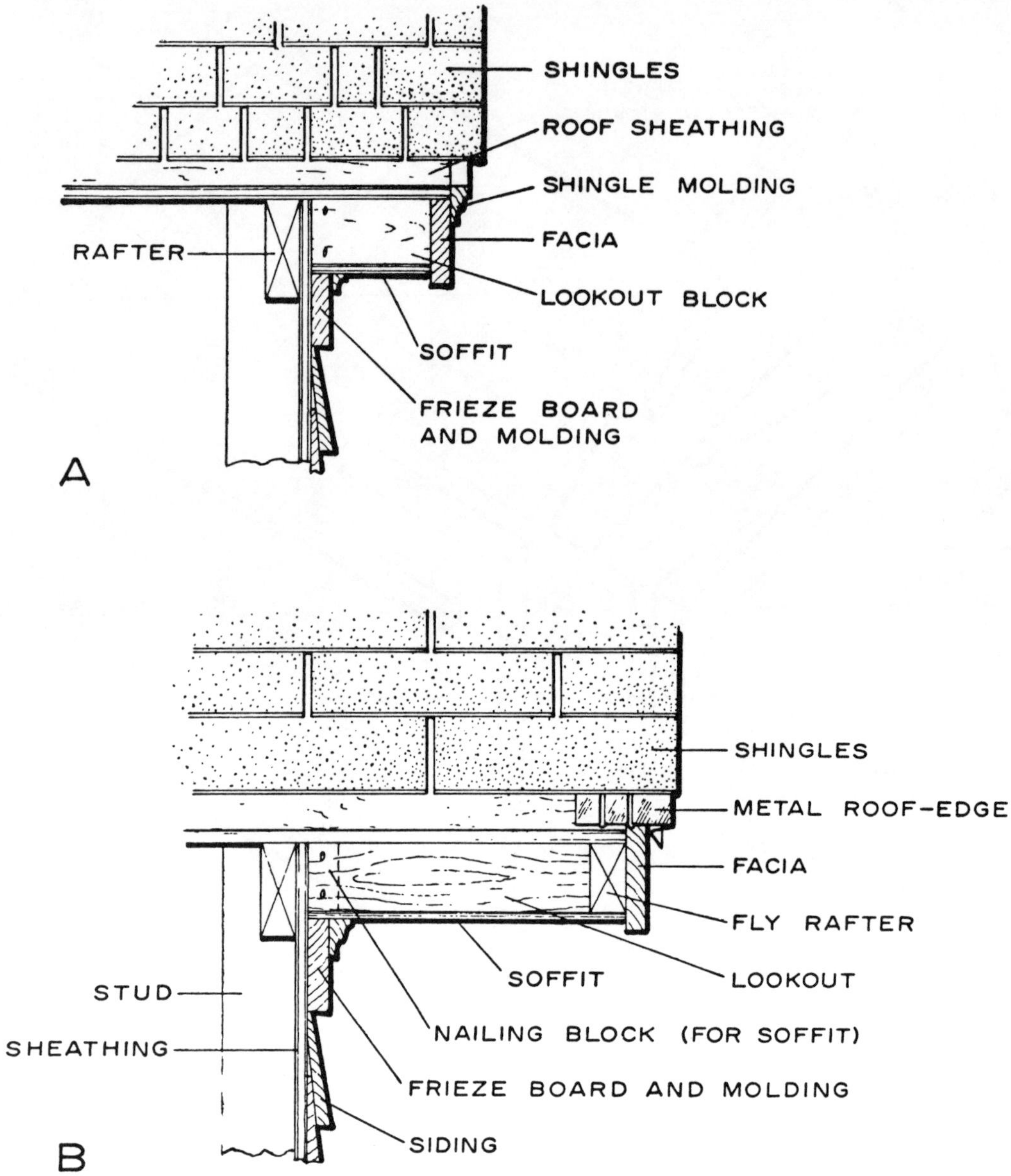

Figure 65.—Normal gable-end extensions: *A*, Narrow overhang; *B*, moderate overhang.

A narrow box cornice often used in houses with Cape Cod or colonial details has a boxed return when the rake section has some projection (fig. 67,*A*). The facia board and shingle molding of the cornice are carried around the corner of the rake projection.

When a wide box cornice has no horizontal lookout members (fig. 63), the soffit of the gable-end overhang is at the same slope and coincides with the cornice soffit (fig. 67,*B*). This is a simple system and is often used when there are wide overhangs at both sides and ends of the house.

A close rake (a gable end with little projection) may be used with a narrow box cornice or a close cornice. In this type, the frieze board of the gable end, into which the siding butts, joints the frieze board or facia of the cornice (fig. 67,*C*).

While close rakes and cornices with little overhang are lower in cost, the extra material and labor required for good gable and cornice overhangs are usually justified. Better sidewall protection and lower paint maintenance costs are only two of the benefits derived from good roof extensions.

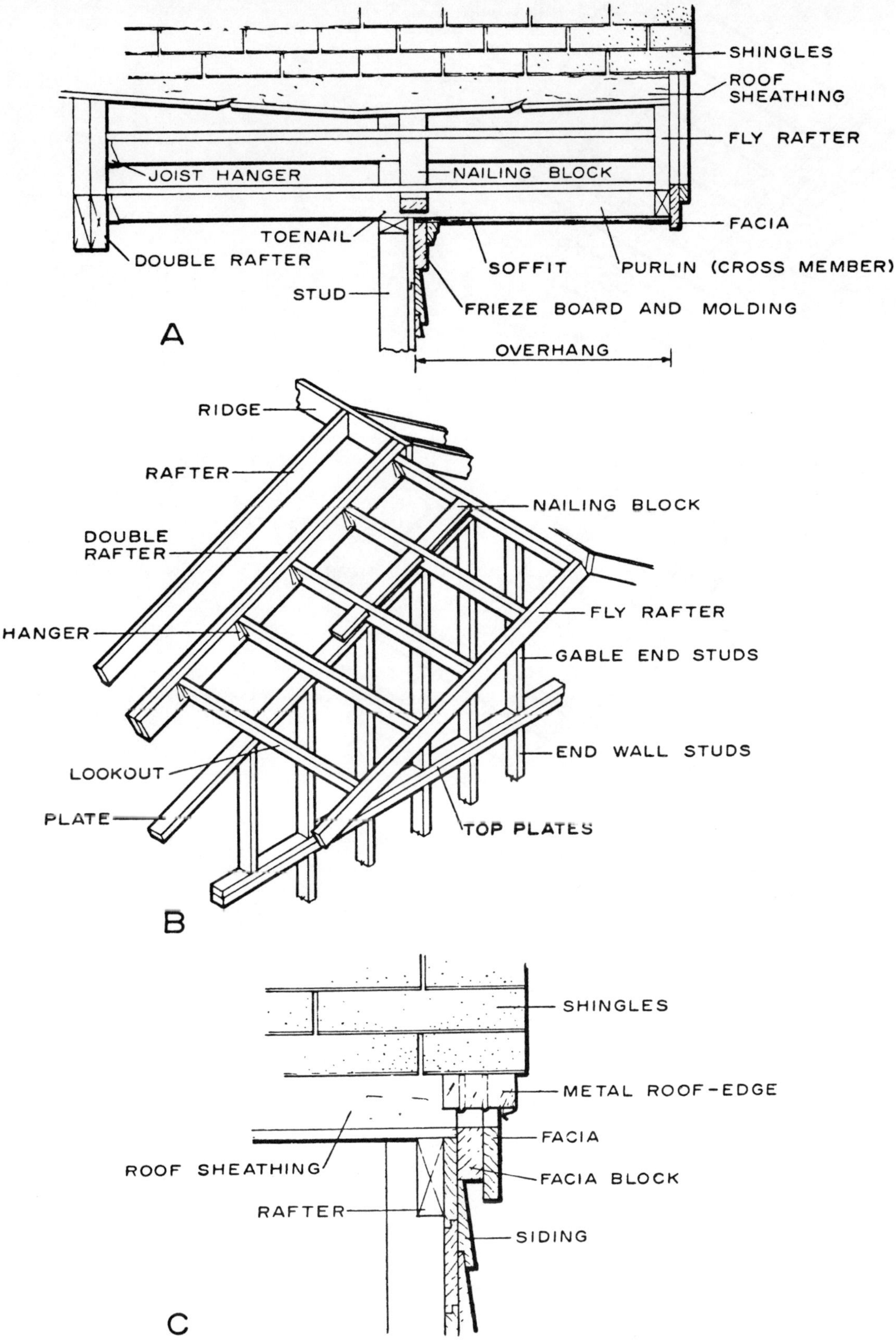

Figure 66.—Special gable-end extensions: *A*, Extra wide overhang; *B*, ladder framing for wide overhang; C, close rake.

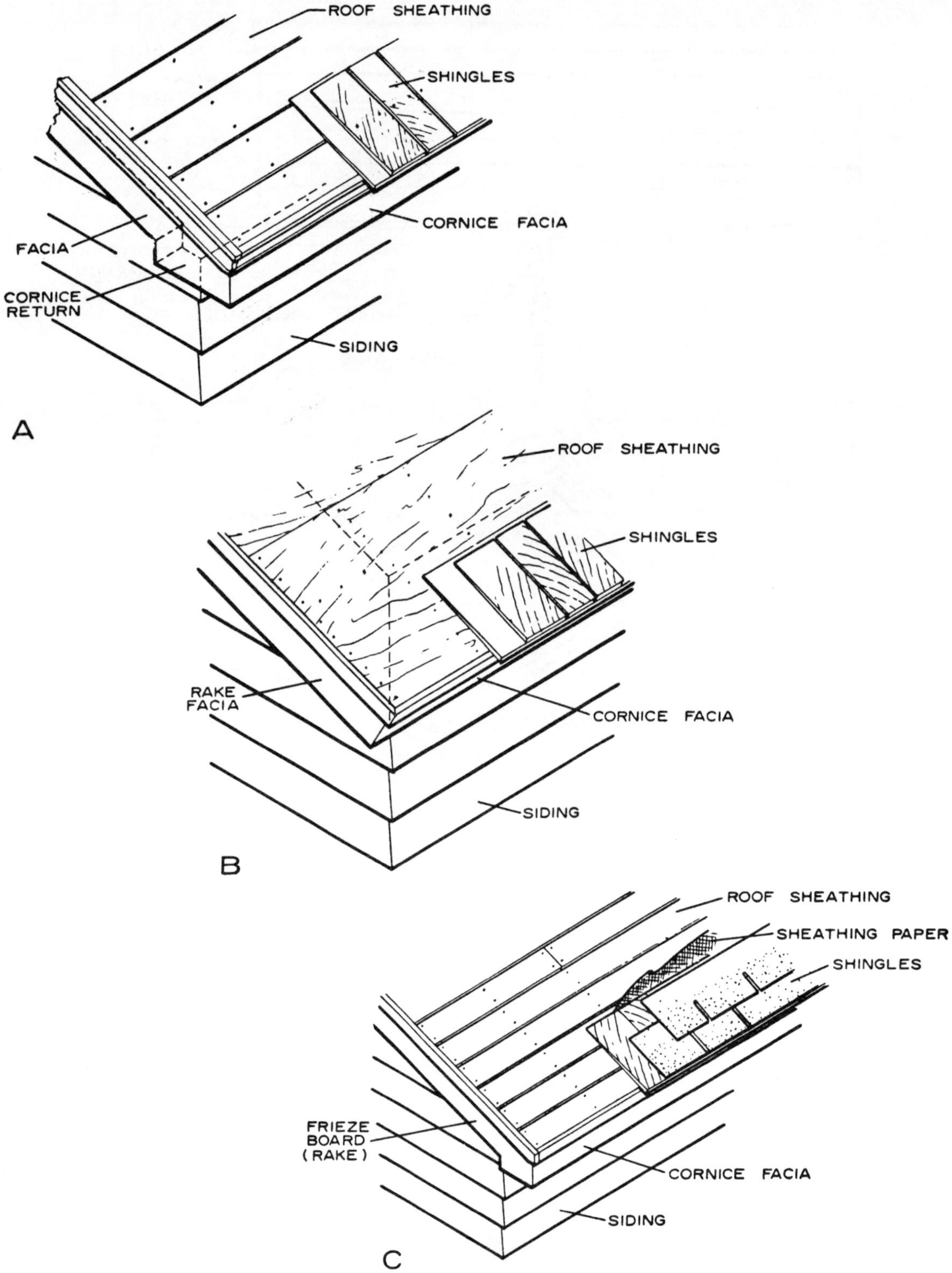

Figure 67.—Cornice returns: *A*, Narrow cornice with boxed return; *B*, wide overhang at cornice and rake; *C*, narrow box cornice and close rake.

CHAPTER 11

ROOF COVERINGS

Roof coverings should provide a long-lived waterproof finish that will protect the building and its contents from rain, snow, and wind. Many materials have withstood the test of time and have proved satisfactory under given service conditions.

Materials

Materials used for pitched roofs are wood, asphalt, and asbestos shingles, and also tile and slate. Sheet materials such as roll roofing, galvanized iron, aluminum, copper, and tin are also used. Perhaps the most common covering for flat or low-pitched roofs is the built-up roof with a gravel topping or cap sheet. Plastic films, often backed with an asbestos sheet, are also being applied on low-slope roofs. While these materials are relatively new, it is likely that their use will increase, especially for roofs with unusual shapes. However, the choice of roofing materials is usually influenced by first cost, local code requirements, house design, or preferences based on past experience.

In shingle application, the exposure distance is important and the amount of exposure generally depends on the roof slope and the type of material used. This may vary from a 5-inch exposure for standard size asphalt and wood shingles on a moderately steep slope to about 3½ inches for flatter slopes. However, even flatter slopes can be used for asphalt shingles with double underlay and triple shingle coverage. Built-up construction is used mainly for flat or low-pitched roofs but can be adapted to steeper slopes by the use of special materials and methods.

Roof underlay material usually consists of 15- or 30-pound asphalt-saturated felt and should be used in moderate and lower slope roofs covered with asphalt, asbestos, or slate shingles, or tile roofing. It is not commonly used for wood shingles or shakes. In areas where moderate to severe snowfalls occur, cornices without proper protection will often be plagued with ice dams (fig. 68,*A*). These are formed when snow melts, runs down the roof, and freezes at the colder cornice area. Gradually, the ice forms a dam that backs up water under the shingles. Under these conditions, it is good practice to use an undercourse (36-in. width) of 45-pound or heavier smooth-surface roll roofing along the eave line as a flashing (fig. 68,*B*). This will minimize the chance of water backing up and entering the wall. However, good attic ventilation and sufficient ceiling insulation are of primary importance in eliminating this harmful nuisance. These details are described in Chapter 16, "Ventilation."

Metal roofs (tin, copper, galvanized iron, or aluminum) are sometimes used on flat decks of dormers, porches, or entryways. Joints should be watertight and the deck properly flashed at the juncture with the house. Nails should be of the same metal as that used on the roof, except that with tin roofs, steel nails may be used. All exposed nailheads in tin roofs should be soldered with a rosin-core solder.

Wood Shingles

Wood shingles of the types commonly used for house roofs are No. 1 grade. Such shingles (*5*) are all-heartwood, all-edgegrain, and tapered. Second grade shingles make good roofs for secondary buildings as well as excellent sidewalls for primary buildings. Western redcedar and redwood are the principal commercial shingle woods, as their heartwood has high decay resistance and low shrinkage.

Four bundles of 16-inch shingles laid 5 inches "to the weather" will cover 100 square feet. Shingles are of random widths, the narrower shingles being in the lower grades. Recommended exposures for the standard shingle sizes are shown in table 3.

TABLE 3.—*Recommended exposure for wood shingles*[1]

Shingle length	Shingle thickness (Green)	Maximum exposure: Slope less[2] than 4 in 12	Maximum exposure: Slope 5 in 12 and over
In.		*In.*	*In.*
16	5 butts in 2 in.	3¾	5
18	5 butts in 2¼ in.	4¼	5½
24	4 butts in 2 in.	5¾	7½

[1] As recommended by the Red Cedar Shingle and Handsplit Shake Bureau.

[2] Minimum slope for main roofs—4 in 12.
Minimum slope for porch roofs—3 in 12.

Figure 69 illustrates the proper method of applying a wood-shingle roof. Underlay or roofing felt is not required for wood shingles except for protection in ice-dam areas. Spaced roof boards under wood shingles are most common, although spaced or solid sheathing is optional.

The following general rules should be followed in the application of wood shingles:

1. Shingles should extend about 1½ inches beyond the eave line and about ¾ inch beyond the rake (gable) edge.

2. Use two rust-resistant nails in each shingle; space them about ¾ inch from the edge and 1½ inches above the butt line of the next course. Use threepenny

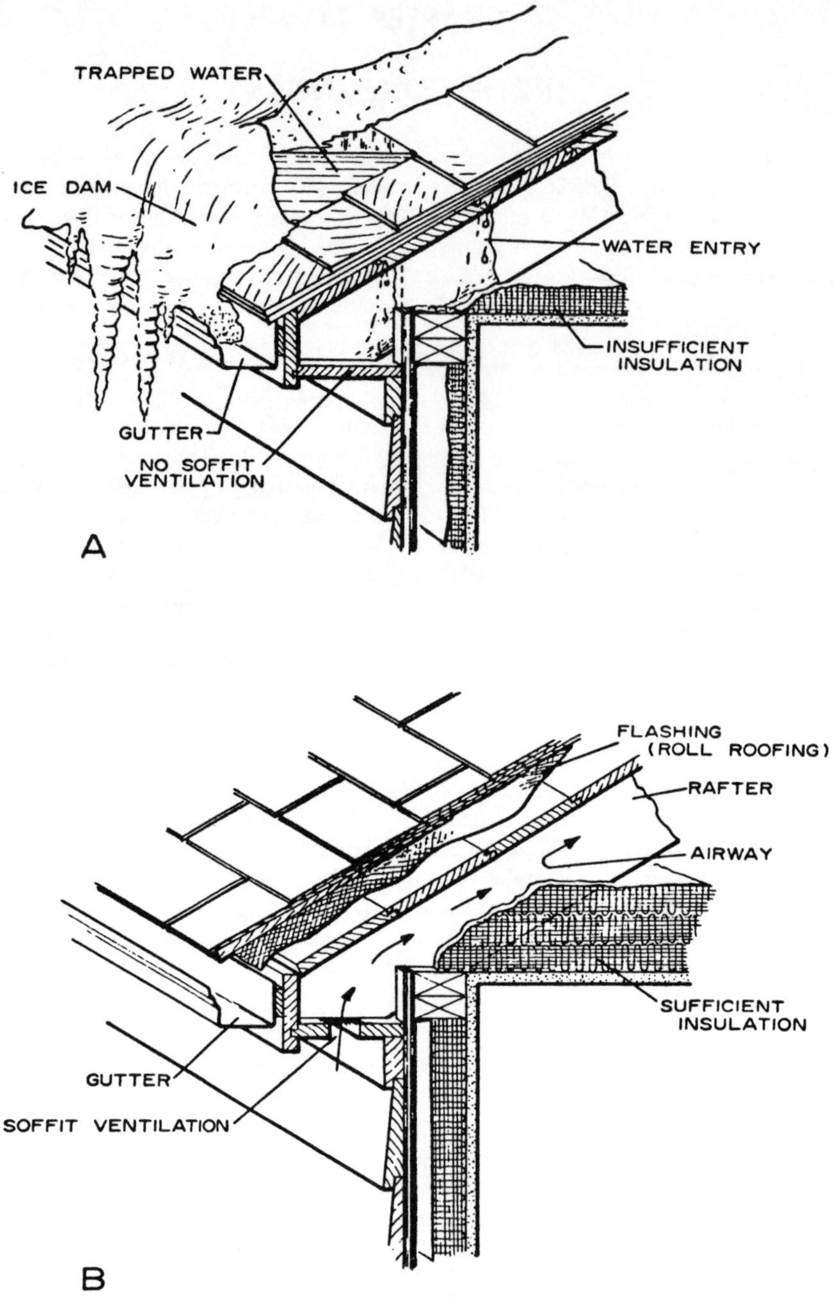

Figure 68.—Snow and ice dams: *A*, Ice dams often build up on the overhang of roofs and in gutters, causing melting snow water to back up under shingles and under the facia board of closed cornices. Damage to ceilings inside and to paint outside results. *B*, Eave protection for snow and ice dams. Lay smooth-surface 45-pound roll roofing on roof sheathing over the eaves extending upward well above the inside line of the wall.

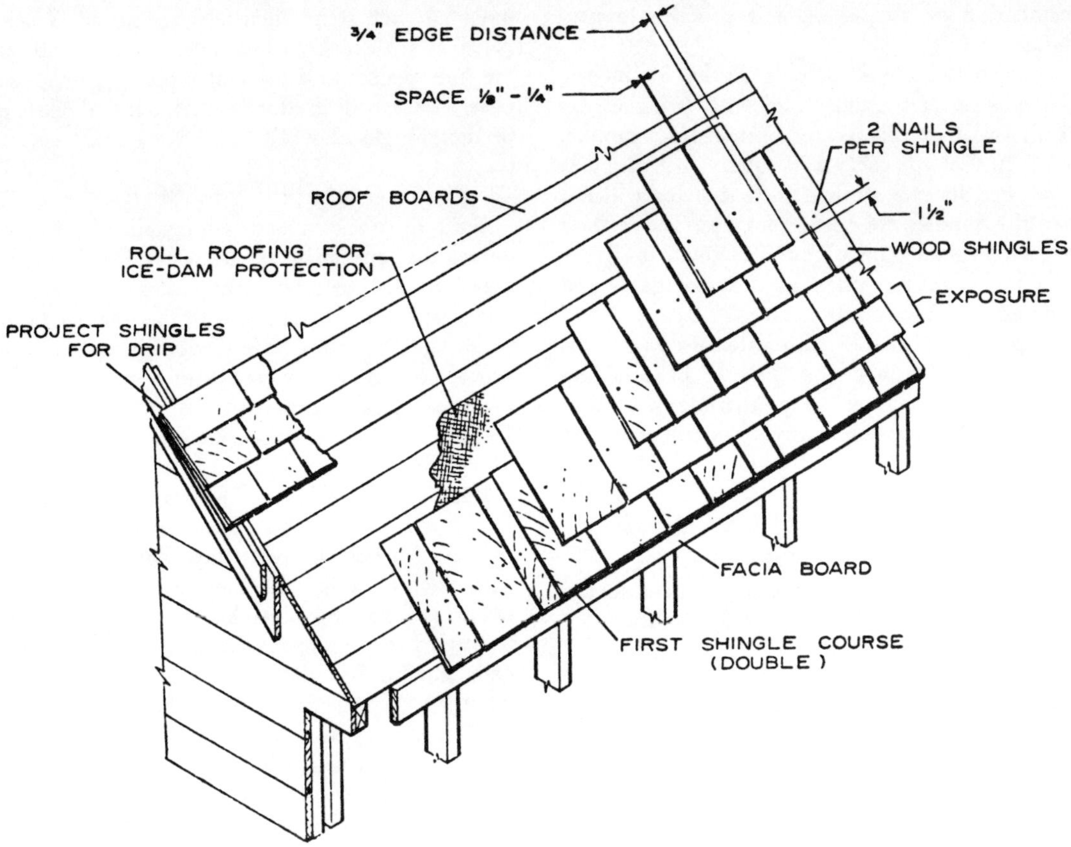

Figure 69.—Installation of wood shingles.

nails for 16- and 18-inch shingles and fourpenny for 24-inch shingles in new construction. A *ring-shank nail* (threaded) is often recommended for plywood roof sheathing less than 1/2 inch thick.

3. The first course of shingles should be doubled. In all courses, allow 1/8- to 1/4-inch space between each shingle for expansion when wet. The joints between shingles should be offset at least 1 1/2 inches from the joints between shingles in the course below. Further, the joints in succeeding courses should be spaced so that they do not directly line up with joints in the second course below.

4. When valleys are present, shingle away from the valleys, selecting and precutting wide valley shingles.

5. A metal edging along the gable end will aid in guiding the water away from the sidewalls.

6. In laying No. 1 all-heartwood edge-grain shingles no splitting of wide shingles is necessary.

Wood shakes are applied much the same as wood shingles. Because shakes are much thicker (longer shakes have the thicker butts), long galvanized nails are used. To create a rustic appearance, the butts are often laid unevenly. Because shakes are longer than shingles, they have a greater exposure. Exposure distance is usually 7 1/2 inches for 18-inch shakes, 10 inches for 24-inch shakes, and 13 inches for 32-inch shakes. Shakes are not smooth on both faces, and because wind-driven snow might enter, it is essential to use an underlay between each course. An 18-inch-wide layer of 30-pound asphalt felt should be used between each course with the bottom edge positioned above the butt edge of the shakes a distance equal to double the weather exposure. A 36-inch wide starting strip of the asphalt felt is used at the eave line. Solid sheathing should be used when wood shakes are used for roofs in areas where wind-driven snow is experienced.

Asphalt Shingles

The usual minimum recommended weight for asphalt shingles is 235 pounds for square-butt strip shingles. This may change in later years, as 210 pounds (weight per square) was considered a minimum several years ago. Strip shingles with a 300-pound weight per square are available, as are lock-type and other shingles weighing 250 pounds and more. Asphalt shingles are also available with seal-type tabs for wind resistance. Many contractors apply a small spot of asphalt roof cement under each tab after installation of regular asphalt shingles to provide similar protection.

The square-butt strip shingle is 12 by 36 inches, has three tabs, and is usually laid with 5 inches exposed to the weather. There are 27 strips in a bundle, and three bundles will cover 100 square feet. Bundles should be piled flat for storage so that strips will not curl when the bundles are opened for use. The method of laying an asphalt-shingle roof is shown in figure 70,*A*. A metal edging is often used at the gable end to provide additional protection (fig. 70,*B*).

Data such as that in table 4 are often used in determining the need for and the method of applying *underlayment* for asphalt shingles on roofs of various slopes. Underlayment is commonly 15-pound saturated felt.

TABLE 4.—*Underlayment requirements for asphalt shingles*

(Headlap for single coverage of underlayment should be 2 inches and for double coverage 19 inches.)

Underlayment	Minimum roof slope	
	Double coverage [1] shingles	Triple coverage [1] shingles
Not required	7 in 12	[2] 4 in 12
Single	[2] 4 in 12	[3] 3 in 12
Double	2 in 12	2 in 12

[1] Double coverage for a 12- by 36-in. shingle is usually an exposure of about 5 in. and about 4 in. for triple coverage.
[2] May be 3 in 12 for porch roofs.
[3] May be 2 in 12 for porch roofs.

An asphalt-shingle roof can also be protected from ice dams by adding an initial layer of 45-pound or heavier roll roofing, 36 inches wide, and insuring good ventilation and insulation within the attic space (fig. 68,*B*).

A course of wood shingles or a metal edging should be used along the eave line before application of the asphalt shingles. The first course of asphalt shingles is doubled; or, if desired, a starter course may be used under the first asphalt-shingle course. This first course should extend downward beyond the wood shingles (or edging) about ½ inch to prevent the water from backing up under the shingles. A ½-inch projection should also be used at the rake.

Several chalklines on the underlay will help aline the shingles so that tab notches will be in a straight line for good appearance. Each shingle strip should be fastened securely according to the manufacturer's directions. The use of six 1-inch galvanized roofing nails for each 12- by 36-inch strip is considered good practice in areas of high winds. A sealed tab or the use of asphalt sealer will also aid in preventing wind damage during storms. Some contractors use four nails for each strip when tabs are sealed. When a nail penetrates a crack or knothole, it should be removed, the hole sealed, and the nail replaced in sound wood; otherwise, it will gradually work out and cause a hump in the shingle above it.

Built-up Roofs

Built-up roof coverings are installed by roofing companies that specialize in this work. Roofs of this type may have 3, 4, or 5 layers of roofer's felt, each mopped down with tar or asphalt, with the final surface coated with asphalt and covered with gravel embedded in asphalt or tar, or covered with a cap sheet. For convenience, it is customary to refer to built-up roofs as 10-, 15-, or 20-year roofs, depending upon the method of application.

For example, a 15-year roof over a wood deck (fig. 71,*A*) may have a base layer of 30-pound saturated roofer's felt laid dry, with edges lapped and held down with roofing nails. All nailing should be done with either (a) roofing nails having ⅜-inch heads driven through 1-inch-diameter tin caps or (b) special roofing nails having 1-inch-diameter heads. The dry sheet is intended to prevent tar or asphalt from entering the rafter spaces. Three layers of 15-pound saturated felt follow, each of which is mopped on with hot tar rather than being nailed. The final coat of tar or asphalt may be covered with roofing gravel or a cap sheet of roll roofing.

The cornice or eave line of projecting roofs is usually finished with metal edging or flashing, which acts as a drip. A metal gravel strip is used in conjunction with the flashing at the eaves when the roof is covered with gravel (fig. 71,*B*). Where built-up roofing is finished against another wall, the roofing is turned up on the wall sheathing over a cant strip and is often also flashed with metal (fig. 71,*C*). This flashing is generally extended up about 4 inches above the bottom of the siding.

Other Roof Coverings

Other roof coverings, including asbestos, slate, tile, metal and others, many of which require specialized applicators, are perhaps less commonly used than wood or asphalt shingles and built-up roofs. Several new materials, such as plastic films and coatings, are showing promise for future moderate-cost roof coverings. However, most of them are more expensive than the materials now commonly being used for houses. These newer materials, however, as well as other new products, are likely to come into more general use during the next decade.

Finish at the Ridge and Hip

The most common type of ridge and hip finish for wood and asphalt shingles is known as the *Boston ridge*. Asphalt-shingle squares (one-third of a 12- by

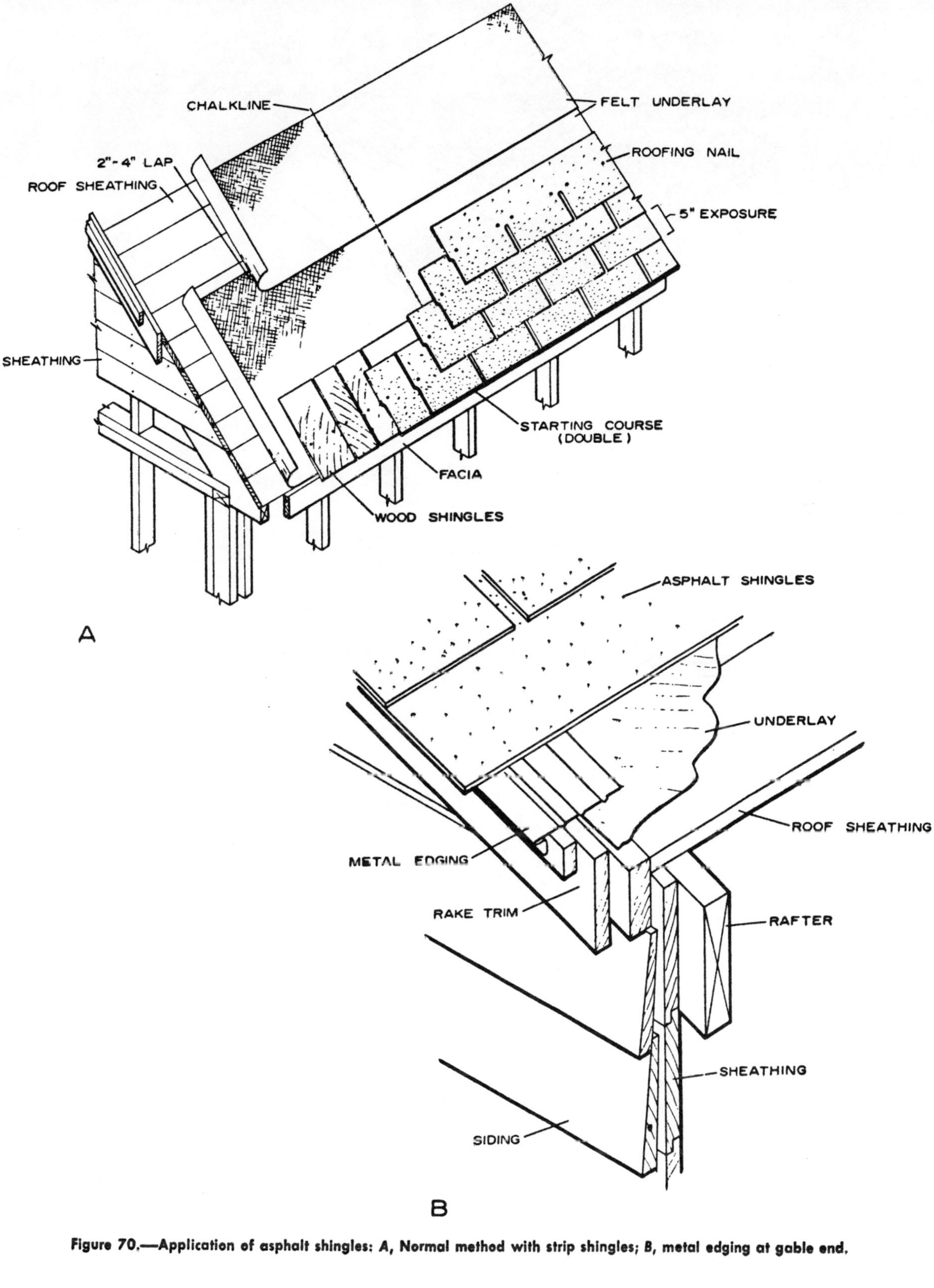

Figure 70.—Application of asphalt shingles: *A*, Normal method with strip shingles; *B*, metal edging at gable end.

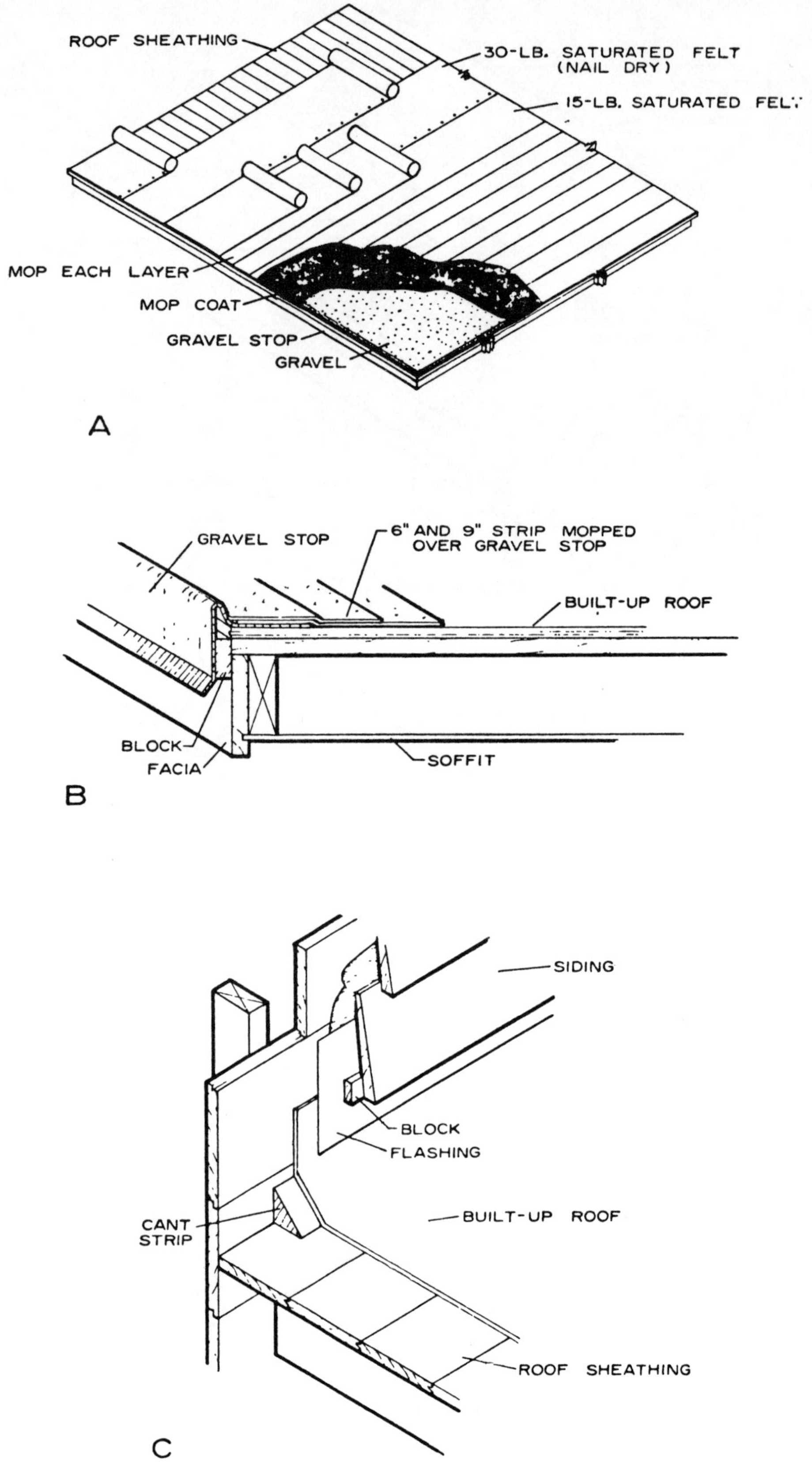

Figure 71.—Built-up roof: *A*, Installation of roof; *B*, gravel stop; C, flashing at building line.

36-inch strip) are used over the ridge and blind-nailed (fig. 72,*A*). Each shingle is lapped 5 to 6 inches to give double coverage. In areas where driving rains occur, it is well to use metal flashing under the shingle ridge. The use of a ribbon of asphalt roofing cement under each lap will also greatly reduce the chance of water penetration.

A wood-shingle roof (fig. 72,*B*) also should be finished in a Boston ridge. Shingles 6 inches wide are alternately lapped, fitted, and blind-nailed. As shown, the shingles are nailed in place so that exposed trimmed edges are alternately lapped. Pre-assembled hip and ridge units are available and save both time and money.

A metal ridge roll can also be used on asphalt-shingle or wood-shingle roofs (fig. 72,*C*). This ridge is formed to the roof slope and should be copper, galvanized iron, or aluminum. Some metal ridges are formed so that they provide an outlet ventilating area. However, the design should be such that it prevents rain or snow blowing in.

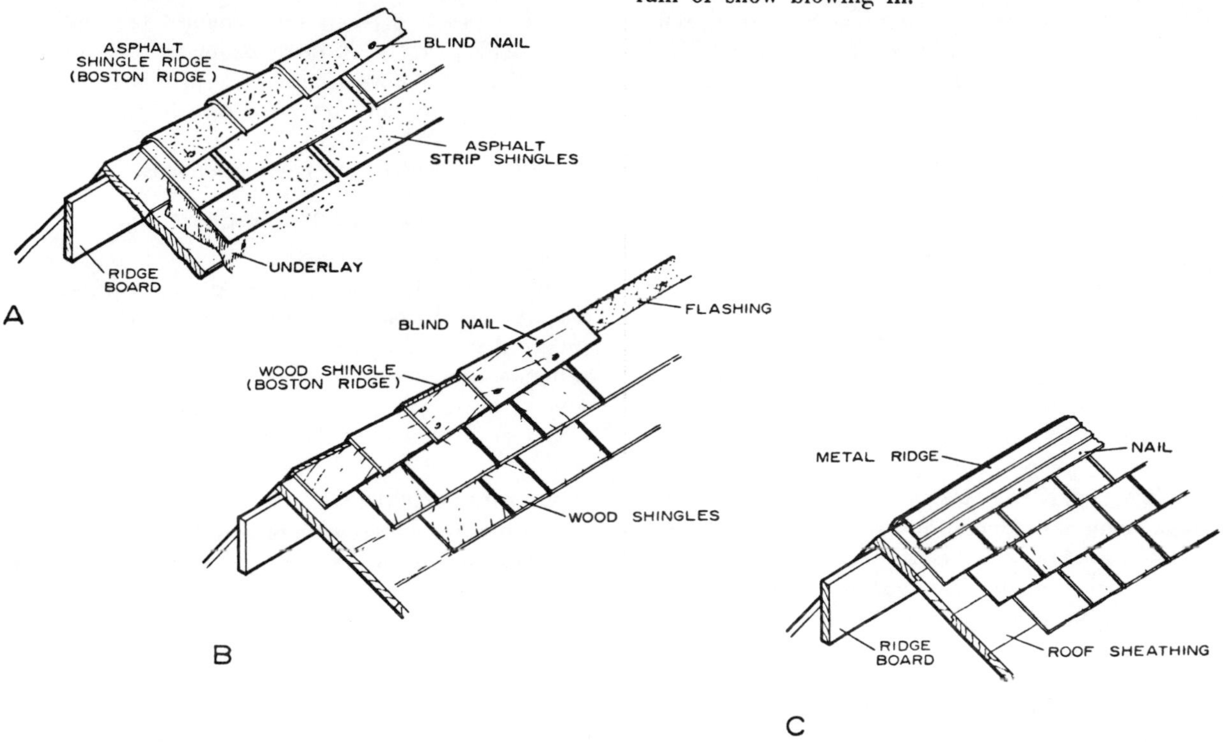

Figure 72.—Finish at ridge: *A*, Boston ridge with asphalt shingles; *B*, Boston ridge with wood shingles; *C*, metal ridge.

CHAPTER 12

EXTERIOR FRAMES, WINDOWS, AND DOORS

Windows, doors, and their frames are millwork items that are usually fully assembled at the factory. Window units, for example, often have the sash fitted and weatherstripped, frame assembled, and exterior casing in place. Standard combination storms and screens or separate units can also be included. Door frames are normally assembled ready for use in the building. All such wood components are treated with a water-repellent preservative at the factory to provide protection before and after they are placed in the walls.

Windows are mainly to allow entry of light and air, but may also be an important part of the architectural design. Some variation may occur, but normally in habitable rooms the glass area should be not less than 10 percent of the floor area. Natural ventilation should be not less than 4 percent of the floor area in a habitable room unless a complete air-conditioning system is used.

Types of Windows

Windows are available in many types, each having advantages. The principal types are double-hung, casement, stationary, awning, and horizontal sliding.

They may be made of wood or metal. Heat loss through metal frames and sash is much greater than through similar wood units. Glass blocks are sometimes used for admitting light in places where transparency or ventilation is not required.

Insulated glass, used both for stationary and moveable sash, consists of two or more sheets of spaced glass with hermetically-sealed edges. This type has more resistance to heat loss than a single thickness and is often used without a storm sash.

Wood sash and door and window frames should be made from a clear grade of all-heartwood stock of a decay-resistant wood species or from wood which is given a preservative treatment. Species commonly used include ponderosa and other pines, the cedars, cypress, redwood, and the spruces.

Tables showing glass size, sash size, and rough opening size are available at lumber dealers, so that the wall openings can be framed accordingly. Typical openings for double-hung windows are shown in the chapter "Wall Framing."

Double-hung Windows

The double-hung window is perhaps the most familiar window type. It consists of an upper and lower sash that slide vertically in separate grooves in the side jambs or in full-width metal weatherstripping (fig. 73). This type of window provides a maximum face opening for ventilation of one-half the total window area. Each sash is provided with springs, balances, or *compression weatherstripping* to hold it in place in any location. Compression weatherstripping, for example, prevents air infiltration, provides tension, and acts as a counterbalance; several types allow the sash to be removed for easy painting or repair.

The *jambs* (sides and top of the frames) are made of nominal 1-inch lumber; the width provides for use with dry-wall or plastered interior finish. Sills are made from nominal 2-inch lumber and sloped at about 3 in 12 for good drainage (fig. 73,*D*). Sash are normally 1⅜ inches thick and wood combination storm and screen windows are usually 1⅛ inches thick.

Sash may be divided into a number of lights by small wood members called *muntins.* A ranch-type house may provide the best appearance with top and bottom sash divided into two horizontal lights. A colonial or Cape Code house usually has each sash divided into six or eight lights. Some manufacturers provided preassembled dividers which snap in place over a single light, dividing it into six or eight lights. This simplifies painting and other maintenance.

Assembled frames are placed in the rough opening over strips of building paper put around the perimeter to minimize air infiltration. The frame is plumbed and nailed to side studs and header through the casings or the blind stops at the sides. Where nails are exposed, such as on the casing, use the corrosion-resistant type.

Hardware for double-hung windows includes the sash lifts that are fastened to the bottom rail, although they are sometimes eliminated by providing a finger groove in the rail. Other hardware consists of sash locks or fasteners located at the meeting rail. They not only lock the window, but draw the sash together to provide a "windtight" fit.

Double-hung windows can be arranged in a number of ways—as a single unit, doubled (or mullion) type, or in groups of three or more. One or two double-hung windows on each side of a large stationary insulated window are often used to effect a window wall. Such large openings must be framed with headers large enough to carry roofloads.

Casement Windows

Casement windows consist of side-hinged sash, usually designed to swing outward (fig. 74) because this type can be made more weathertight than the in-swinging style. Screens are located inside these out-swinging windows and winter protection is obtained with a storm sash or by using insulated glass in the sash. One advantage of the casement window over the double-hung type is that the entire window area can be opened for ventilation.

Weatherstripping is also provided for this type of window, and units are usually received from the factory entirely assembled with hardware in place. Closing hardware consists of a rotary operator and sash lock. As in the double-hung units, casement sash can be used in a number of ways—as a pair or in combinations of two or more pairs. Style variations are achieved by divided lights. Snap-in muntins provided a small, multiple-pane appearance for traditional styling.

Metal sash are sometimes used but, because of low insulating value, should be installed carefully to prevent condensation and frosting on the interior surfaces during cold weather. A full storm-window unit is sometimes necessary to eliminate this problem in cold climates.

Stationary Windows

Stationary windows used alone or in combination with double-hung or casement windows usually consist of a wood sash with a large single light of insulated glass. They are designed to provide light, as well as for attractive appearance, and are fastened permanently into the frame (fig. 75). Because of their size, (sometimes 6 to 8 feet wide) 1¾-inch-thick sash is used to provide strength. The thickness is usually required because of the thickness of the insulating glass.

Other types of stationary windows may be used

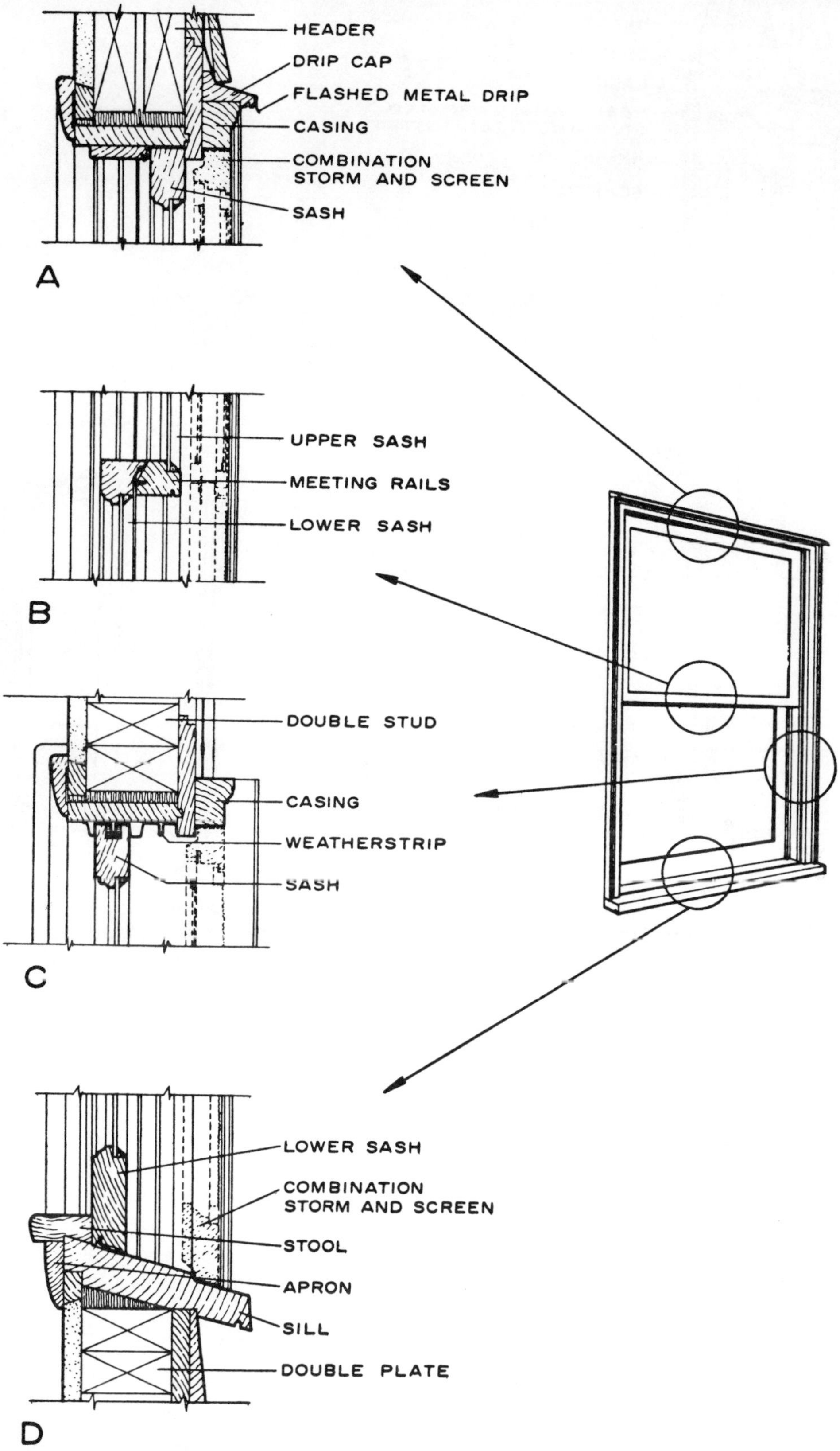

Figure 73.—Double-hung windows. Cross sections: *A*, Head jamb; *B*, meeting rails; *C*, side jambs; *D*, sill.

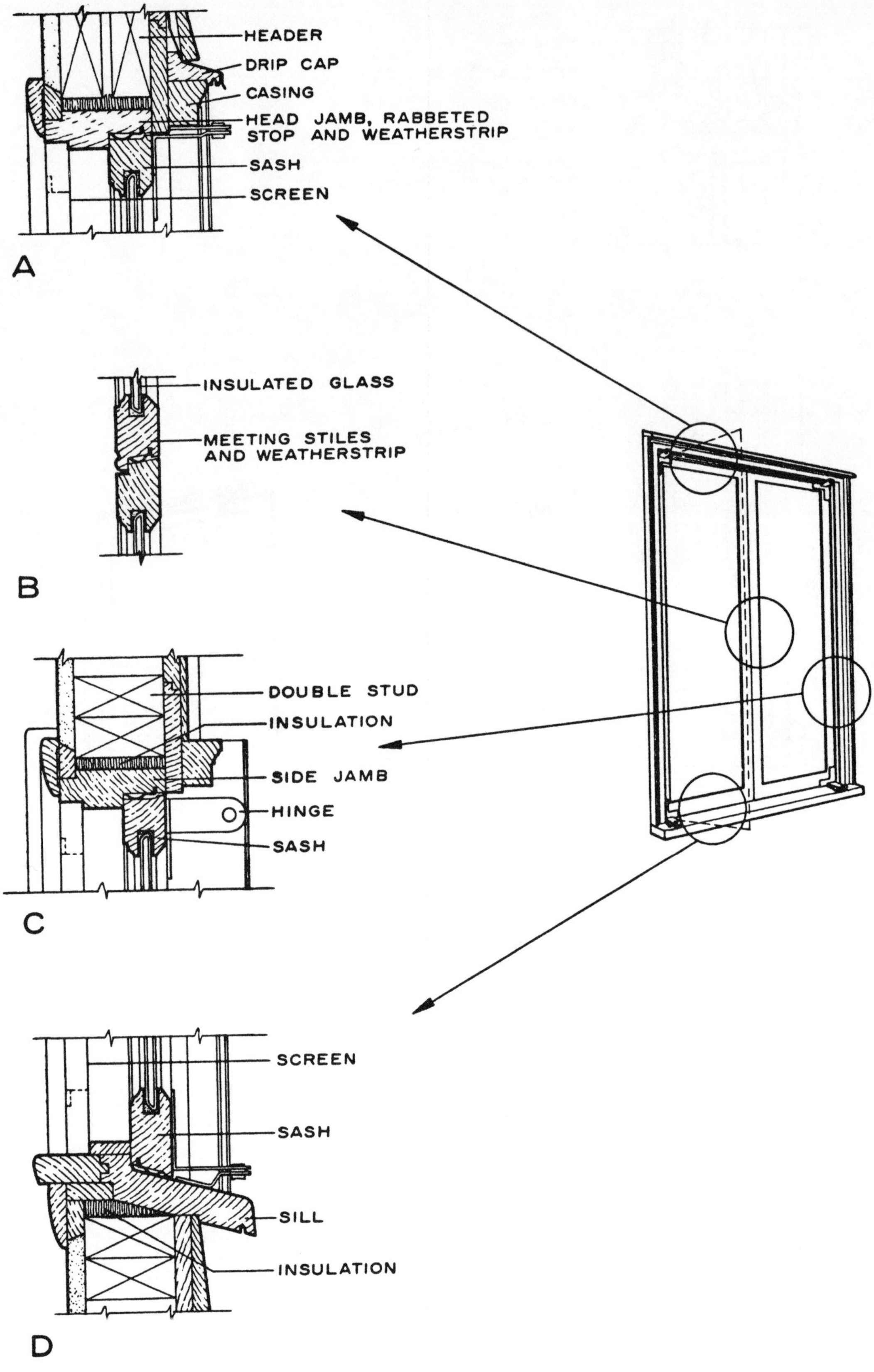

Figure 74.—Outswinging casement sash. Cross sections: *A,* Head jamb; *B,* meeting stiles; C, side jambs· *D,* sill.

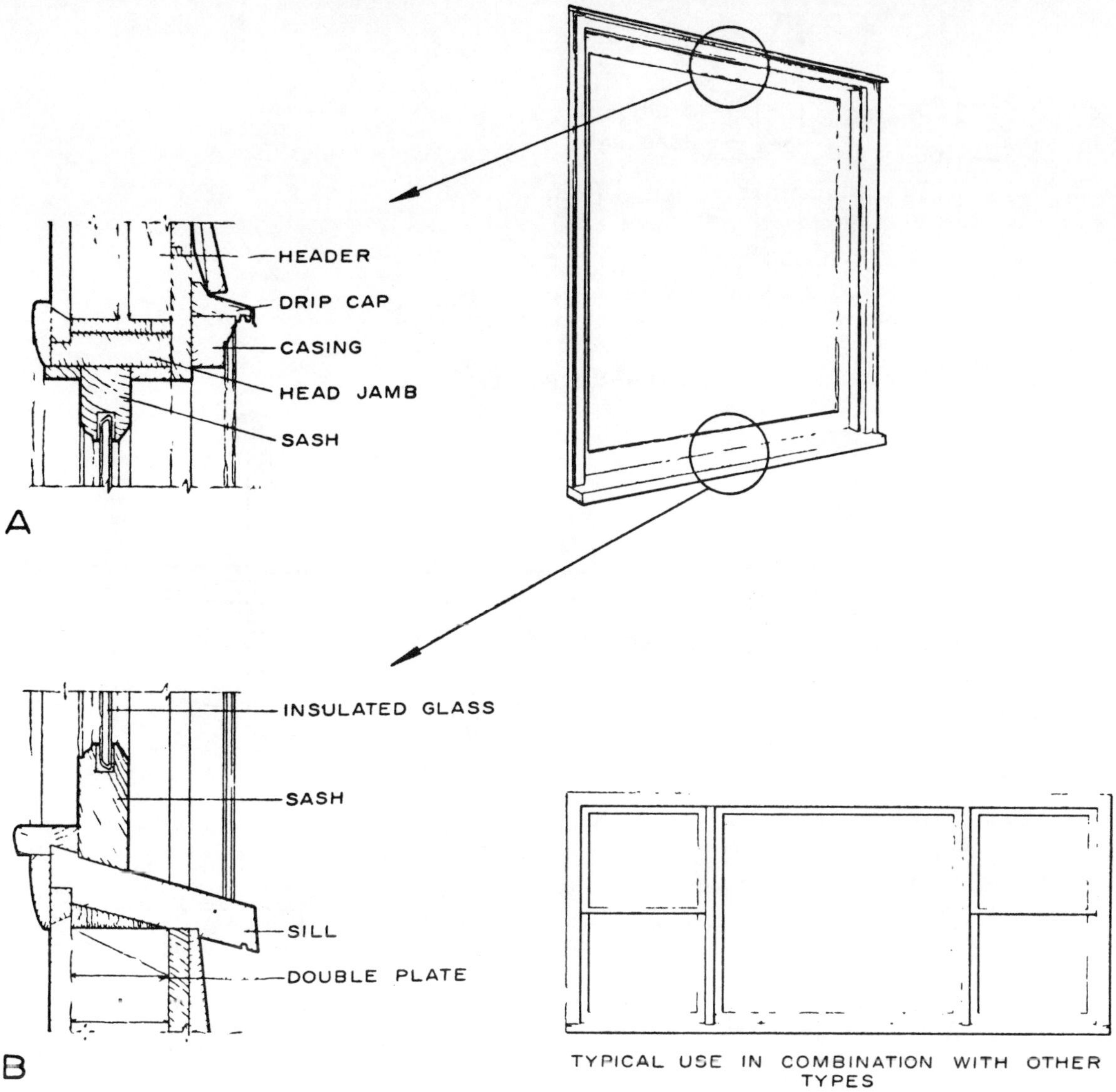

Figure 75—Stationary window. Cross sections: *A*, Head jamb; *B*, sill.

without a sash. The glass is set directly into rabbeted frame members and held in place with stops. As with all window-sash units, back puttying and face puttying of the glass (with or without a stop) will assure moisture-resistance.

Awning Windows

An awning window unit consists of a frame in which one or more operative sash are installed (fig. 76). They often are made up for a large window wall and consist of three or more units in width and height

Sash of the awning type are made to swing outward at the bottom. A similar unit, called the hopper type, is one in which the top of the sash swings inward. Both types provide protection from rain when open.

Jambs are usually $1\frac{1}{16}$ inches or more thick because they are rabbeted, while the sill is at least $1\frac{5}{16}$ inches thick when two or more sash are used in a complete frame. Each sash may also be provided with an individual frame, so that any combination in width and height can be used. Awning or hopper window units may consist of a combination of one or more fixed sash with the remainder being the operable type. Operable sash are provided with hinges, pivots, and sash supporting arms.

Weatherstripping and storm sash and screens are usually provided. The storm sash is eliminated when the windows are glazed with insulated glass.

Horizontal-sliding Window Units

Horizontal-sliding windows appear similar to casement sash. However, the sash (in pairs) slide hori-

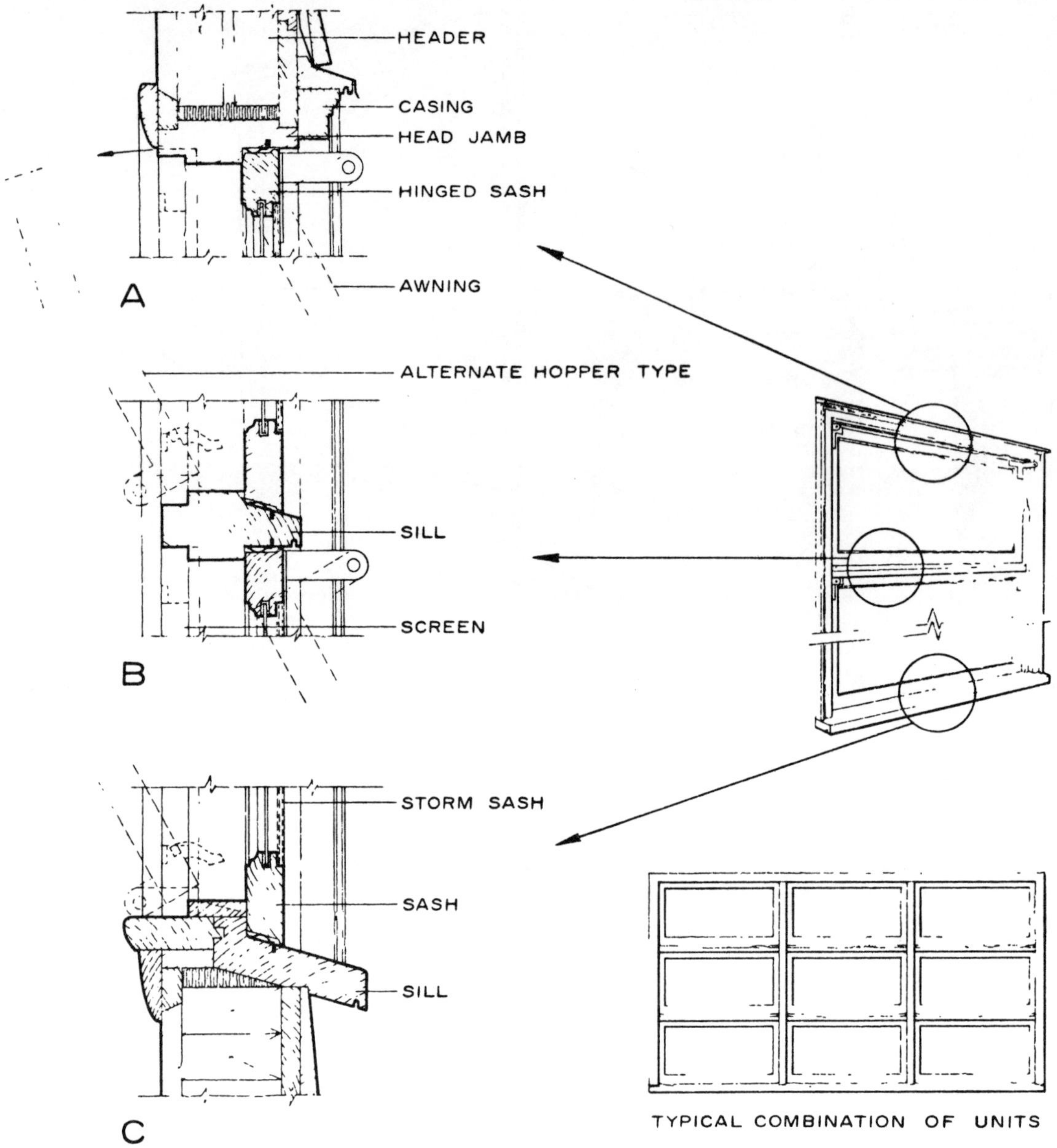

Figure 76.—Awning window. Cross sections: *A*, Head jamb; *B*, horizontal mullion; *C*, sill.

zontally in separate tracks or guides located on the sill and head jamb. Multiple window openings consist of two or more single units and may be used when a window-wall effect is desired. As in most modern window units of all types, weatherstripping, water-repellent preservative treatments, and sometimes hardware are included in these fully factory-assembled units.

Exterior Doors and Frames

Exterior doors are 1¾ inches thick and not less than 6 feet 8 inches high. The main entrance door is 3 feet wide and the side or rear service door 2 feet 8 inches wide.

The frames for these doors are made of 1⅛-inch or thicker material, so that rabbeting of side and head jambs provides stops for the main door (fig. 77). The wood sill is often oak for wear resistance, but when softer species are used, a metal nosing and wear strips are included. As in many of the window units, the outside casings provide space for the 1⅛-inch combination or screen door.

The frame is nailed to studs and headers of the rough opening through the outside casing. The sill

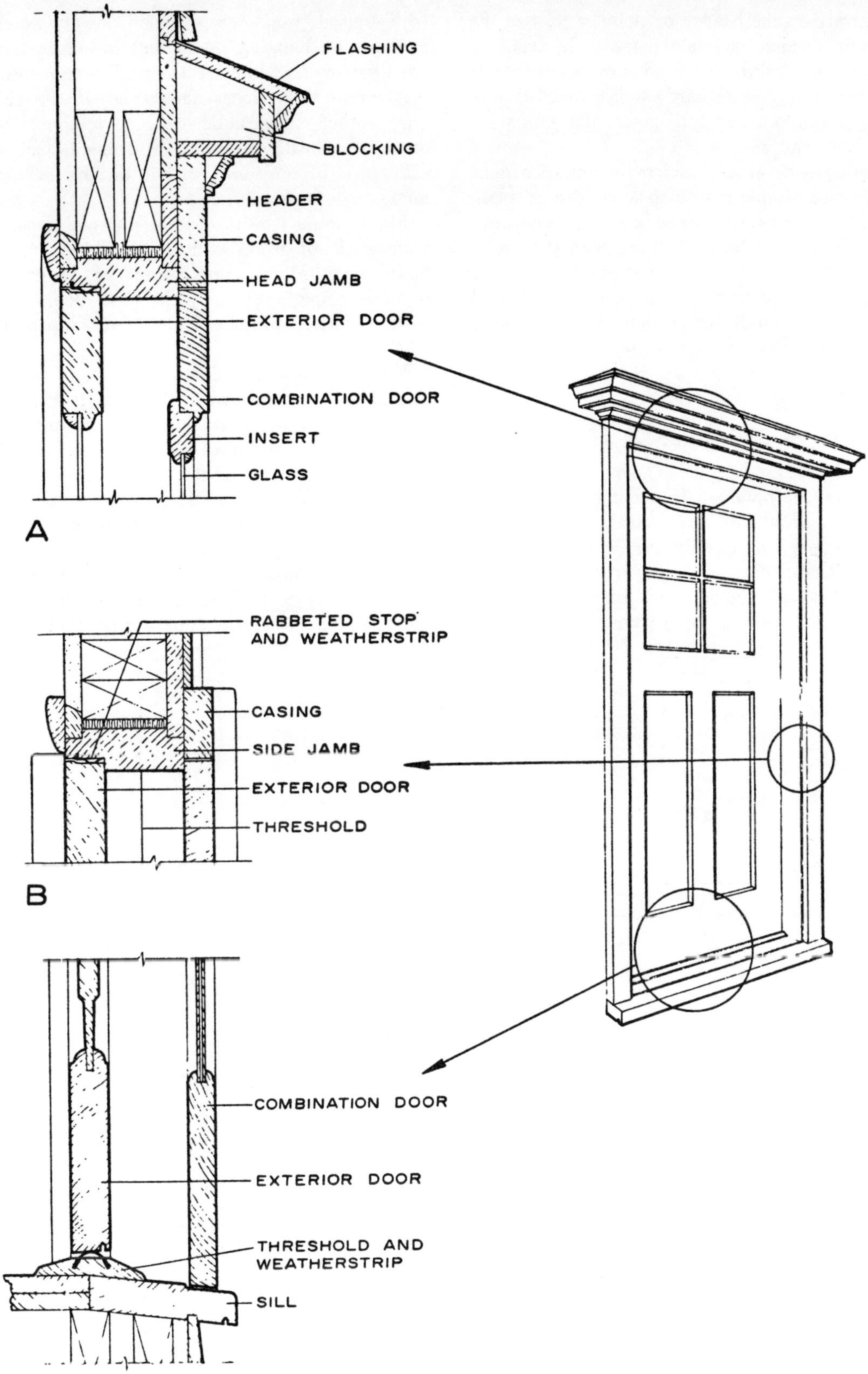

Figure 77.—Exterior door and frame. Exterior-door and combination-door (screen and storm) cross sections: *A*, Head jamb; *B*, side jamb; C, sill.

must rest firmly on the header or stringer joist of the floor framing, which commonly must be trimmed with a saw and hand ax or other means. After finish flooring is in place, a hardwood or metal threshold with a plastic weatherstop covers the joints between the floor and sill.

The exterior trim around the main entrance door can vary from a simple casing to a molded or plain pilaster with a decorative head casing. Decorative designs should always be in keeping with the architecture of the house. Many combinations of door and entry designs are used with contemporary houses, and manufacturers have millwork which is adaptable to this and similar styles. If there is an entry hall, it is usually desirable to have glass included in the main door if no other light is provided.

Types of Exterior Doors

Exterior doors and outside combination and storm doors can be obtained in a number of designs to fit the style of almost any house. Doors in the traditional pattern are usually the *panel type* (fig. 78,*A*). They consist of *stiles* (solid vertical members), *rails* (solid cross members), and *filler panels* in a number of designs. Glazed upper panels are combined with raised wood or plywood lower panels. For methods of hanging doors and installing hardware, see Chapter 21. "Interior Doors, Frames, and Trim."

Exterior flush doors should be of the solid-core type rather than hollow-core to minimize warping during the heating season. (Warping is caused by a difference in moisture content on the exposed and unexposed faces.)

Flush doors consist of thin plywood faces over a framework of wood with a woodblock or particle board core. Many combinations of designs can be obtained, ranging from plain flush doors to others with a variety of panels and glazed openings (fig. 78,*B*).

Wood combination doors (storm and screen) are available in several styles (fig. 78,*C*). Panels which include screen and storm inserts are normally located in the upper portion of the door. Some types can be obtained with self-storing features, similar to window combination units. Heat loss through metal combination doors is greater than through similar type wood doors.

Weatherstripping of the $1\frac{3}{4}$-inch-thick exterior door will reduce both air infiltration and frosting of the glass on the storm door during cold weather.

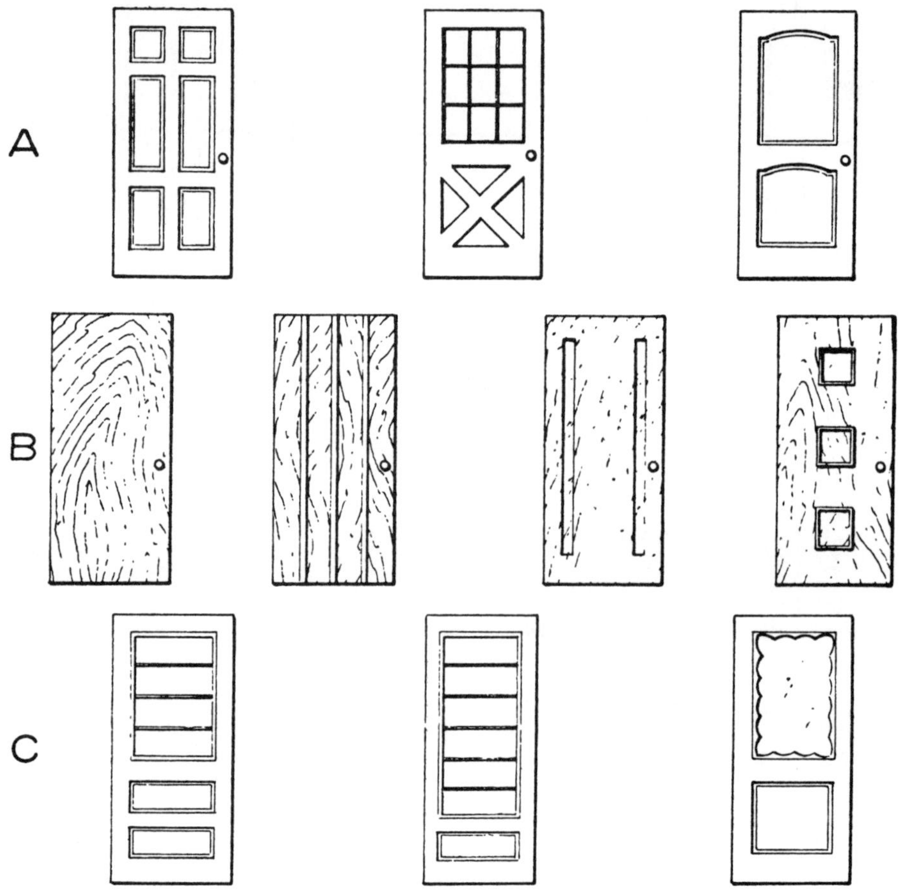

Figrue 78.—Exterior doors: *A*, Traditional panel; *B*, flush; *C*, combination.

CHAPTER 13

EXTERIOR COVERINGS

Because siding and other types of coverings used for exterior walls have an important influence on the appearance as well as on the maintenance of the house, a careful selection of the pattern should be made. The homeowner now has a choice of many wood and wood-base materials which may be used to cover exterior walls. Masonry, veneeers, metal or plastic siding, and other nonwood materials are additional choices. Wood siding can be obtained in many different patterns and can be finished naturally, stained, or painted. Wood shingles, plywood, wood siding or paneling, fiberboard, and hardboard are some of the types and as exterior coverings. Many prefinished sidings are available, and the coatings and films applied to several types of base materials presumably eliminate the need of refinishing for many years.

Wood Siding

One of the materials most characteristic of the exteriors of American houses is *wood siding.* The essential properties required for siding are good painting characteristics, easy working qualities, and freedom from warp. Such properties are present to a *high* degree in the cedars, eastern white pine, sugar pine, western white pine, cypress, and redwood; to a *good* degree in western hemlock, ponderosa pine, the spruces, and yellow-poplar; and to a *fair* degree in Douglas-fir, western larch, and southern pine (5).

Material used for exterior siding which is to be painted, should preferably be of a high grade and free from knots, pitch pockets, and waney edges. Vertical grain and mixed grain (both vertical and flat) are available in some species such as redwood and western redcedar.

The moisture content at the time of application should be that which it would attain in service. This would be approximately 10 to 12 percent except in the dry Southwestern States where the moisture content should average about 8 to 9 percent. To minimize seasonal movement due to changes in moisture content, vertical-grain (edge-grain) siding is preferred. While this is not as important for a stained finish, the use of edge-grain siding for a paint finish will result in longer paint life. A 3-minute dip in a water-repellent preservative (Federal Specification TT-W-572) before siding is installed will not only result in longer paint life, but also will resist moisture entry and decay. Some manufacturers supply siding with this treatment. Freshly cut ends should be brush-treated on the job.

Horizontal Sidings

Some wood siding patterns are used only horizontally and others only vertically. Some may be used in either manner if adequate nailing areas are provided. Following are descriptions of each of the general types.

Bevel Siding

Plain bevel siding can be obtained in sizes from ½ by 4 inches to ½ by 8 inches, and also in sizes of ¾ by 8 inches and ¾ by 10 inches (fig. 79). "Anzac" siding (fig. 79) is ¾ by 12 inches in size. Usually the finished width of bevel siding is about ½ inch less than the size listed. One side of bevel siding has a smooth planed surface, while the other has a rough resawn surface. For a stained finish, the rough or sawn side is exposed because wood stain is most successful and longer lasting on rough wood surfaces.

Dolly Varden Siding

Dolly Varden siding is similar to true bevel siding except that shiplap edges are used, resulting in a constant exposure distance (fig. 79). Because is lies flat against the studs, it is sometimes used for garages and similar buildings without sheathing. Diagonal bracing is then needed to provide racking resistance to the wall.

Other Horizontal Sidings

Regular *drop sidings* can be obtained in several patterns, two of which are shown in figure 79. This siding, with matched or shiplap edges, can be obtained in 1- by 6- and 1- by 8-inch sizes. This type is commonly used for lower cost dwellings and for garages, usually without benefit of sheathing. Tests conducted at the Forest Products Laboratory have shown that the tongued-and-grooved (matched) patterns have greater resistance to the penetration of wind-driven rain than the shiplap patterns, when both are treated with a water-repellent preservative.

Fiberboard and *hardboard sidings* are also available in various forms. Some have a backing to provide rigidity and strength while others are used directly over sheathing. Plywood horizontal lap siding, with medium density overlaid surface, is also avaiable as an exterior covering material. It is usually ⅜ inch thick and 12 and 16 inches wide. It is applied in much the same manner as wood siding, except that a shingle wedge is used behind each vertical joint.

Sidings for Horizontal or Vertical Applications

A number of siding or paneling patterns can be used horizontally or vertically (fig. 79). These are manufactured in nominal 1-inch thicknesses and in

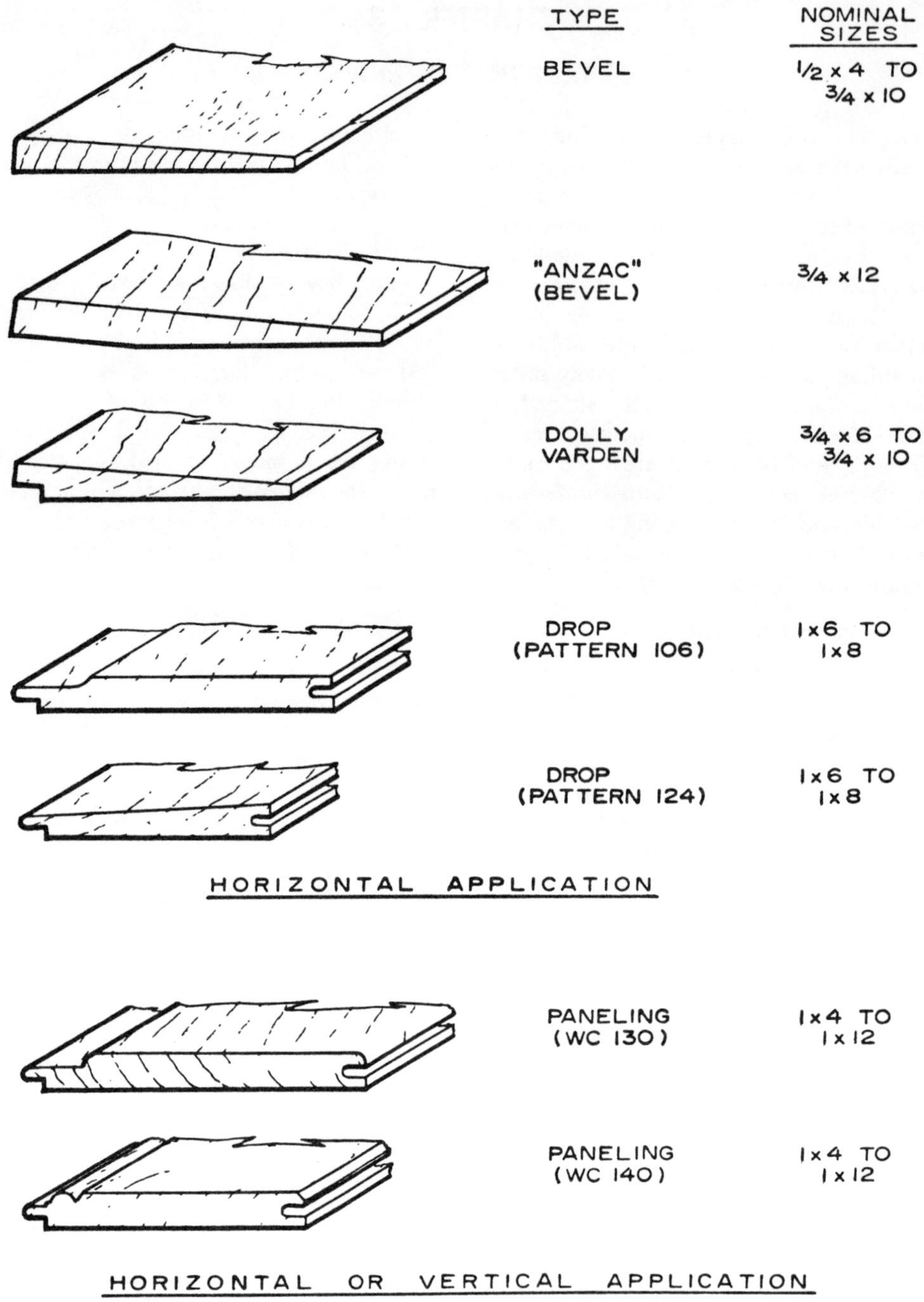

Figure 79.—Wood siding types.

widths from 4 to 12 inches. Both dressed and matched and shiplapped edges are available. The narrow and medium-width patterns will likely be more satisfactory when there are moderate moisture content changes. Wide patterns are more successful if they are vertical grain to keep shrinkage to a minimum. The correct moisture content is also important when tongued and grooved siding is wide, to prevent shrinkage to a point where the tongue is exposed.

Treating the edges of both drop and the matched and shiplapped sidings with water-repellent preservative usually prevents wind-driven rain from penetrating the joints if exposed to weather. In areas under wide overhangs, or in porches or other protected sections, this treatment is not as important. Some manufacturers provide siding with this treatment applied at the factory.

Sidings for Vertical Application

A method of siding application, popular for some architectural styles, utilizes rough-sawn boards and battens applied vertically. These boards can be arranged in several ways: (a) Board and batten, (b) batten and board, and (c) board and board (fig. 80). As in the vertical application of most siding materials, nominal 1-inch sheathing boards or plywood sheathing ⅝ or ¾ inch thick should be used for nailing surfaces. When other types of sheathing materials or thinner plywoods are used, nailing blocks between studs commonly provide the nailing areas. Nailers of 1 by 4 inches, laid horizontally and spaced from 16 to 24 inches apart vertically, can be used over nonwood sheathing. However, special or thicker casing is sometimes required around doors and window frames when this system is used. It is good practice to use a building paper over the sheathing before applying the vertical siding.

Sidings with Sheet Materials

A number of sheet materials are now available for use as siding. These include plywood in a variety of face treatments and species, paper-overlaid plywood, and hardboard. Plywood or paper-overlaid plywood is sometimes used without sheathing and is known as panel siding with ⅜-inch often considered the minimum thickness for such use for 16-inch stud spacing. However, from the standpoint of stiffness and strength, better performance is usually obtained by using ½- or ⅝-inch thickness.

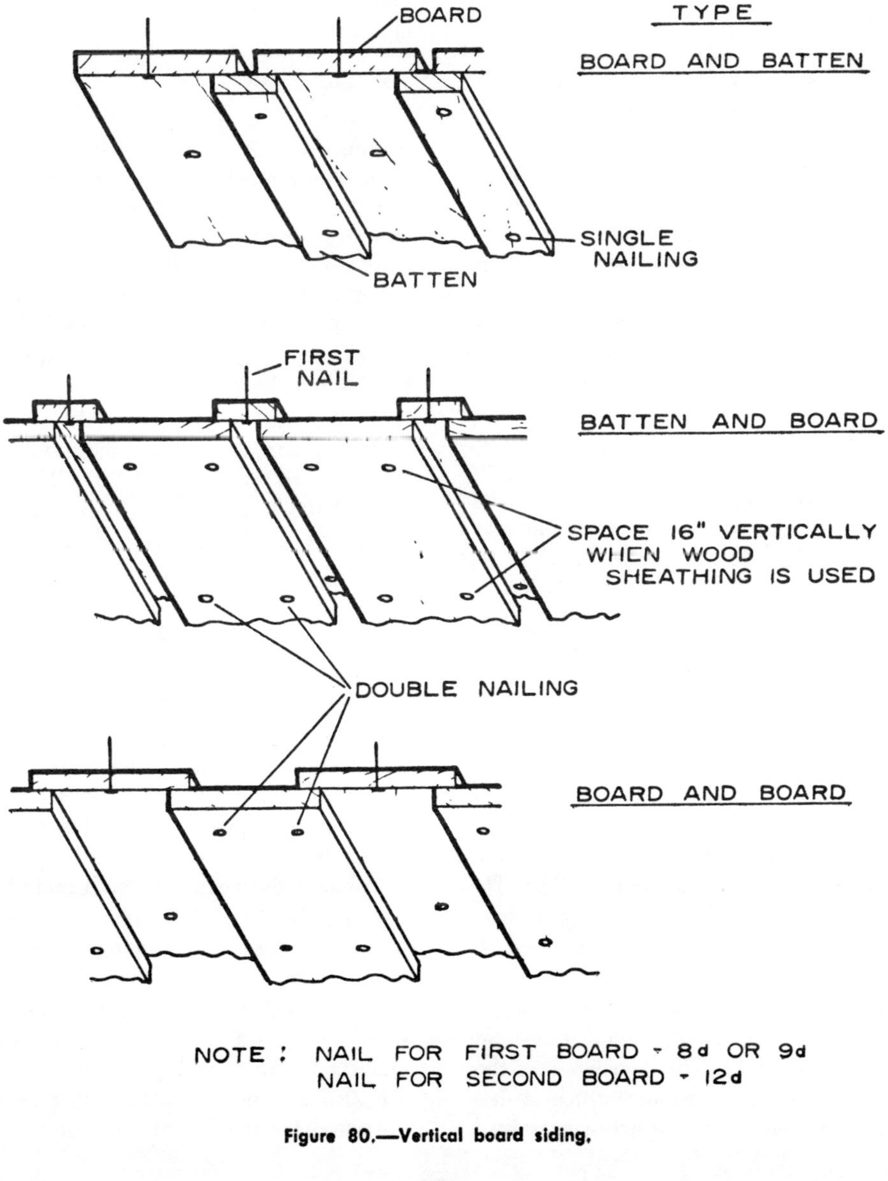

Figure 80.—Vertical board siding.

These 4- by 8-foot and longer sheets must be applied vertically with intermediate and perimeter nailing to provide the desired rigidity. Most other methods of applying sheet materials require some type of sheathing beneath. When horizontal joints are necessary, they should be protected by a simple flashing.

An exterior-grade plywood should always be used for siding, and can be obtained in such surfaces as grooved, brushed, and saw-textured. These surfaces are usually finished with some type of stain. If shiplap or matched edges are not provided, some method of providing a waterproof joint should be used. This often consists of calking and a batten at each joint and a batten at each stud if closer spacing is desired for appearance. An edge treatment of water-repellent preservative will also aid in reducing moisture penetration. Allow 1/16-inch edge and end spacing when installing plywood in sheet form.

Exterior grade particleboard might also be considered for panel siding. FHA Material Use Bulletin No. 32 (9) lists the requirements when this material is used. Normally 5/8-inch thickness is required for 16-inch stud spacing and 3/4 inch for 24-inch stud spacing. The finish must be with an approved paint, and the stud wall behind must have corner bracing.

Paper-overlaid plywood has many of the advantages of plywood with the addition of providing a very satisfactory base for paint. A medium-density, overlaid plywood is most commonly used.

Hardboard sheets used for siding are applied the same way as plywood, that is, by using battens at vertical points and at intermediate studs. Medium-density fiberboards might also be used in some areas as exterior coverings over certain types of sheathing.

Many of these sheet materials resist the passage of water vapor. Hence, when they are used, it is important that a good vapor barrier, well installed, be employed on the warm side of the insulated walls. These factors are described in Chapter 15, "Thermal Insulation and Vapor Barriers."

Wood Shingles and Shakes

Grades and Species

Wood shingles and shakes are desirable for sidewalls in many styles of houses. In Cape Cod or Colonial houses, shingles may be painted or stained. For ranch or contemporary designs, wide exposures of shingles or shakes often add a desired effect. They are easily stained and thus provide a finish which is long-lasting on those species commonly used for shingles.

Western redcedar is perhaps the most available species, although northern white-cedar, baldcypress, and redwood are also satisfactory. The heartwood of these species has a natural decay resistance which is desirable if shingles are to remain unpainted or unstained.

Western redcedar shingles can be obtained in three grades. The first-grade (No. 1) is all heartwood, edge grain, and knot free; it is primarily intended for roofs but is desirable in double-course sidewall application where much of the face is exposed.

Second-grade shingles (No. 2) are most often used in single-course application for sidewalls, since only three-fourths of the shingle length is blemish-free. A 1-inch width of sapwood and mixed vertical and flat grain are permissible.

The third-grade shingle (No. 3) is clear for 6 inches from the butt. Flat grain is acceptable, as are greater widths of sapwood. Third-grade shingles are likely to be somewhat thinner than the first and second grades; they are used for secondary buildings and sometimes as the undercourse in double-course application.

A lower grade than the third grade, known as under-coursing shingle, is used only as the under and completely covered course in double-course sidewall application.

Shingle Sizes

Wood shingles are available in three standard lengths—16, 18, and 24 inches. The 16-inch length is perhaps the most popular, having five butt thicknesses per 2 inches when green (designated a 5/2). These shingles are packed in bundles with 20 courses on each side. Four bundles will cover 100 square feet of wall or roof with an exposure of 5 inches. The 18- and 24-inch-length shingles have thicker butts, five in 2¼ inches for the 18-inch shingles and four in 2 inches for the 24-inch lengths.

Shakes are usually available in several types, the most popular being the split-and-resawn. The sawed face is used as the back face. The butt thickness of each shake ranges between ¾ and 1½ inches. They are usually packed in bundles (20 sq. ft.), five bundles to the square.

Other Exterior Finish

Nonwood materials, such as asbestos-cement siding and shingles, metal sidings, and the like are available and are used in some types of architectural design. Stucco or a cement plaster finish, preferably over a wire mesh base, is most often seen in the Southwest and the West Coast areas. Masonry veneers may be used effectively with wood siding in various finishes to enhance the beauty of both materials.

Some homebuilders favor an exterior covering which requires a minimum of maintenance. While some of the nonwood materials are chosen for this reason, developments by the paint industry are providing comparable long-life coatings for wood-base materials. Plastic films on wood siding or plywood are also promising, so that little or no refinishing is indicated for the life of the building.

Installation of Siding

One of the important factors in successful performance of various siding materials is the type of fasteners used. Nails are the most common of these, and it is poor economy indeed to use them sparingly. Corrosion-resistant nails, galvanized or made of aluminum, stainless steel, or similar metals, may cost more, but their use will insure spot-free siding under adverse conditions.

Two types of nails are commonly used with siding, the finishing nail having a small head and the siding nail having a moderate-size flat head. The small-head finishing nail is set (driven with a nail set) about 1/16 inch below the face of the siding, and the hole is filled with putty after the prime coat of paint is applied. The flathead siding nail, most commonly used, is driven flush with the face of the siding and the head later covered with paint.

Ordinary steel-wire nails tend to rust in a short time and cause a disfiguring stain on the face of the siding. In some cases, the small-head nails will show rust spots through the putty and paint. Noncorrosive nails that will not cause rust are readily available.

Siding to be "natural finished" with a water-repellent preservative or stain should be fastened with stainless steel or aluminum nails. In some types of prefinished sidings, nails with color-matched heads are supplied.

In recent years, nails with modified shanks have become quite popular. These include the *annularly* threaded shank nail and the *helically* threaded shank nail. Both have greater withdrawal resistance than the smooth shank nail and, for this reason, a shorter nail is often used.

Exposed nails in siding should be driven just flush with the surface of the wood. Overdriving may not only show the hammer mark, but may also cause objectionable splitting and crushing of the wood. In sidings with prefinished surfaces or overlays, the nails should be driven so as not to damage the finished surface.

Bevel Siding

The minimum lap for bevel siding should not be less than 1 inch. The average exposure distance is usually determined by the distance from the underside of the window sill to the top of the drip cap (fig. 81). From the standpoint of weather resistance and appearance, the butt edge of the first course of siding above the window should coincide with the top of the window drip cap. In many one-story houses with an overhang, this course of siding is often replaced with a frieze board (fig. 62). It is also desirable that the bottom of a siding course be flush with the underside of the window sill. However, this may not always be possible because of varying window heights and types that might be used in a house.

One system used to determine the siding exposure width so that it is about equal both above and below the window sill is described below:

Divide the overall height of the window frame by the approximate recommended exposure distance for the siding used (4 for 6-inch-wide siding, 6 for 8-inch siding, 8 for 10-inch siding, and 10 for 12-inch siding). This will result in the number of courses between the top and bottom of the window. For example, the overall height of our sample window from top of the drip cap to the bottom of the sill is 61 inches. If 12-inch siding is used, the number of courses would be $61/10 = 6.1$ or six courses. To obtain the exact exposure distance, divide 61 by 6 and the result would be 10⅙ inches. The next step is to determine the exposure distance from the bottom of the sill to just below the top of the foundation wall. If this is 31 inches, three courses at 10⅓ inches each would be used. Thus, the exposure distance above and below the window would be almost the same (fig. 81).

When this system is not satisfactory because of big differences in the two areas, it is preferable to use an equal exposure distance for the entire wall height and notch the siding at the window sill. The fit should be tight to prevent moisture entry.

Siding may be installed starting with the bottom course. It is normally blocked out with a starting strip the some thickness as the top of the siding board (fig. 81). Each succeeding course overlaps the upper edge of the lower course. Siding should be nailed to each stud or on 16-inch centers. When plywood or wood sheathing or spaced wood nailing strips are used over nonwood sheathing, sevenpenny or eightpenny nails (2¼ and 2½ in. long) may be used for ¾-inch-thick siding. However, if gypsum or fiberboard sheathing is used, the tenpenny nail is recommended to penetrate into the stud. For ½-inch-thick siding, nails may be ¼ inch shorter than those used for ¾-inch siding.

The nails should be located far enough up from the butt to miss the top of the lower siding course (fig. 82). This clearance distance is usually ⅛ inch. This allow for slight movement of the siding due to moisture changes without causing splitting. Such an allowance is especially required for the wider sidings of 8 to 12 inches wide.

It is good practice to avoid butt joints whenever possible. Use the longer sections of siding under windows and other long stretches and utilize the shorter lengths for areas between windows and doors. If unavoidable, butt joints should be made over a stud and staggered between courses as much as practical (fig. 81).

Siding should be *square-cut* to provide a good joint at window and door casings and at butt joints. Open joints permit moisture to enter, often leading to paint deterioration. It is good practice to brush

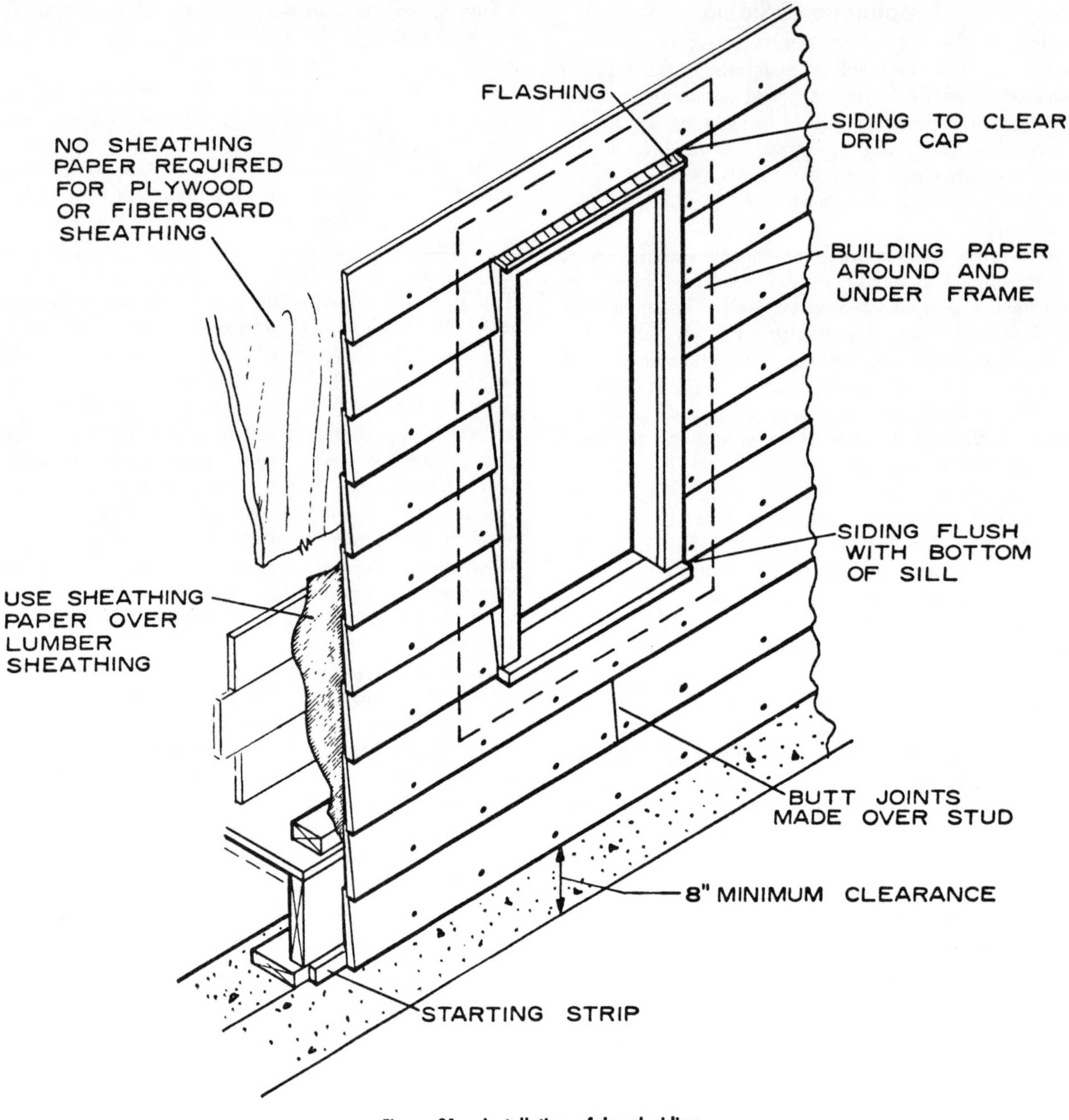

Figure 81.—Installation of bevel siding.

or dip the fresh-cut ends of the siding in a water-repellent preservative before boards are nailed in place. Using a small finger-actuated oil can to apply the water-repellent preservative at end and butt joints after siding is in place is also helpful.

Drop and Similar Sidings

Drop siding is installed much the same as lap siding except for spacing and nailing. Drop, Dolly Varden, and similar sidings have a constant exposure distance. This face width is normally 5¼ inches for 1- by 6-inch siding and 7¼ inches for 1- by 8-inch siding. Normally, one or two eightpenny or ninepenny nails should be used at each stud crossing, depending on the width (fig. 82). Length of nail depends on type of sheathing used, but penetration into the stud or through the wood backing should be at least 1½ inches.

Horizontally applied matched paneling in narrow widths should be blind-nailed at the tongue with a corrosion-resistant finishing nail (fig. 82). For widths greater than 6 inches. an additional nail should be used as shown.

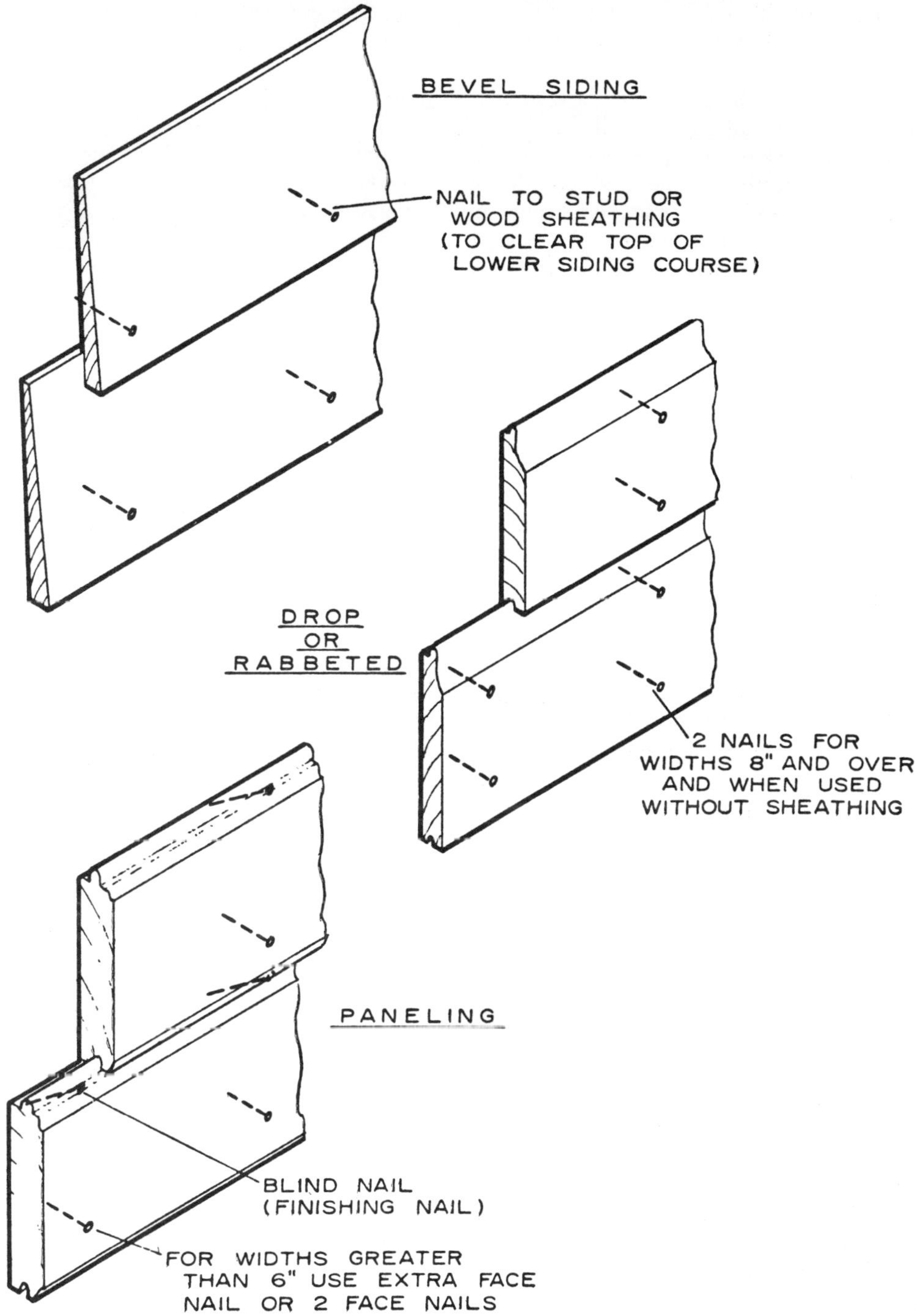

Figure 82.—Nailing of siding.

Other materials such as plywood, hardboard, or medium-density fiberboard, which are used horizontally in widths up to 12 inches, should be applied in the same manner as lap or drop siding, depending on the pattern. Prepackaged siding should be applied according to manufacturers' directions.

Vertical Sidings

Vertically applied matched and similar sidings having interlapping joints are nailed in the same manner as when applied horizontally. However, they should be nailed to blocking used between studs or to wood or plywood sheathing. Blocking is spaced from 16 to

24 inches apart. With plywood or nominal 1-inch board sheathing, nails should be spaced on 16-inch centers.

When the various combinations of boards and battens are used, they should also be nailed to blocking spaced from 16 to 24 inches apart between studs, or closer for wood sheathing. The first boards or battens should be fastened with one eightpenny or ninepenny nail at each blocking, to provide at least 1½-inch penetration. For wide under-boards, two nails spaced about 2 inches apart may be used rather than the single row along the center (fig. 80). The second or top boards or battens should be nailed with twelvepenny nails. Nails of the top board or batten should always miss the under-boards and not be nailed through them (fig. 80). In such applications, double nails should be spaced closely to prevent splitting if the board shrinks. It is also good practice to use a sheathing paper, such as 15 pound asphalt felt, under vertical siding.

Plywood and Other Sheet Siding

Exterior grade plywood, paper-overlaid plywood, and similar sheet materials used for siding are usually applied vertically. When used over sheathing, plywood should be at least ¼ inch thick, although 5/16 and ⅜ inch will normally provide a more even surface. Hardboard should be ¼ inch thick and materials such as medium-density fiberboard should be ½ inch.

All nailing should be over studs and total effective penetration into wood should be at least 1½ inches. For example, ⅜-inch plywood siding over ¾-inch wood sheathing would require about a sevenpenny nail, which is 2¼ inches long. This would result in a 1⅛-inch-penetration into the stud, but a total effective penetration of 1⅞ inches into wood.

Plywood should be nailed at 6-inch intervals around the perimeter and 12 inches at intermediate members. Hardboard siding should be nailed at 4- and 8-inch intervals. All types of sheet material should have a joint calked with mastic unless the joints are of the interlapping or matched type or battens are installed. A strip of 15-pound asphalt felt under uncalked joints is good practice.

Corner Treatment

The method of finishing wood siding or other materials at exterior corners is often influenced by the overall design of the house. A mitered corner effect on horizontal siding or the use of corner boards are perhaps the most common methods of treatment.

Mitering corners (fig. 83,*A*) of bevel and similar sidings, unless carefully done to prevent openings, is not always satisfactory. To maintain a good joint, it is necessary that the joint fit tightly the full depth of the miter. It is also good practice to treat the ends with a water-repellent preservative prior to nailing.

Metal corners (fig. 83,*B*) are perhaps more commonly used than the mitered corner and give a mitered effect. They are easily placed over each corner as the siding is installed. The metal corners should fit tightly without openings and be nailed on each side to the sheathing or corner stud beneath. If made of galavanized iron, they should be cleaned with a mild acid wash and primed with a metal primer before the house is painted to prevent early peeling of the paint. Weathering of the metal will also prepare it for the prime paint coat.

Corner boards of various types and sizes may be used for horizontal sidings of all types (fig. 83,*C*). They also provide a satisfactory termination for plywood and similar sheet materials. Vertical applications of matched paneling or of boards and battens are terminated by lapping one side and nailing into the edge of this member, as well as to the nailing members beneath. Corner boards are usually 1⅛- or 1⅜-inch material and for a distinctive appearance might be quite narrow. Plain outside casing commonly used for window and door frames can be adapted for corner boards.

Prefinished shingle or shake exteriors sometimes are used with color-matched metal corners. They can also be lapped over the adjacent corner shingle, alternating each course. This is called "lacing." This type of corner treatment usually requires that some kind of flashing be used beneath.

When siding returns against a roof surface, such as at a dormer, there should be a clearance of about 2 inches (fig. 83,*D*). Siding cut tight against the shingles retains moisture after rains and usually results in paint peeling. Shingle flashing extending well up on the dormer wall will provide the necessary resistance to entry of wind-driven rain. Here again, a water-repellent preservative should be used on the ends of the siding at the roofline.

Interior corners (fig. 83,*E*) are butted against a square corner board of nominal 1¼- or 1⅜-inch size, depending on the thickness of the siding.

Material Transition

At times, the materials used in the gable ends and in the walls below differ in form and application. The details of construction used at the juncture of the two materials should be such that good drainage is assured. For example, if vertical boards and battens are used at the gable end and horizontal siding below, a drip cap or similar molding might be used (fig. 84). Flashing should be used over and above the drip cap so that moisture will clear the gable material.

Another method of material transition might also be used. By extending the plate and studs of the gable end out from the wall a short distance, or by

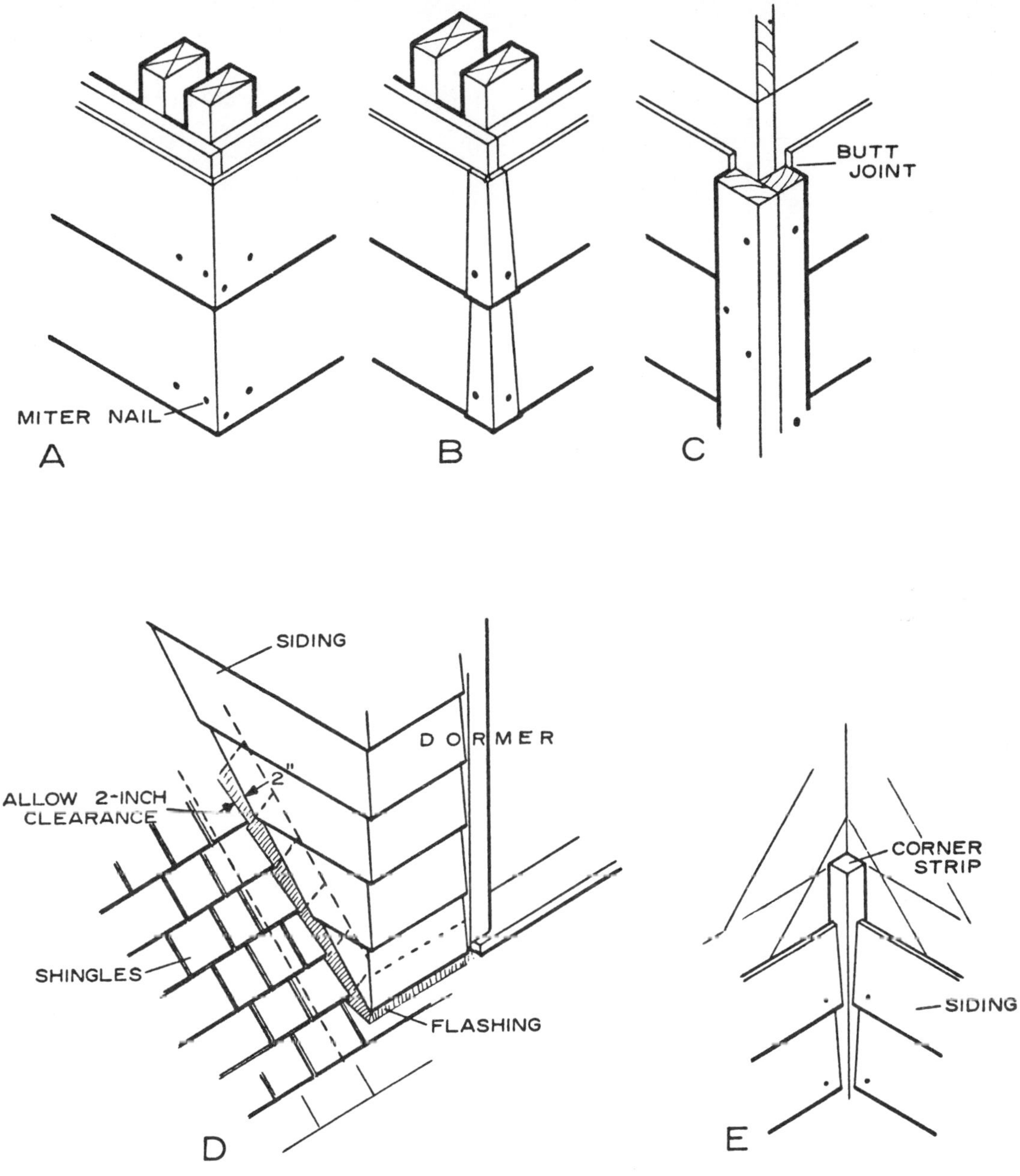

Figure 83.—Siding details: *A,* Miter corner; *B,* metal corners; *C,* corner boards; *D,* siding return at roof; *E,* interior corner.

the use of furring strips, the gable siding will project beyond the wall siding and provide good drainage (fig. 85).

Installation of Wood Shingles and Shakes

Wood shingles and shakes are applied in a single- or double-course pattern. They may be used over wood or plywood sheathing. If sheathing is $\frac{3}{8}$-inch plywood, use threaded nails. For nonwood sheathing, 1- by 3- or 1- by 4-inch wood nailing strips are used as a base. In the single-course method, one course is simply laid over the other as lap siding is applied. The shingles can be second grade because only one-half or less of the butt portion is exposed (fig. 86). Shingles should not be soaked before application but should usually be laid up with about $\frac{1}{8}$ to $\frac{1}{4}$ inch space between adjacent shingles to allow for expansion during rainy weather. When a "siding effect" is desired, shingles should be laid up so that they are only lightly in contact. Prestained or treated shingles provide the best results for this system.

In a double-course system, the undercourse is ap-

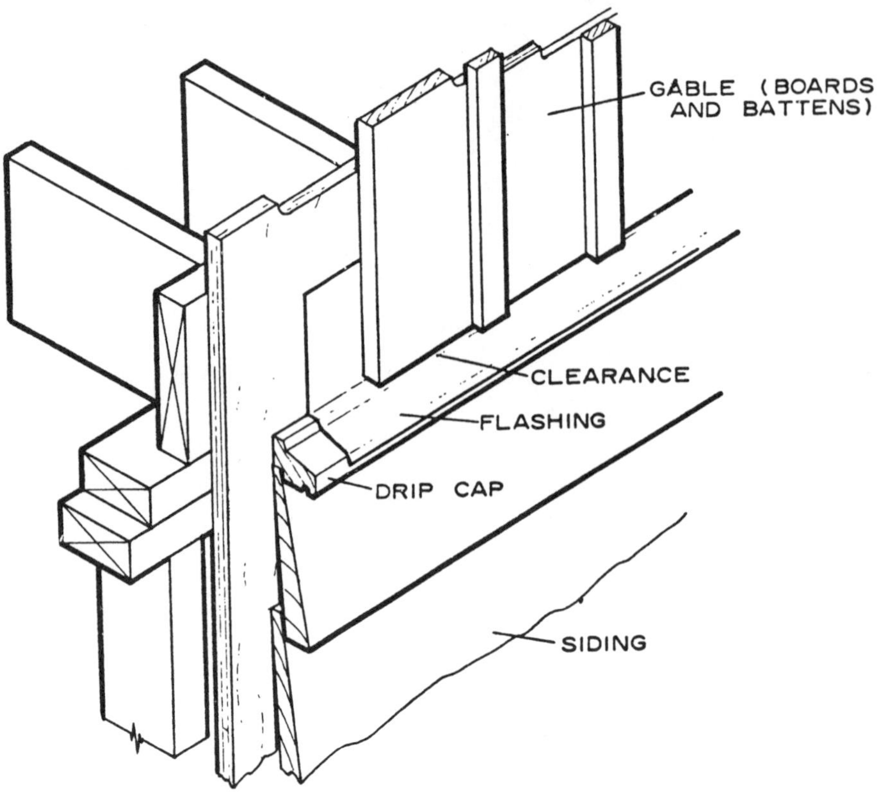

Figure 84.—Gable-end finish (material transition).

plied over the wall and the top course nailed directly over a ¼- to ½-inch projection of the butt (fig. 87). The first course should be nailed only enough to hold it in place while the outer course is being applied. The first shingles can be a lower quality, such as third grade or the undercourse grade. The top course, because much of the shingle length is exposed, should be first-grade shingles.

Exposure distance for various length shingles and shakes can be guided by the recommendations in table 5.

TABLE 5.—*Exposure distances for wood shingles and shakes on sidewalls*

Material	Length	Maximum exposure		
		Single coursing	Double coursing	
			No. 1 grade	No. 2 grade
	In.	*In.*	*In.*	*In.*
Shingles	16	7½	12	10
	18	8½	14	11
	24	11½	16	14
Shakes (hand split and resawn)	18	8½	14	--------
	24	11½	20	--------
	32	15	--------	--------

As in roof shingles, joints should be "broken" so that butt joints of the upper shingles are at least 1½ inches from the under-shingle joints.

Closed or open joints may be used in the application of shingles to sidewalls at the discretion of the builder (fig. 86). Spacing of ¼ to ⅜ inch produces an individual effect, while close spacing produces a shadow line similar to bevel siding.

Shingles and shakes should be applied with rust-resistant nails long enough to penetrate into the wood backing strips or sheathing. In single coursing, a threepenny or fourpenny zinc-coated "shingle" nail is commonly used. In double coursing, where nails are exposed, a fivepenny zinc-coated nail with a small flat head is used for the top course and threepenny or fourpenny size for the undercourse. Use building paper over lumber sheathing.

Nails should be placed in from the edge of the shingle a distance of ¾ inch (fig. 86). Use two nails for each shingle up to 8 inches wide and three nails for shingles over 8 inches. In single-course applications, nails should be placed 1 inch above the butt line of the next higher course. In double coursing, the use of a piece of shiplap sheathing as a guide allows the outer course to extend ½ inch below the undercourse, producing a shadow line (fig. 87). Nails should be placed 2 inches above the bottom of the single or shake. Rived or fluted processed shakes,

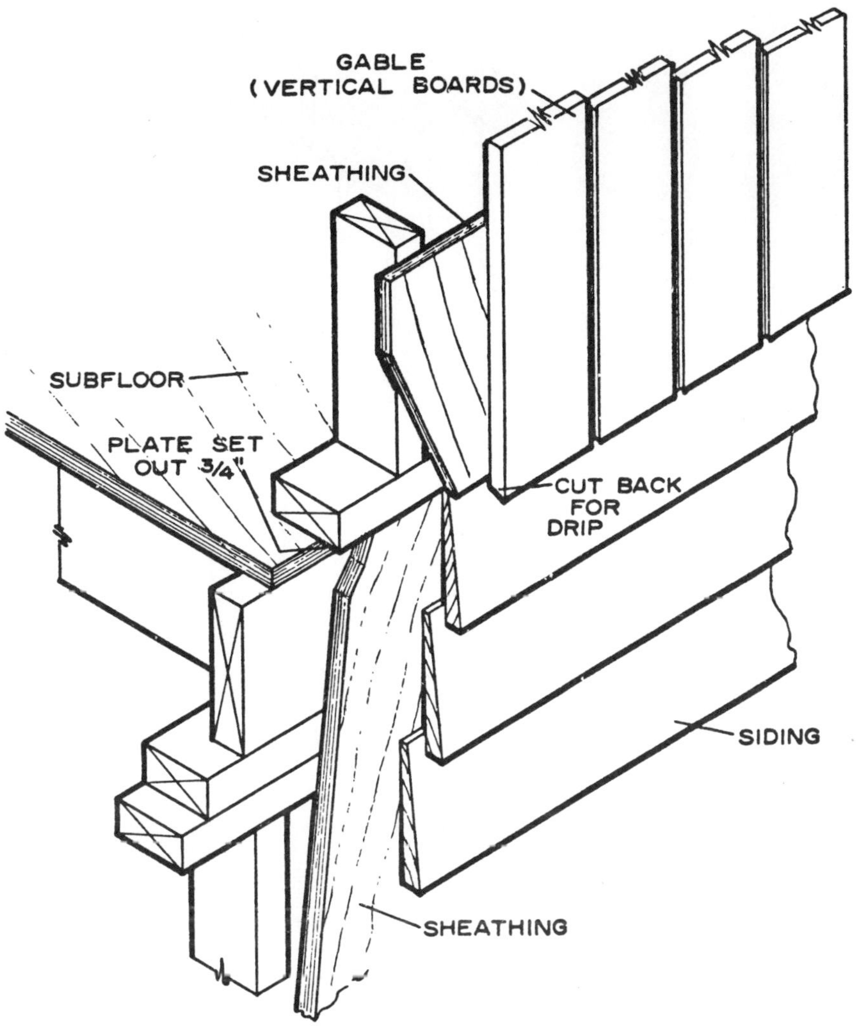

Figure 85.—Gable-end projection (material transition).

usually factory-stained, are available and have a distinct effect when laid with closely fitted edges in a double-course pattern.

Nonwood Coverings

Asbestos-Cement Shingles

Asbestos-cement shingles and similar nonwood exterior coverings should be applied in accordance with the manufacturer's directions. They are used over wood or plywood sheathing or over spaced nailing strips. Nails are of the noncorrosive type and usually are available to match the color of the shingles. Manufacturers also supply matching color corners.

Cement-Plaster Finish

Stucco and similar cement-mortar finishes, most commonly used in the Southwest, are applied over a coated expanded-metal lath and, usually, over some type of sheathing. However, in some areas where local building regulations permit, such a finish is applied to metal lath fastened directly to the braced wall framework. Waterproof paper is used over the studs before the metal lath is applied.

When a plastered exterior is applied to two-story houses, balloon framing is recommended (fig. 35). If platform framing is used for one-story houses (fig. 31), shrinkage of joists and sills may cause an unsightly bulge or break in the cement-plaster at those points unless joists have reached moisture equilibrium. This stresses the need for proper moisture content of the framing members when this type of finish is used.

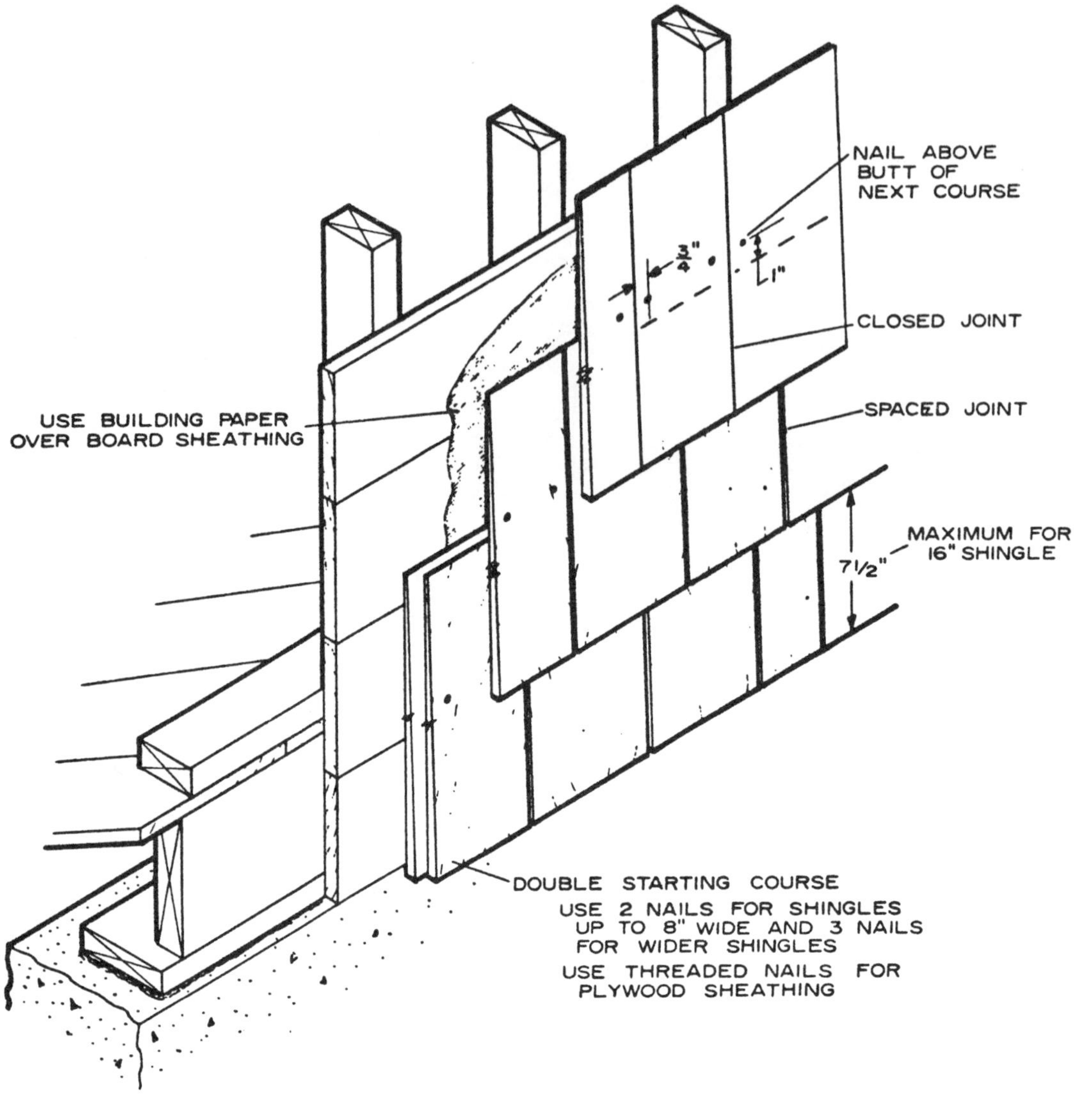

Figure 86.—Single-coursing of sidewalls (wood shingles - shakes)

Masonry Veneer

Brick or stone veneer is used for all or part of the exterior wall finish for some styles of architecture. The use of balloon framing for brick-veneered two-story houses will prevent cracks due to shrinkage of floor joists. It is good practice, when possible, to delay applying the masonry finish over platform framing until the joists and other members reach moisture equilibrium. The use of a waterproof paper backing and sufficient wall ties is important. Details on the installation of masonry veneer are shown in figure 12.

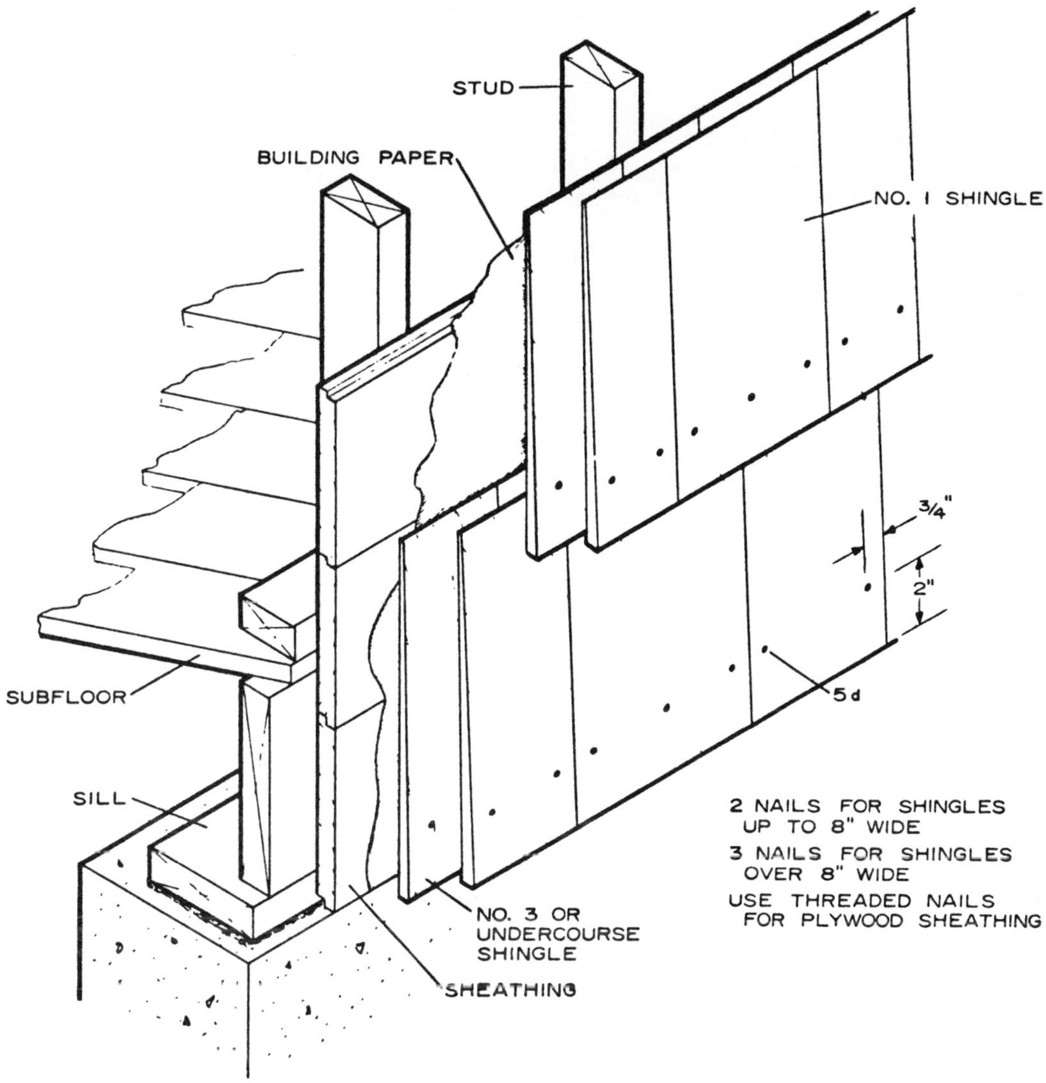

Figure 87.—Double-coursing of sidewalls (wood shingles-shakes).

CHAPTER 14

FRAMING DETAILS FOR PLUMBING, HEATING, AND OTHER UTILITIES

It is desirable, when framing a house, to limit cutting of framing members for installation of plumbing lines and other utilities. A little planning before framing is started will reduce the need for cutting joists and other members. This is more easily accomplished in one-story houses, however, than in two-story houses. In a single-story house, many of the connections are made in the basement area; in two-story houses they must be made between the first-floor ceiling joists. Thus, it is sometimes necessary to cut or notch joists, but this should be done in a manner least detrimental to their strength (*12*).

Plumbing Stack Vents

One wall of the bath, kitchen, or utility room is normally used to carry the water, vent, and drainage lines. This is usually the wall behind the water closet where connections can be easily made to the tub or shower and to the lavatory. When 4-inch cast-iron bell pipe is used in the soil and vent stack, it is necessary to use 2- by 6- or 2- by 8-inch plates to provide space for the pipe and the connections. Some contractors use a double row of studs placed flatwise so that no drilling is required for the horizontal runs (fig. 88,*A*).

Building regulations in some areas allow the use of 3-inch pipe for venting purposes in one-story houses. When this size is used, 2- by 4-inch plates and studs may be employed. However, it is then necessary to reinforce the top plates, which have been cut, by using a *double scab* (fig. 88,*B*). Scabs are well nailed on each side of the stack and should extend over two studs. Small angle irons can also be used.

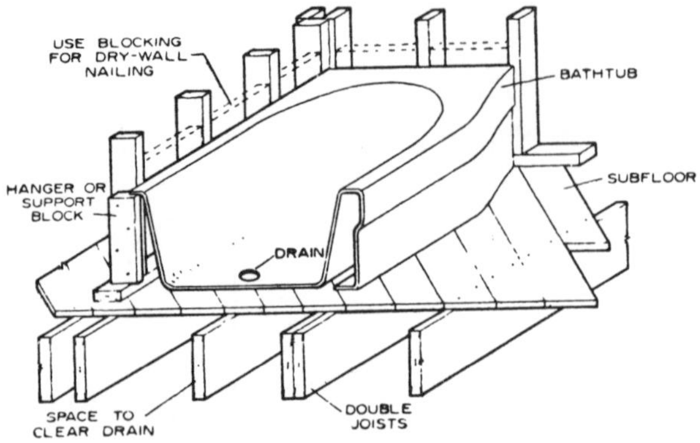

Figure 89.—Framing for bathtub.

Bathtub Framing

A bathtub full of water is heavy; so floor joists must be arranged to carry the load without excessive deflection. Too great a deflection will sometimes cause an opening above the edge of the tub. Joists should be doubled at the outer edge (fig. 89). The intermediate joist should be spaced to clear the drain. Metal hangers or wood blocking support the inner edge of the tub at the wall line.

Cutting Floor Joists

Floor joists should be cut, notched, or drilled only where the effect on strength is minor. While it is always desirable to prevent cutting joists whenever possible, sometimes such alterations are required. Joists or other structural members should then be reinforced by nailing a reinforcing scab to each side or by adding an additional member. Well-nailed plywood scabs on one or both sides of altered joists also provide a good method of reinforcing these members.

Notching the top or bottom of the joist should only be done in the end quarter of the span and not more than one-sixth of the depth. When greater alterations are required, headers and tail beams should be added around the altered area. This may occur where the closet bend must cross the normal joist locations. In other words, it should be framed out similar to a stair opening (fig. 28).

When necessary, holes may be bored in joists if the diameter is no greater than 2 inches and the edge of the hole is not less than 2½ to 3 inches from the top or bottom edge of the joists (fig. 90). This usually limits the joist size to a 2 by 8 or larger member.

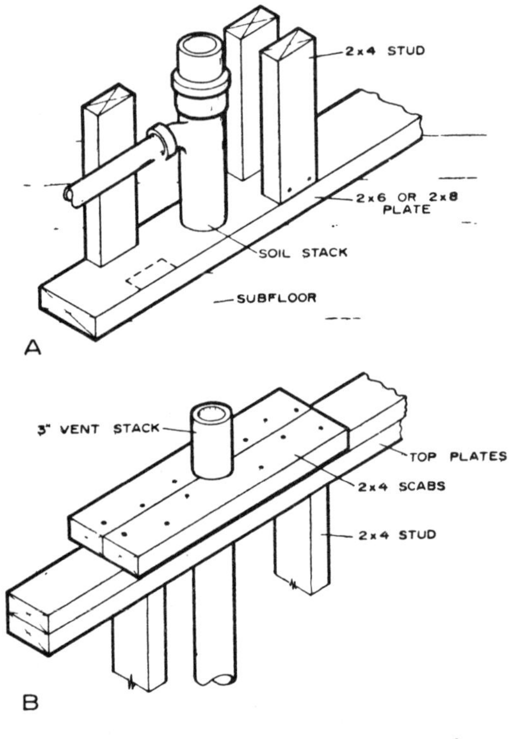

Figure 88.—Plumbing stacks: *A*, 4-inch cast-iron stack; *B*, 3-inch pipe for vent.

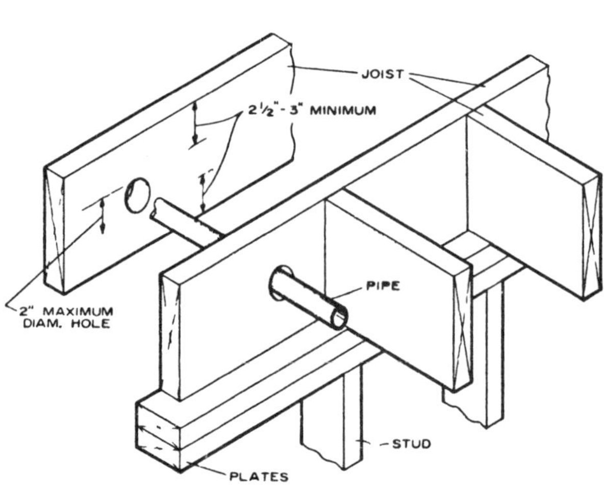

Figure 90.—Drilled holes in joists.

Such a method of installation is suitable where joist direction changes and the pipe can be inserted from the long direction, such as from the plumbing wall to a tub on the second floor. Connections for first-floor plumbing can normally be made without cutting or drilling of joists.

Alterations for Heating Ducts

A number of systems are used to heat a house, from a multi-controlled hot-water system to a simple floor or wall furnace. Central air conditioning combined with the heating system is becoming a normal part of house construction. Ducts and piping should be laid out so that framing or other structural parts can be adjusted to accommodate them. However, the system which requires heat or cooling ducts and return lines is perhaps the most important from the standpoint of framing changes required.

Supply and Cold Air Return Ducts

The installation of ducts for a forced-warm-air or air-conditioning system usually requires the removal of the soleplate and the subfloor at the duct location. Supply ducts are made to dimensions that permit them to be placed between studs. When the same duct system is used for heating and cooling, the duct sizes are generally larger than when they are designed for heating alone. Such systems often have two sets of registers; one near the floor for heat and one near the ceiling for more efficient cooling. Both are furnished with dampers for control.

Walls and joists are normally located so that they do not have to be cut when heating ducts are installed. This is especially true when partitions are at right angles to the floor joists.

When a load-bearing partition requires a doubled parallel floor joist as well as a warm-air duct, the joists can be spaced apart to allow room for the duct (fig. 91). This will eliminate the need for excessive cutting of framing members or the use of intricate pipe angles.

Cold-air returns are generally located in the floor between joists or in the walls at floor level (fig. 92). They are sometimes located in outside walls, in which case they should be lined with metal. *Unlined ducts* in exterior walls have been known to be responsible for exterior-wall paint failures, especially those from a second-floor room.

The elbow from the return duct below the floor is usually placed between floor joists. The space between floor joists, when enclosed with sheet metal, serves as a cold-air return. Other cold-air returns may connect with the same joist-space return duct.

Framing for Convectors

Convectors and hot-water or steam radiators are sometimes recessed partly into the wall to provide

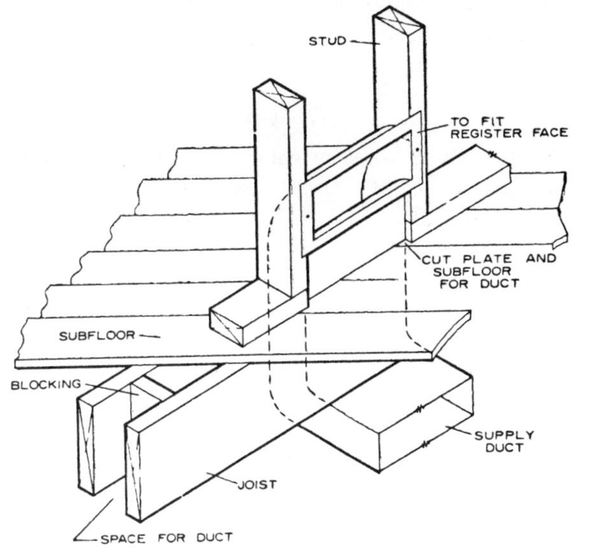

Figure 91.—Spaced joists for supply ducts.

more usable space in the room and improve appearance by the installation of a decorative grill. Such framing usually requires the addition of a doubled header to carry the wall load from the studs above (fig. 93). Size of the headers depends on the span and should be designed the same as those for window or door openings. The sizes in the tabulation listed under window and door framing in Chapter 6, "Wall Framing", should be used to determine the correct

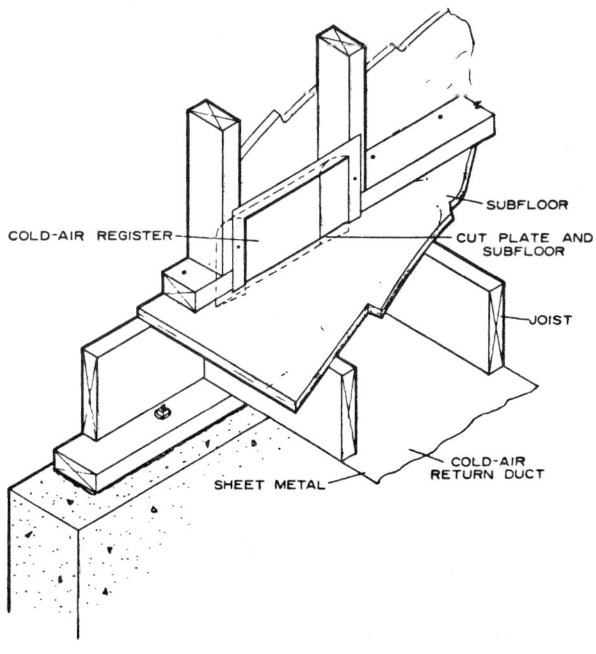

Figure 92.—Cold-air return.

sizes. Because only $1\frac{5}{8}$ inches of space in the wall is available for insulation, a highly efficient insulation (one with a low "k" value) is sometimes used.

Wiring

House wiring for electrical services is usually started some time after the house has been closed in. The initial phase, of it, termed "roughing in," includes the installation of conduit or cable and the location of switch, light, and outlet boxes with wires ready to connect. This roughing-in work is done before the plaster base or dry-wall finish is applied, and before the insulation is placed in the walls or ceilings. The placement of the fixtures, the switches, and switch plates is done after plastering.

Framing changes for wiring are usually of a minor nature and, for the most part, consist of holes drilled in the studs for the flexible conduit. Although these holes are small in diameter, they should comply with locations shown in figure 90. Perhaps the only area which requires some planning to prevent excessive cutting or drilling is the location of wall switches at entrance door frames. By spacing the doubled framing studs to allow for location of multiple switch boxes, little cutting will be required.

Switches or convenience outlet boxes on exterior walls must be sealed to prevent water vapor movement. Sealing of the vapor barrier around the box is important and will be discussed further in Chapter 15, "Thermal Insulation and Vapor Barriers."

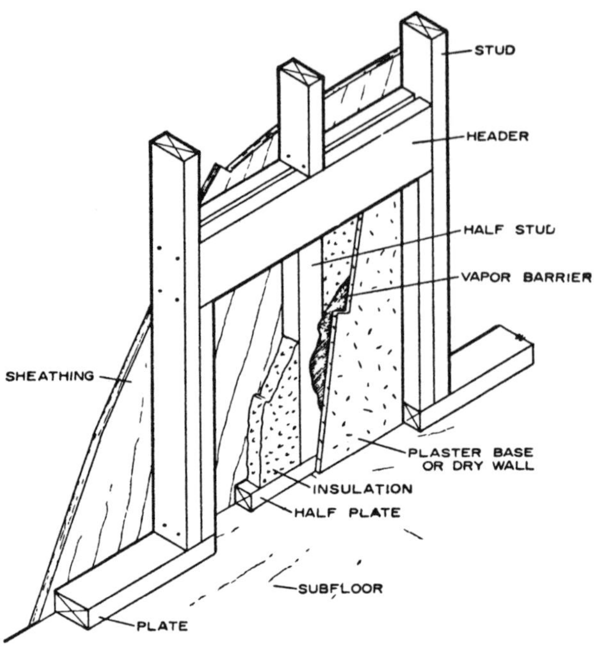

Figure 93.—Framing for a convector recess.

CHAPTER 15

THERMAL INSULATION AND VAPOR BARRIERS

Most materials used in houses have some insulating value. Even air spaces between studs resist the passage of heat. However, when these stud spaces are filled or partially filled with a material high in resistance to heat transmission, namely thermal insulation, the stud space has many times the insulating value of the air alone.

The inflow of heat through outside walls and roofs in hot weather or its outflow during cold weather have important effects upon (a) the comfort of the occupants of a building and (b) the cost of providing either heating or cooling to maintain temperatures at acceptable limits for occupancy. During cold weather, high resistance to heat flow also means a saving in fuel. While the wood in the walls provides good insulation, commercial insulating materials are usually incorporated into exposed walls, ceilings, and floors to increase the resistance to heat passage. The use of insulation in warmer climates is justified with air conditioning, not only because of reduced operating costs but also because units of smaller capacity are required. Thus, whether from the standpoint of thermal insulation alone in cold climates or whether for the benefit of reducing cooling costs, the use of 2 inches or more of insulation in the walls can certainly be justified.

Average winter low-temperature zones of the United States are shown in figure 94. These data are used in determining the size of heating plant required after calculating heat loss. This information is also useful in selecting the amount of insulation for walls, ceilings, and floors.

Insulating Materials

Commercial insulation is manufactured in a variety of forms and types, each with advantages for specific uses. Materials commonly used for insulation may be grouped in the following general classes: (1) Flexible insulation (blanket and batt); (2) loose-fill insulation; (3) reflective insulation; (4) rigid insulation (struc-

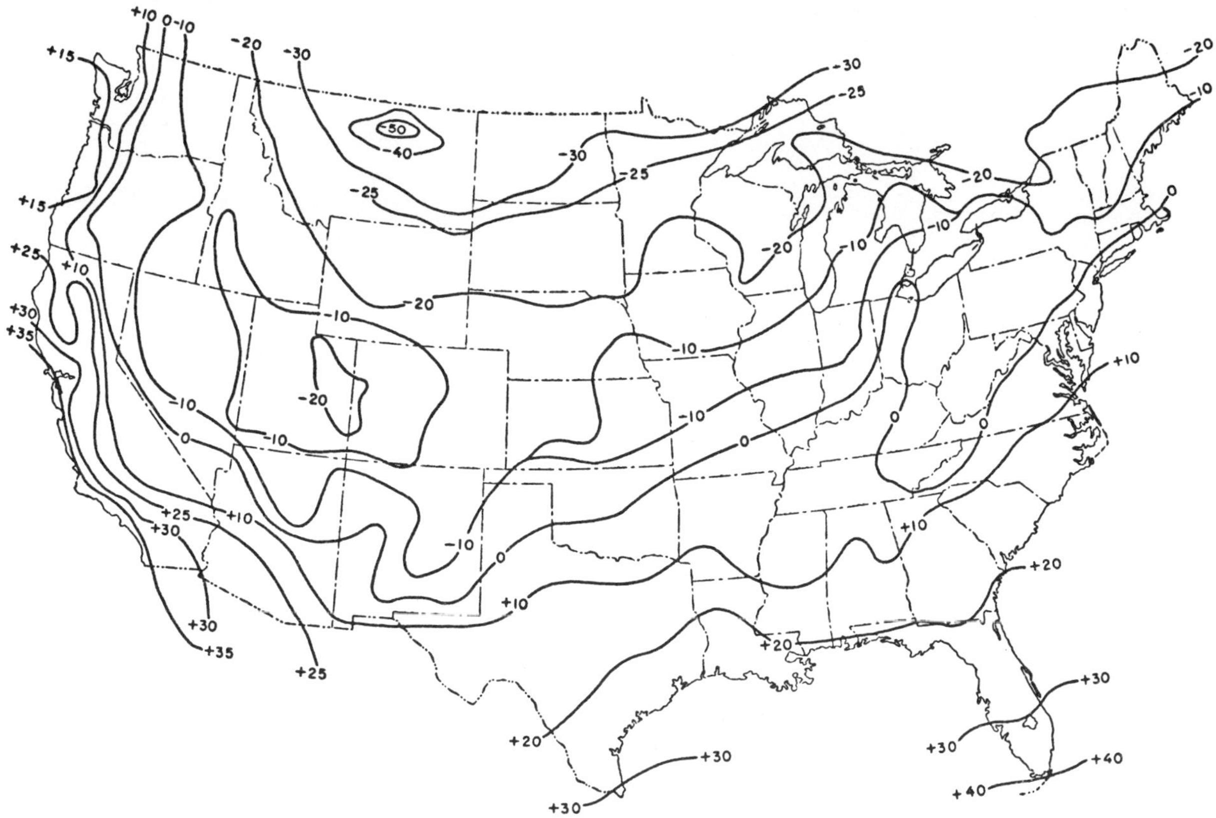

Figure 94.—Average outside design temperature zones of the United States.

tural and nonstructural); and (5) miscellaneous types.

The thermal properties of most building materials are known, and the rate of heat flow or coefficient of transmission for most combinations of construction can be calculated (4). This coefficient, or *U-value*, is a measure of heat transmission between air on the warm side and air on the cold side of the construction unit. The insulating value of the wall will vary with different types of construction, with materials used in construction, and with different types and thickness of insulation. Comparisons of U-values may be made and used to evaluate different combinations of materials and insulation based on overall heat loss, potential fuel savings, influence on comfort, and installation costs.

Air spaces add to the total resistance of a wall section to heat transmission, but an air space is not as effective as it would be if filled with an insulating material. Great importance is frequently given to dead-air spaces in speaking of a wall section. Actually, the air in never dead in cells where there are differences in temperature on opposite sides of the space, because the difference causes convection currents.

Information regarding the calculated U-values for typical constructions with various combinations of insulation may be found in "Thermal Insulation from Wood for Buildings: Effects of Moisture and Its Control" (*15*).

Flexible Insulation

Flexible insulation is manufactured in two types, *blanket* and *batt.* Blanket insulation (fig. 95,*A*) is furnished in rolls or packages in widths suited to 16- and 24-inch stud and joist spacing. Usual thicknesses are $1\frac{1}{2}$, 2, and 3 inches. The body of the blanket is made of felted mats of mineral or vegetable fibers, such as rock or glass wool, wood fiber, and cotton. Organic insulations are treated to make them resistant to fire, decay, insects, and vermin. Most blanket insulation is covered with paper or other sheet material with tabs on the sides for fastening to studs or joists. One covering sheet serves as a vapor barrier to resist movement of water vapor and should always face the warm side of the wall. Aluminum foil or asphalt or plastic laminated paper are commonly used as barrier materials.

Batt insulation (fig. 95,*B*) is also made of fibrous material preformed to thicknesses of 4 and 6 inches for 16- and 24-inch joist spacing. It is supplied with or without a vapor barrier. One friction type of fibrous

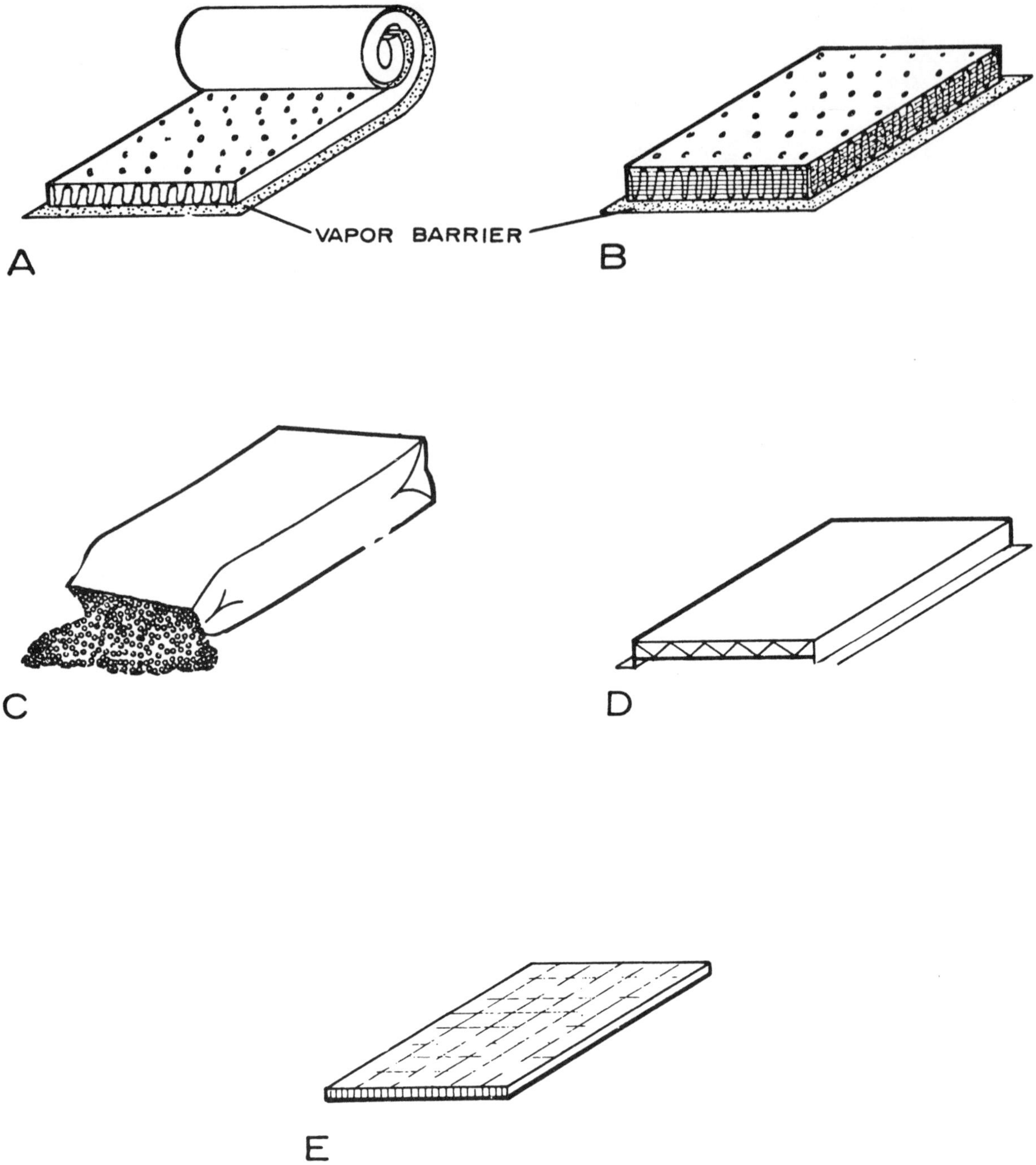

Figure 95.—Types of Insulation: *A*, Blanket; *B*, batt; *C*, fill; *D*, reflective (one type), *E*, rigid.

glass batt is supplied without a covering and is designed to remain in place without the normal fastening methods.

Loose Fill Insulation

Loose fill insulation (fig. 95,*C*) is usually composed of materials used in bulk form, supplied in bags or bales, and placed by pouring, blowing, or packing by hand. This includes rock or glass wool, wood fibers, shredded redwood bark, cork, wood pulp products, vermiculite, sawdust, and shavings.

Fill insulation is suited for use between first-floor ceiling joists in unheated attics. It is also used in sidewalls of existing houses that were not insulated during construction. Where no vapor barrier was installed during construction, suitable paint coatings, as described later in this chapter, should be used for vapor barriers when blown insulation is added to an existing house.

Reflective Insulation

Most materials reflect some radiant heat, and some materials have this property to a very high degree (*4*). Materials high in reflective properties include aluminum foil, sheet metal with tin coating, and paper products coated with a reflective oxide composition. Such materials can be used in enclosed stud spaces, in attics, and in similar locations to retard heat transfer by radiation. These reflective insulations are effective only when used where the reflective surface faces an air space at least ¾ inch or more deep. Where a reflective surface contacts another material, the reflective properties are lost and the material has little or no insulating value.

Reflective insulations are equally effective regardless of whether the reflective surface faces the warm or cold side. However, there is a decided difference in the equivalent conductance and the resistance to heat flow. The difference depends on (a) the orientation of the reflecting material and the dead air space, (b) the direction of heat flow (horizontal, up, or down), and (c) the mean summer or winter temperatures. Each possibility requires separate consideration. However, reflective insulation is perhaps more effective in preventing summer heat flow through ceilings and walls. It should likely be considered more for use in the southern portion of the United States than in the northern portion.

Reflective insulation of the foil type is sometimes applied to blankets and to the stud-surface side of gypsum lath. Metal foil suitably mounted on some supporting base makes an excellent vapor barrier. The type of reflective insulation shown in figure 95,*D* includes reflective surfaces and air spaces between the outer sheets.

Rigid Insulation

Rigid insulation is usually a fiberboard material manufactured in sheet and other forms (fig. 95,*E*). However, rigid insulations are also made from such materials as inorganic fiber and glass fiber, though not commonly used in a house in this form. The most common types are made from processed wood, sugarcane, or other vegetable products. Structural insulating boards, in densities ranging from 15 to 31 pounds per cubic foot, are fabricated in such forms as building boards, roof decking, sheathing, and wallboard. While they have moderately good insulating properties, their primary purpose is structural.

Roof insulation is nonstructural and serves mainly to provide thermal resistance to heat flow in roofs. It is called "slab" or "block" insulation and is manufactured in rigid units ½ to 3 inches thick and usually 2 by 4 feet in size.

In house construction, perhaps the most common forms of rigid insulation are sheathing and decorative coverings in sheets or in tile squares. Sheathing board is made in thicknesses of ½ and 25/32 inch. It is coated or impregnated with an asphalt compound to provide water resistance. Sheets are made in 2- by 8-foot size for horizontal application and 4- by 8-feet or longer for vertical application.

Miscellaneous Insulation

Some insulations do not fit in the classifications previously described, such as insulation blankets made up of multiple layers of corrugated paper. Other types, such as lightweight vermiculite and perlite aggregates, are sometimes used in plaster as a means of reducing heat transmission.

Other materials are foamed-in-place insulations. which include sprayed and plastic foam types. Sprayed insulation is usually inorganic fibrous material blown against a clean surface which has been primed with an adhesive coating. It is often left exposed for acoustical as well as insulating properties.

Expanded *polystyrene* and *urethane* plastic foams may be molded or foamed-in-place. Urethane insulation may also be applied by spraying. Polystyrene and urethane in board form can be obtained in thicknesses from ½ to 2 inches.

Values in table 6 will provide some comparison of the insulating value of the various materials. These are expressed as "k" values or heat conductivity and are defined as the amount of heat, in British thermal units, that will pass in 1 hour through 1 square foot of material 1 inch thick per 1° F. temperature differ-

TABLE 6.—*Thermal conductivity values of some insulating materials*

Insulation group		"k" range (conductivity)
General	Specific type	
Flexible		0.25 – 0.27
Fill	Standard materials	.28 – .30
	Vermiculite	.45 – .48
Reflective (2 sides)		[1]
Rigid	Insulating fiberboard	.35 – .36
	Sheathing fiberboard	.42 – .55
Foam	Polystyrene	.25 – .29
	Urethane	.15 – .17
Wood	Low density	.60 – .65

[1] Insulating value is equal to slightly more than 1 inch of flexible insulation. (Resistance, "R" = 4.3)

ence between faces of the material. Simply expressed, "k" represents heat loss; the lower this numerical value, the better the insulating qualities.

Insulation is also rated on its resistance or "R" value, which is merely another expression of its insulating value. The "R" value is usually expressed as the total resistance of the wall or of a thick insulating blanket or batt, whereas "k" is the rating per inch of thickness. For example, a "k" value of 1 inch of insulation is 0.25. Then the resistance, "R" is $\frac{1}{0.25}$ or 4.0. If there is three inches of this insulation, the total "R" is three times 4.0, or 12.0.

The "U" value is the overall heat-loss value of all materials in the wall. The lower this value, the better the insulating value. Specific insulating values for various materials are also available (*4, 15*). For comparison with table 6, the "U" value of window glass is:

Glass	*U value*
Single	1.13
Double	
Insulated, with ¼-inch air space	.61
Storm sash over single glazed window	.53

Where to Insulate

To reduce heat loss from the house during cold weather in most climates, all walls, ceilings, roofs, and floors that separate heated from unheated spaces should be insulated.

Insulation should be placed on all outside walls and in the ceiling (fig. 96,*A*). In houses involving unheated crawl spaces, it should be placed between the floor joists or around the wall perimeter. If a flexible type of insulation (blanket or batt) is used, it should be well-supported between joists by slats and a galvanized wire mesh, or by a rigid board with the vapor barrier installed toward the subflooring. Press-fit or friction insulations fit tightly between joists and require only a small amount of support to hold them in place. Reflective insulation is often used for crawl spaces, but only one dead-air space should be assumed in calculating heat loss when the crawl space is ventilated. A ground cover of roll roofing or plastic film such as polyethylene should be placed on the soil of crawl spaces to decrease the moisture content of the space as well as of the wood members.

In 1½-story houses, insulation should be placed along all walls, floors, and ceilings that are adjacent to unheated areas (fig. 96,*B*). These include stairways, dwarf (knee) walls, and dormers. Provisions should be made for ventilation of the unheated areas.

Where attic space is unheated and a stairway is included, insulation should be used around the stairway as well as in the first-floor ceiling (fig. 96,*C*). The door leading to the attic should be weather-stripped to prevent heat loss. Walls adjoining an unheated garage or porch should also be insulated.

In houses with flat or low-pitched roofs (fig. 96,*D*), insulation should be used in the ceiling area with sufficient space allowed above for clear unobstructed ventilation between the joists. Insulation should be used along the perimeter of houses built on slabs. A vapor barrier should be included under the slab.

In the summer, outside surfaces exposed to the direct rays of the sun may attain temperatures of 50° F. or more above shade temperatures and, of course, tend to transfer this heat toward the inside of the house. Insulation in the walls and in attic areas retards the flow of heat and, consequently, less heat is transferred through such areas, resulting in improved summer comfort conditions.

Where air-conditioning systems are used, insulation should be placed in all exposed ceilings and walls in the same manner as insulating against cold-weather heat loss. Shading of glass against direct rays of the sun and the use of insulated glass will aid in reducing the air-conditioning load.

Ventilation of attic and roof spaces is an important adjunct to insulation. Without ventilation, an attic space may become very hot and hold the heat for many hours. (See Chapter 16, "Ventilation.") Obviously, more heat will be transmitted through the ceiling when the attic temperature is 150° F. than if it is 100° to 120° F. Ventilation methods suggested for protection against cold-weather condensation apply equally well to protection against excessive hot-weather roof temperatures.

The use of storm windows or insulated glass will greatly reduce heat loss. Almost twice as much heat loss occurs through a single glass as through a window glazed with insulated glass or protected by a storm sash. Furthermore, double glass will normally prevent surface condensation and frost forming on inner glass surfaces in winter. When excessive condensation persists, paint failures or even decay of the sash rail or other parts can occur.

How to Install Insulation

Blanket insulation or batt insulation with a vapor barrier should be placed between framing members so that the tabs of the barrier lap the edge of the studs as well as the top and bottom plates. This method is not often popular with the contractor because it is more difficult to apply the dry wall or rock lath (plaster base). However, it assures a minimum amount of vapor loss compared to the loss when tabs are stapled to the sides of the studs. To protect the head and soleplate as well as the headers over openings, it is good practice to use narrow strips of vapor barrier material along the top and bottom of the wall (fig. 97,*A*). Ordinarily, these areas are not covered too well by the barrier on the blanket or

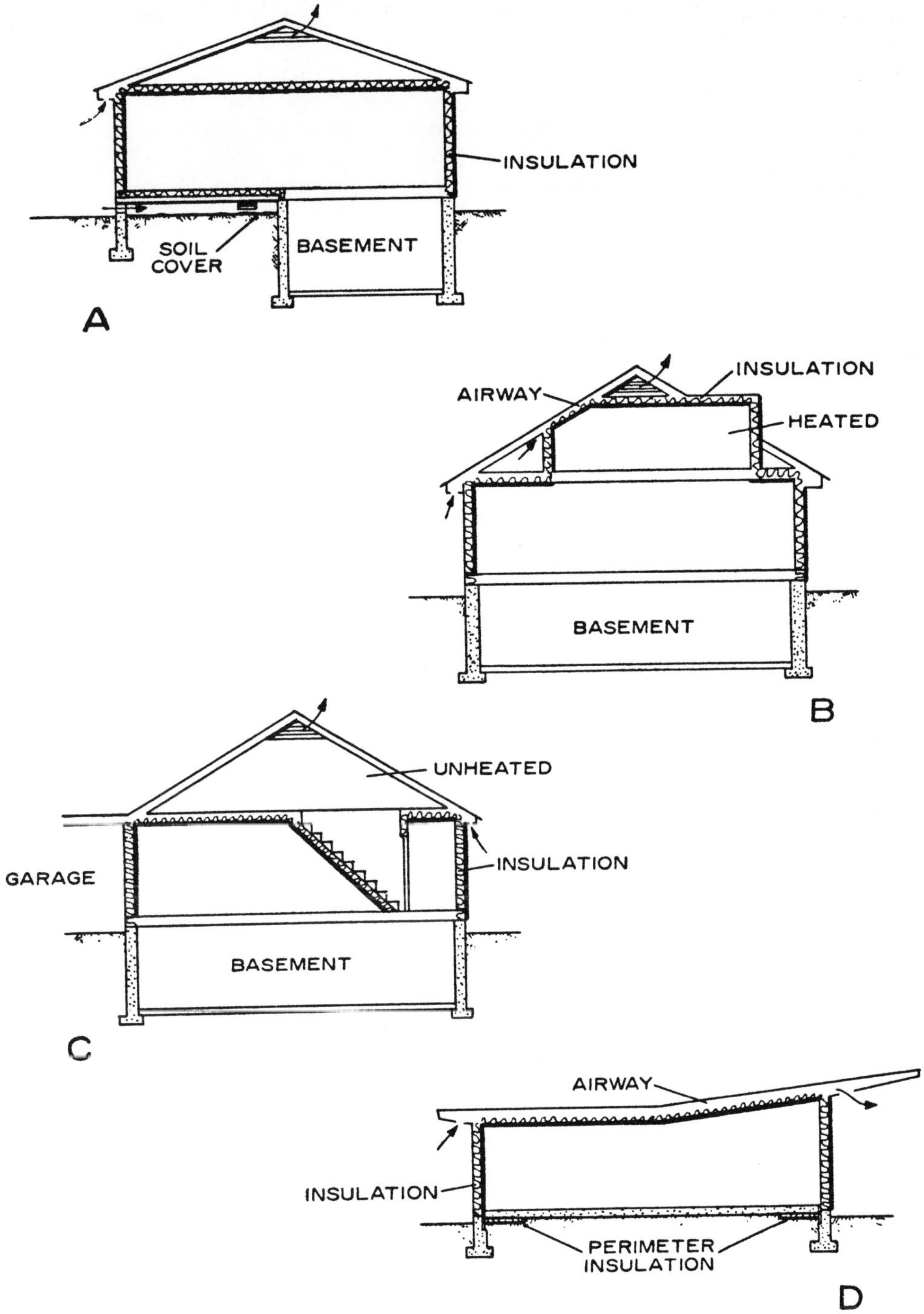

Figure 96.—Placement of insulation: *A*, In walls, floor, and ceiling; *B*, in 1½-story house; *C*, at attic door; *D*, in flat roof.

batt. A hand stapler is commonly used to fasten the insulation and the barriers in place.

For insulation without a barrier (press-fit or friction type), a plastic film vapor barrier such as 4-mil polyethylene is commonly used to envelop the entire exposed wall and ceiling (fig. 97,*B*). It covers the openings as well as window and door headers and edge studs. This system is one of the best from the standpoint of resistance to vapor movement. Furthermore, it does not have the installation inconveniences encountered when tabs of the insulation are stapled over the edges of the studs. After the dry wall is installed or plastering is completed, the film is trimmed around the window and door openings.

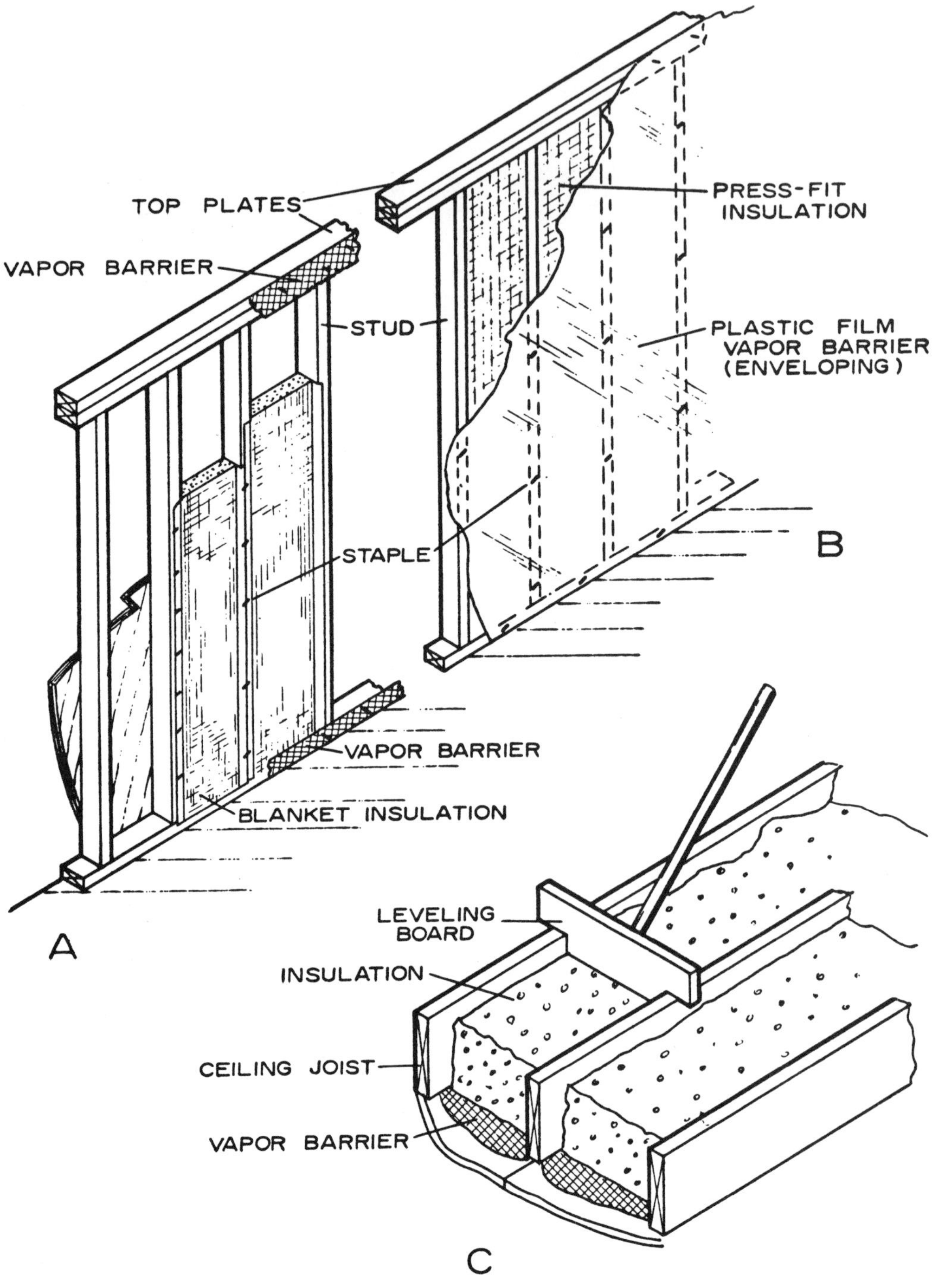

Figure 97.—Application of insulation: *A*, Wall section with blanket type; *B*, wall section with "press-fit" insulation; C, ceiling with full insulation.

Reflective insulation, in a single-sheet form with two reflective surfaces, should be placed to divide the space formed by the framing members into two approximately equal spaces. Some reflective insulations include air spaces and are furnished with nailing tabs. This type is fastened to the studs to provide at least a ¾-inch space on each side of the reflective surfaces.

Fill insulation is commonly used in ceiling areas and is poured or blown into place (fig. 97,*C*). A vapor barrier should be used on the warm side (the bottom, in case of ceiling joists) before insulation is placed. A leveling board (as shown) will give a constant insulation thickness. Thick batt insulation is also used in ceiling areas. Batt and fill insulation might also be combined to obtain the desired thickness with the vapor barrier against the back face of the ceiling finish. Ceiling insulation 6 or more inches thick greatly reduces heat loss in the winter and also provides summertime protection.

Precautions in Insulating

Areas over door and window frames and along side and head jambs also require insulation. Because these areas are filled with small sections of insulation, a vapor barrier must be used around the opening as well as over the header above the openings (fig. 98,*A*). Enveloping the entire wall eliminates the need for this type of vapor barrier installation.

In 1½- and 2-story houses and in basements, the area at the joist header at outside walls should be insulated and protected with a vapor barrier (fig. 98,*B*).

Insulation should be placed behind electrical outlet boxes and other utility connections in exposed walls to minimize condensation on cold surfaces.

Vapor Barriers

Some discussion of vapor barriers has been included in the previous sections because vapor barriers are usually a part of flexible insulation. However, further information is included in the following paragraphs.

Most building materials are permeable to water vapor. This presents problems because considerable water vapor is generated in a house from cooking, dishwashing, laundering, bathing, humidifiers, and other sources. In cold climates during cold weather, this vapor may pass through wall and ceiling materials and condense in the wall or attic space; subsequently, in severe cases, it may damage the exterior paint and interior finish, or even result in decay in structural members. For protection, a material highly resistive to vapor transmission, called a *vapor barrier*, should be used on the warm side of a wall or below the insulation in an attic space.

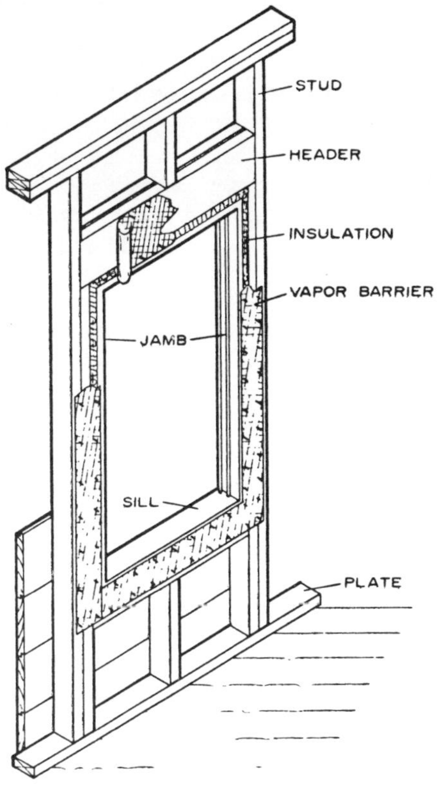

A

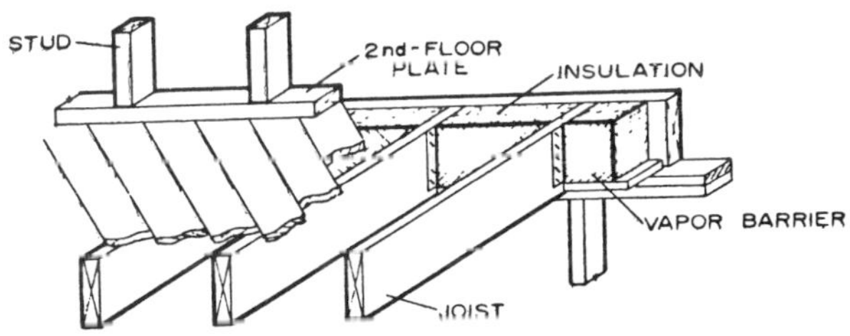

B

Figure 98.—Precautions in insulating: *A*, Around openings; *B*, joist space in outside walls.

Among the effective vapor-barrier materials are asphalt laminated papers, aluminum foil, and plastic films. Most blanket and batt insulations are provided with a vapor barrier on one side, some of them with paper-backed aluminum foil. Foil-backed gypsum lath or gypsum boards are also available and serve as excellent vapor barriers.

The perm values of vapor barriers vary (*17*), but ordinarily it is good practice to use those which have values less than ¼ (0.25) perm. Although a value of ½ perm is considered adequate, aging reduces the effectiveness of some materials.

Some types of flexible blanket and batt insulations

have a barrier material on one side. Such flexible insulations should be attached with the tabs at their sides fastened on the inside (narrow) edges of the studs, and the blanket should be cut long enough so that the cover sheet can lap over the face of the sole-plate at the bottom and over the plate at the top of the stud space. However, such a method of attachment is not the common practice of most installers. When a positive seal is desired, wall-height rolls of plastic-film vapor barriers should be applied over studs, plates, and window and door headers. This system, called "enveloping," is used over insulation having no vapor barrier or to insure excellent protection when used over any type of insulation. The barrier should be fitted tightly around outlet boxes and sealed if necessary. A ribbon of sealing compound around an outlet or switch box will minimize vapor loss at this area. Cold-air returns in outside walls should consist of metal ducts to prevent vapor loss and subsequent paint problems.

Paint coatings on plaster may be very effective as vapor barriers if materials are properly chosen and applied. They do not however, offer protection during the period of construction, and moisture may cause paint blisters on exterior paint before the interior paint can be applied. This is most likely to happen in buildings that are constructed during periods when outdoor temperatures are 25° F. or more below inside temperatures. Paint coatings cannot be considered a substitute for the membrane types of vapor barriers, but they do provide some protection for houses where other types of vapor barriers were not installed during construction.

Of the various types of paint, one coat of *aluminum primer* followed by two decorative coats of *flat wall* or *lead and oil* paint is quite effective. For rough plaster or for buildings in very cold climates, two coats of the aluminum primer may be necessary. A primer and sealer of the pigmented type, followed by decorative finish coats or two coats of rubber-base paint, are also effective in retarding vapor transmission.

Because no type of vapor barrier can be considered 100 percent resistive, and some vapor leakage into the wall may be expected, the flow of vapor to the outside should not be impeded by materials of relatively high vapor resistance on the cold side of the vapor barrier. For example, sheathing paper should be of a type that is waterproof but not highly vapor resistant. This also applies to "permanent" outer coverings or siding. In such cases, the vapor barrier should have an equally low perm value. This will reduce the danger of condensation on cold surfaces within the wall.

CHAPTER 16

VENTILATION

Condensation of moisture vapor may occur in attic spaces and under flat roofs during cold weather. Even where vapor barriers are used, some vapor will probably work into these spaces around pipes and other inadequately protected areas and some through the vapor barrier itself. Although the amount might be unimportant if equally distributed, it may be sufficiently concentrated in some cold spot to cause damage. While wood shingle and wood shake roofs do not resist vapor movement, such roofings as asphalt shingles and built-up roofs are highly resistant. The most practical method of removing the moisture is by adequately ventilating the roof spaces.

A warm attic that is inadequately ventilated and insulated may cause formation of *ice dams* at the cornice. During cold weather after a heavy snowfall, heat causes the snow next to the roof to melt (fig. 68). Water running down the roof freezes on the colder surface of the cornice, often forming an ice dam at the gutter which may cause water to back up at the eaves and into the wall and ceiling. Similar dams often form in roof valleys. Ventilation thus provides part of the answer to the problems. With a well-insulated ceiling and adequate ventilation, attic temperatures are low and melting of snow over the attic space will be greatly reduced (*15*).

In hot weather, ventilation of attic and roof spaces offers an effective means of removing hot air and thereby materially lowering the temperature in these spaces. Insulation should be used between ceiling joists below the attic or roof space to further retard heat flow into the rooms below and materially improve comfort conditions.

It is common practice to install louvered openings in the end walls of gable roofs for ventilation. Air movement through such openings depends primarily on wind direction and velocity, and no appreciable movement can be expected when there is no wind or unless one or more openings face the wind. More positive air movement can be obtained by providing openings in the soffit areas of the roof overhang in addition to openings at the gable ends or ridge. Hip-roof houses are best ventilated by inlet ventilators in the soffit area and by outlet ventilators along the ridge.

The differences in temperature between the attic and the outside will then create an air movement independent of the wind, and also a more positive movement when there is wind.

Where there is a crawl space under house or porch, ventilation is necessary to remove moisture vapor rising from the soil. Such vapor may otherwise condense on the wood below the floor and facilitate decay. A permanent vapor barrier on the soil of the crawl space greatly reduces the amount of ventilating area required.

Tight construction (including storm window and storm doors) and the use of humidifiers have created potential moisture problems which must be resolved through planning of adequate ventilation as well as the proper use of vapor barriers. Blocking of ventilating areas, for example, must be avoided as such practices will prevent ventilation of attic spaces. Inadequate ventilation will often lead to moisture problems which can result in unnecessary costs to correct.

Area of Ventilators

Types of ventilators and minimum recommended sizes have been generally established for various types of roofs. The minimum net area for attic or roof-space ventilators is based on the projected ceiling area of the rooms below (fig. 99). The ratio of ventilator openings as shown are net areas, and the actual area must be increased to allow for any restrictions such as louvers and wire cloth or screen. The screen area should be double the specified net area shown in figures 99 to 101.

To obtain extra area of screen without adding to the area of the vent, use a frame of required size to hold the screen away from the ventilator opening. Use as coarse a screen as conditions permit, not smaller than No. 16, for lint and dirt tend to clog fine-mesh screens. Screens should be installed in such a way that paint brushes will not easily contact the screen and close the mesh with paint.

Gable Roofs

Louvered openings are generally provided in the end walls of gable roofs and should be as close to the ridge as possible (fig. 99,*A*). The net area for the openings should be 1/300 of the ceiling area (fig. 99,*A*). For example, where the ceiling area equals 1,200 square feet, the minimum total net area of the ventilators should be 4 square feet.

As previously explained, more positive air move-

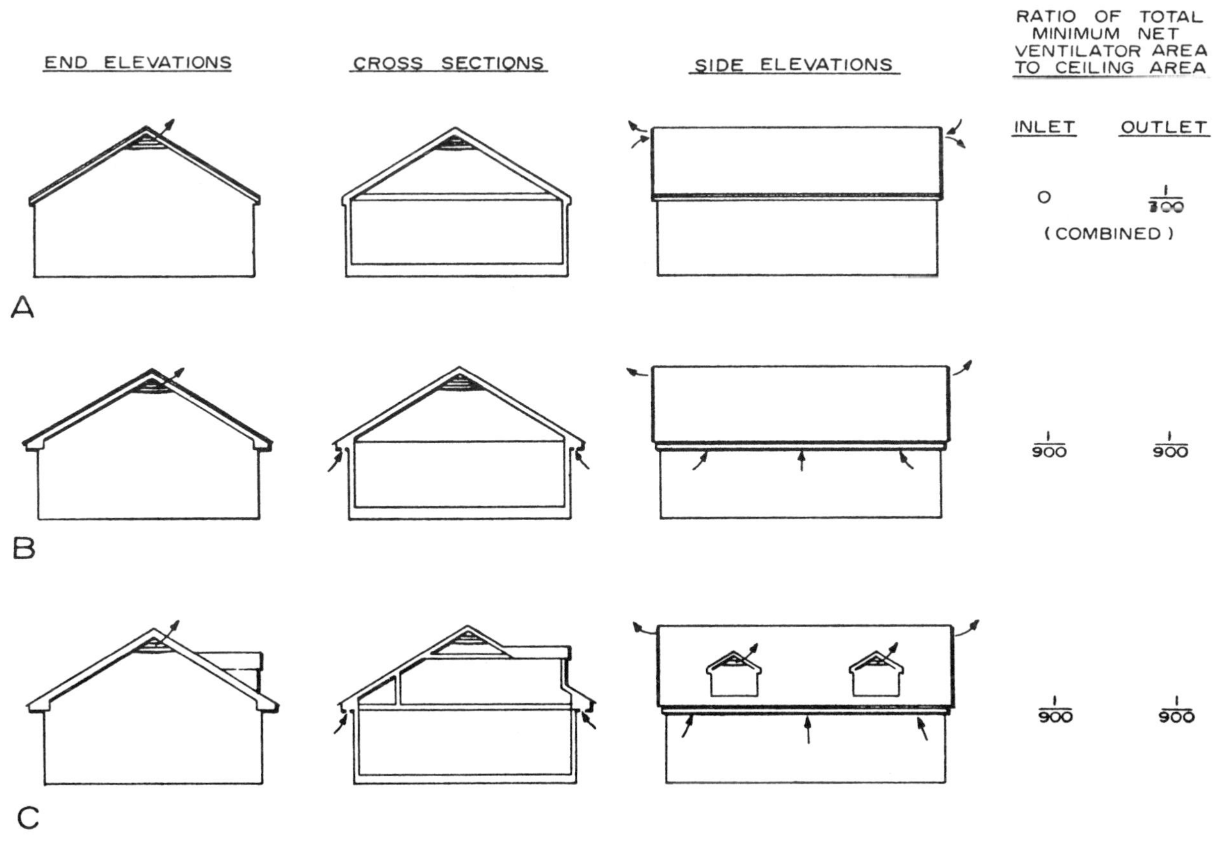

Figure 99.—Ventilating areas of gable roofs: *A*, Louvers in end walls; *B*, louvers in end walls with additional openings in soffit area; *C*, louvers at end walls with additional openings at eaves and dormers. Cross section of C shows free opening for air movement between roof boards and ceiling insulation of attic room.

ment can be obtained if additional openings are provided in the soffit area. The minimum ventilation areas for this method are shown in figure 99,*B*.

Where there are rooms in the attic with sloping ceilings under the roof, the insulation should follow the roof slope and be so placed that there is a free opening of at least 1½ inches between the roof boards and insulation for air movement (fig. 99,*C*).

Hip Roofs

Hip roofs should have air-inlet openings in the soffit area of the eaves and outlet openings at or near the peak. For minimum net areas of openings see figure 100,*A*. The most efficient type of inlet opening is the continuous slot, which should provide a free opening of not less than ¾ inch. The air-outlet opening near the peak can be a globe-type metal ventilator or several smaller roof ventilators located near the ridge. They can be located below the peak on the rear slope of the roof so that they will not be visible from the front of the house. Gabled extensions of a hip-roof house are sometimes used to provide efficient outlet ventilators (fig. 100,*B*).

Flat Roofs

A greater ratio of ventilating area is required in some types of flat roofs than in pitched roofs because the air movement is less positive and is dependent upon wind. It is important that there be a clear open space above the ceiling insulation and below the roof sheathing for free air movement from inlet to outlet openings. Solid blocking should *not* be used for bridging or for bracing over bearing partitions if its use prevents the air circulation.

Perhaps the most common type of flat or low-pitched roof is one in which the rafters extend beyond the wall, forming an overhang (fig. 101,*A*). When soffits are used, this area can contain the combined inlet-outlet ventilators, preferably a continuous slot. When single ventilators are used, they should be distributed evenly along the overhang.

A parapet-type wall and flat roof combination may be constructed with the ceiling joists separate from the roof joists or combined. When members are separate, the space between can be used for an airway (fig. 101,*B*). Inlet and outlet vents are then located as shown, or a series of outlet stack vents can be used along the centerline of the roof in combination with the inlet vents. When ceiling joists and flat rafters are served by one member in parapet construction, vents may be located as shown in figure 101,*C*. Wall inlet ventilators combined with center stack outlet vents is another variable in this type of roof.

Types and Location of Outlet Ventilators

Various styles of gable-end ventilators are available ready for installation. Many are made with metal louvers and frames, while others may be made of wood to fit the house design more closely. However, the most important factors are to have sufficient net ventilating area and to locate ventilators as close to the ridge as possible without affecting house appearance.

One of the types commonly used fits the slope of the roof and is located near the ridge (fig. 102,*A*). It can

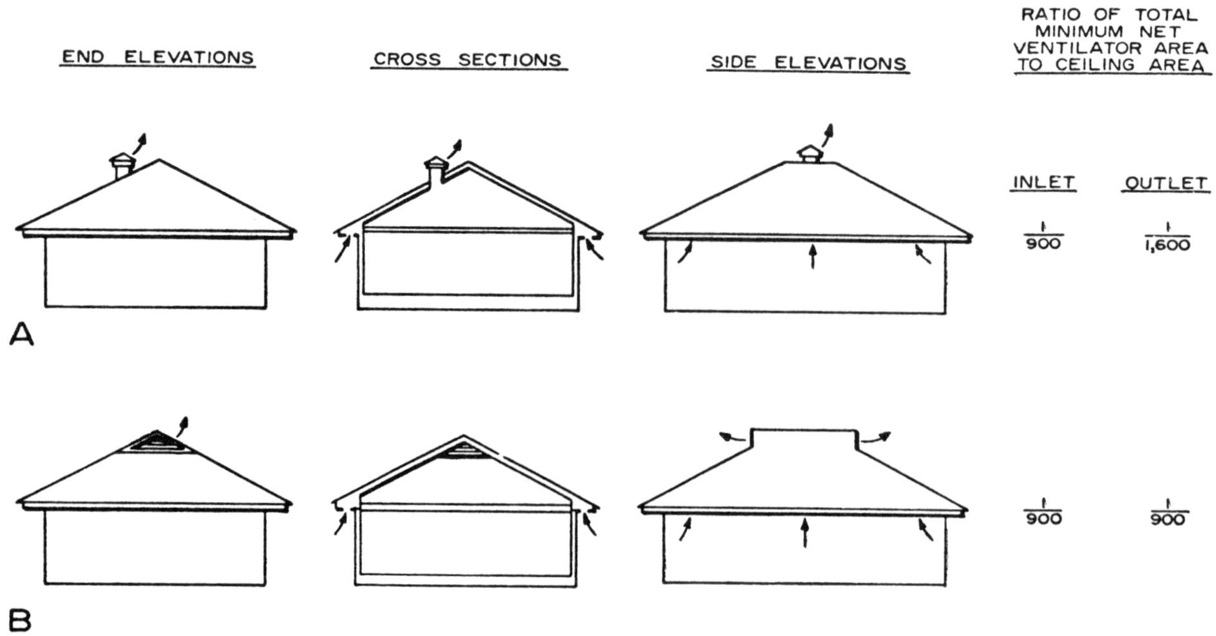

Figure 100.—Ventilating areas of hip roofs: *A*, Inlet openings beneath eaves and outlet vent near peak; *B*, inlet openings beneath eaves and ridge outlets.

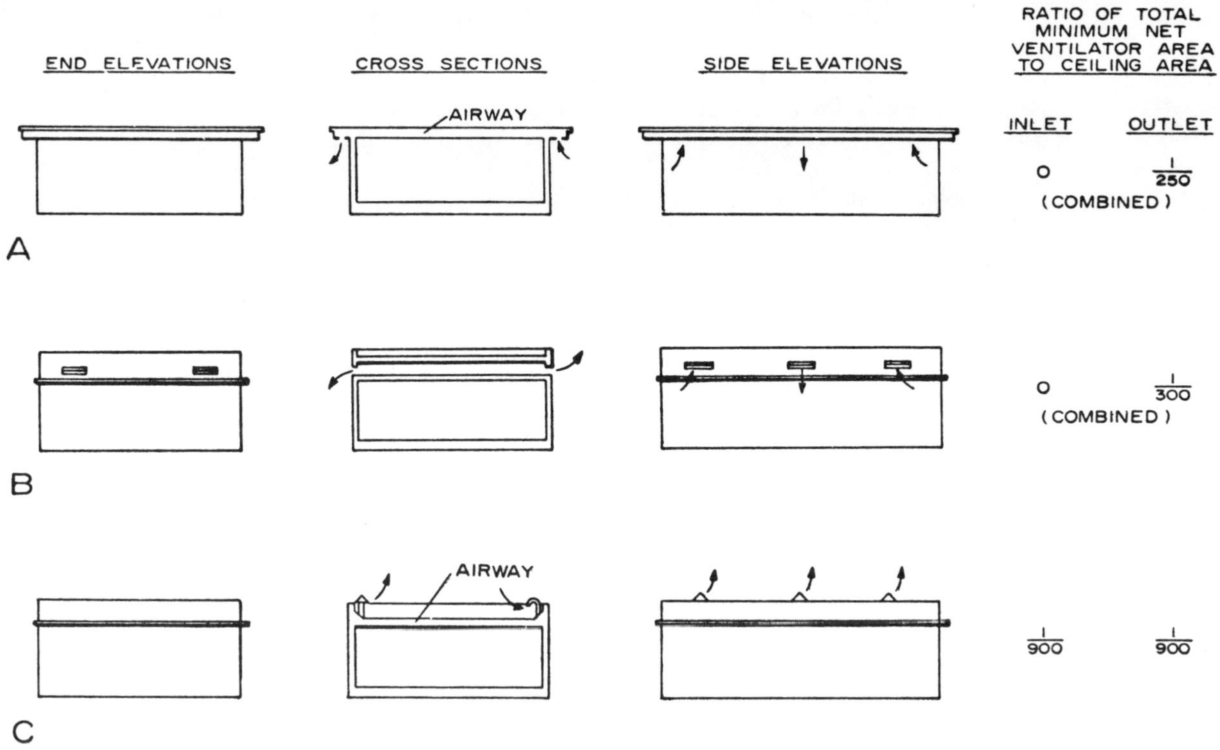

Figure 101.—Ventilating area of flat roofs: *A*, Ventilator openings under overhanging eaves where ceiling and roof joists are combined; *B*, for roof with a parapet where roof and ceiling joists are separate; *C*, for roof with a parapet where roof and ceiling joists are combined.

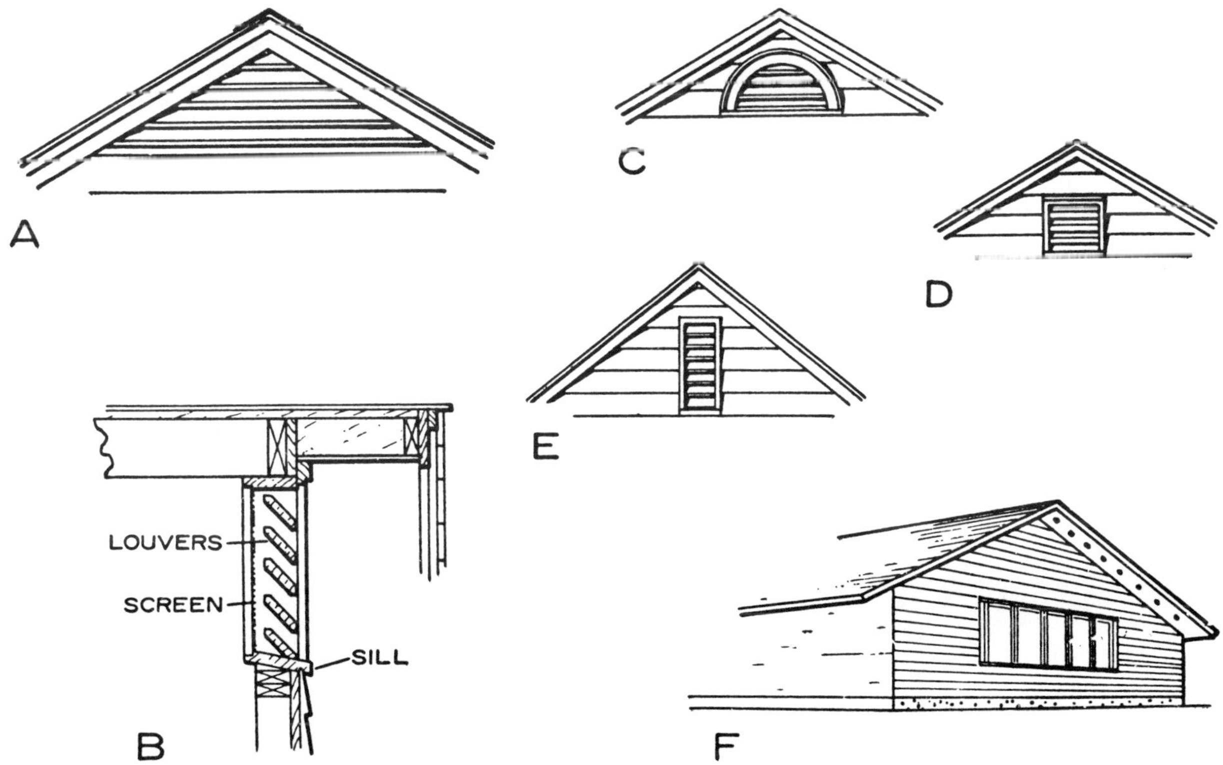

Figure 102.—Outlet ventilators: *A*, Triangular; *B*, typical cross section; *C*, half-circle; *D*, square; *E*, vertical; *F*, soffit.

be made of wood or metal; in metal it is often adjustable to conform to the roof slope. A wood ventilator of this type is enclosed in a frame and placed in the rough opening much as a window frame (fig. 102,*B*). Other forms of gable-end ventilators which might be used are shown in figures 102,*C*, *D*, and *E*.

A system of attic ventilation which can be used on houses with a wide roof overhang at the gable end consists of a series of small vents or a continuous slot located on the underside of the soffit areas (fig. 102,*F*). Several large openings located near the ridge might also be used. This system is especially desirable on low-pitched roofs where standard wall ventilators may not be suitable.

It is important that the roof framing at the wall line does not block off ventilation areas to the attic area. This might be accomplished by the use of a "ladder" frame extension. A flat nailing block used at the wall line will provide airways into the attic (fig. 66,*B*). This can also be adapted to narrower rake sections by providing ventilating areas to the attic.

Types and Location of Inlet Ventilators

Small, well-distributed ventilators or a continuous slot in the soffit provide inlet ventilation. These small

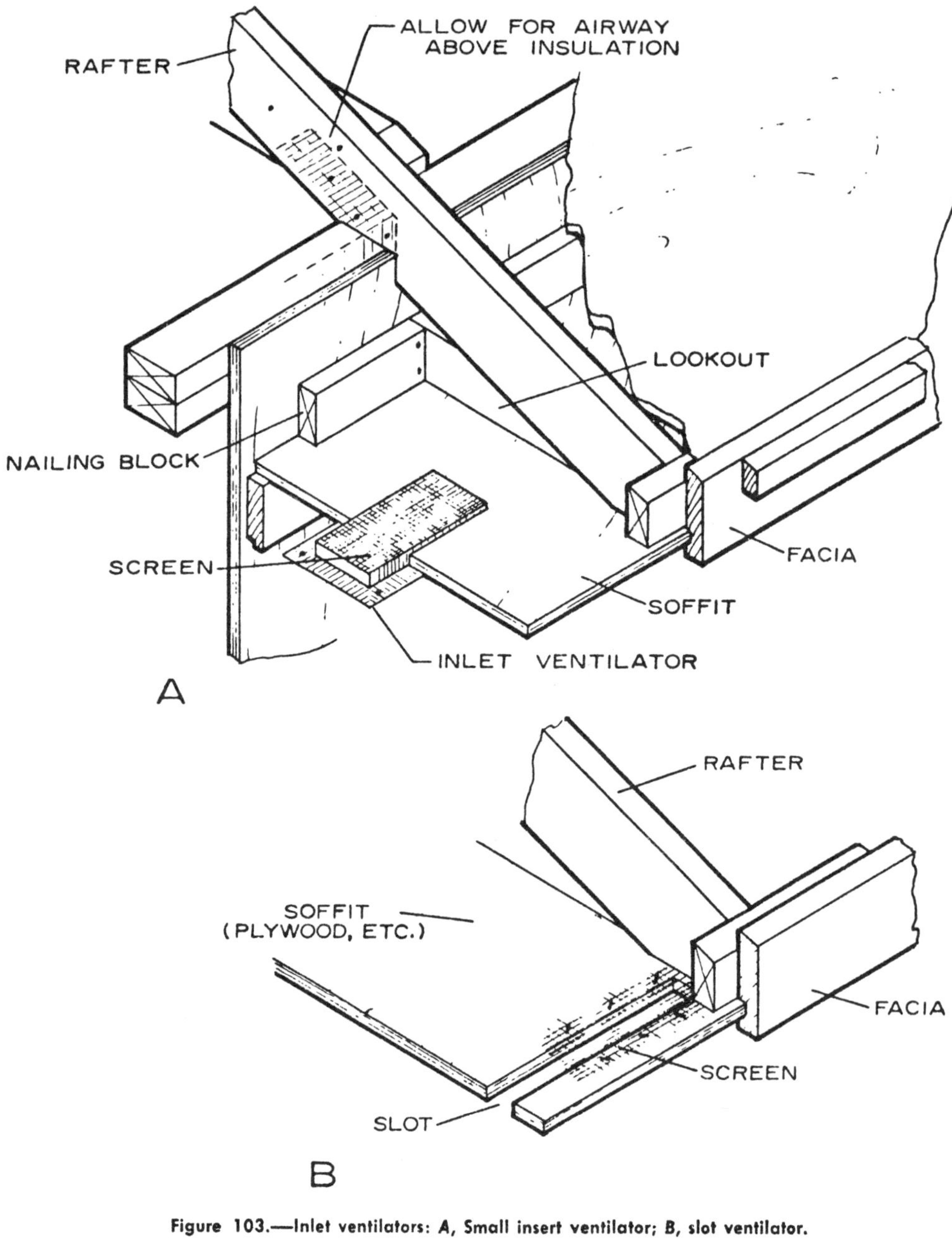

Figure 103.—Inlet ventilators: *A*, Small insert ventilator; *B*, slot ventilator.

louvered and screened vents can be obtained in most lumberyards or hardware stores and are simple to install.

Only small sections need to be cut out of the soffit; these can be sawed out before the soffit is applied. It is more desirable to use a number of smaller well-distributed ventilators than several large ones (fig. 103,*A*). Any blocking which might be required between rafters at the wall line should be installed so as to provide an airway into the attic area.

A continuous screened slot, which is often desirable, should be located near the outer edge of the soffit near the facia (fig. 103,*B*). Locating the slot in this area will minimize the chance of snow entering. This type may also be used on the extension of flat roofs.

Crawl-space Ventilation and Soil Cover

The crawl space below the floor of a basementless house and under porches should be ventilated and protected from ground moisture by the use of a *soil cover* (fig. 104). The soil cover should be a vapor barrier with a perm value of less than 1.0. This includes such barrier materials as plastic films, roll roofing, and asphalt laminated paper. Such protection will minimize the effect of ground moisture on the wood framing members. High moisture content and humidity encourage staining and decay of untreated members.

Where there is a partial basement open to a crawl-space area, no wall vents are required if there is some type of operable window. The use of a soil cover in the crawl space is still important, however. For crawl

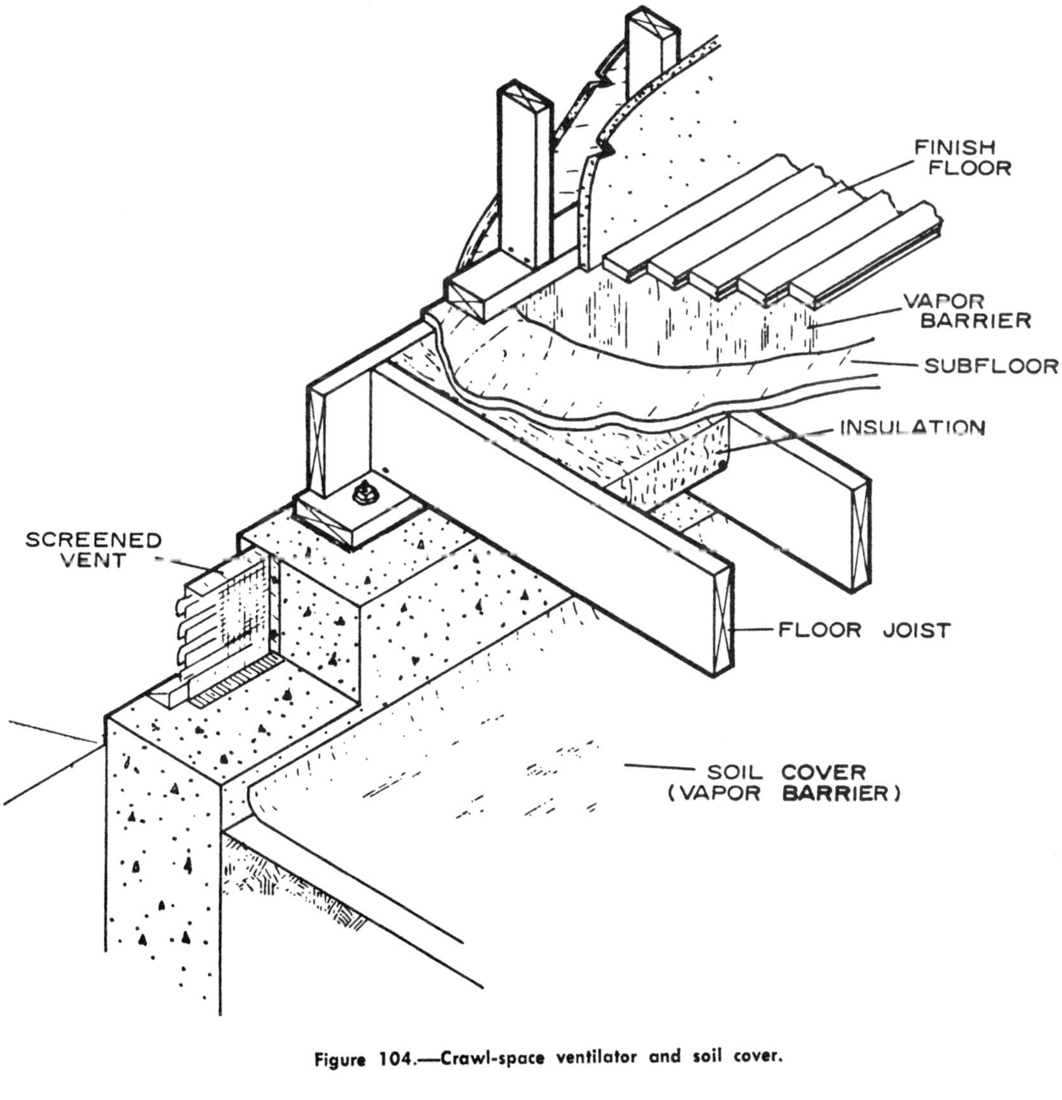

Figure 104.—Crawl-space ventilator and soil cover.

spaces with no basement area, provide at least four foundation-wall vents near corners of the building. The total free (net) area of the ventilators should be equal to 1/160 of the ground area when no soil cover is used. Thus, for a ground area of 1.200 square feet, a total net ventilating area of about 8 square feet is required, or 2 square feet for each of four ventilators. More smaller ventilators having the same net ratio is satisfactory.

When a vapor barrier ground cover is used, the required ventilating area is greatly reduced. The net ventilating area required with a ground cover is 1/1600 of the ground area, or for the 1,200-square-foot house, an area of 0.75 square foot. This should be divided between two small ventilators located on opposite sides of the crawl space. Vents should be covered (fig. 104) with a corrosion-resistant screen of No. 8 mesh.

The use of a ground cover is normally recommended under all conditions. It not only protects wood framing members from ground moisture but also allows the use of small, inconspicuous ventilators.

CHAPTER 17

SOUND INSULATION[5]

Development of the "quiet" home or the need for incorporating sound insulation in a new house is becoming more and more important. In the past, the reduction of sound transfer between rooms was more important in apartments, motels, and hotels than in private homes. However, house designs now often incorporate a family room or "active" living room as well as "quiet" living room. It is usually desirable in such designs to isolate these rooms from the remainder of the house. Sound insulation between the bedroom area and the living area is usually desirable, as is isolation of the bathrooms and lavatories. Isolation from outdoor sounds is also often advisable. Thus, sound control has become a vital part of house design and construction, and will be even more important in the coming years.

How Sound Travels

How does sound travel, and how is it transferred through a wall or floor? Airborne noises inside a house, such as loud conversation or a barking dog, create sound waves which radiate outward from the source through the air until they strike a wall, floor, or ceiling. These surfaces are set in vibration by the fluctuating pressure of the sound wave in the air. Because the wall vibrates, it conducts sound to the other side in varying degrees, depending on the wall construction.

The resistance of a building element, such as a wall, to the passage of airborne sound is rated by its *Sound Transmission Class* (STC). Thus, the higher the number, the better the sound barrier. The approximate effectiveness of walls with varying STC numbers is shown in the following tabulation:

STC No.	*Effectiveness*
25	Normal speech can be understood quite easily
35	Loud speech audible but not intelligible
45	Must strain to hear loud speech
48	Some loud speech barely audible
50	Loud speech not audible

Sound travels readily through the air and also through some materials. When airborne sound strikes a conventional wall, the studs act as sound conductors unless they are separated in some way from the covering material. Electrical switches or convenience outlets placed back-to-back in a wall readily pass sound. Faulty construction, such as poorly fitted doors, often allows sound to travel through. Thus, good construction practices are important in providing sound-resistant walls, as well as those measures commonly used to stop ordinary sounds.

Thick walls of dense materials such as masonry can stop sound. But in the wood-frame house, an interior masonry wall results in increased costs and structural problems created by heavy walls. To provide a satisfactory sound-resistant wall economically has been a problem. At one time, sound-resistant frame construction for the home involved significant additional costs because it usually meant double walls or suspended ceilings. However, a relatively simple system has been developed using sound-deadening insulating board in conjunction with a gypsum board outer covering. This provides good sound-transmission resistance suitable for use in the home with only slight additional cost. A number of combinations are possible with this system, providing different STC ratings.

Wall Construction

As the preceding STC tabulation shows, a wall providing sufficient resistance to airborne sound transfer

[5] Data and information contained in this chapter were obtained in part from literature references (*3, 8, 10, 14*).

likely has an STC rating of 45 or greater. Thus, in construction of such a wall between the rooms of a house, its cost as related to the STC rating should be considered. As shown in figure 105, details *A*, with gypsum wallboard, and *B*, with plastered wall, are those commonly used for partition walls. However, the hypothetical rating of 45 cannot be obtained in this construction. An 8-inch concrete block wall (fig. 105,*C*) has the minimum rating, but this construction is not always practical in a wood-frame house.

Good STC ratings can be obtained in a wood-frame wall by using the combination of materials shown in figure 105,*D* and *E*. One-half-inch sound-deadening board nailed to the studs, followed by a lamination

WALL DETAIL	DESCRIPTION	STC RATING
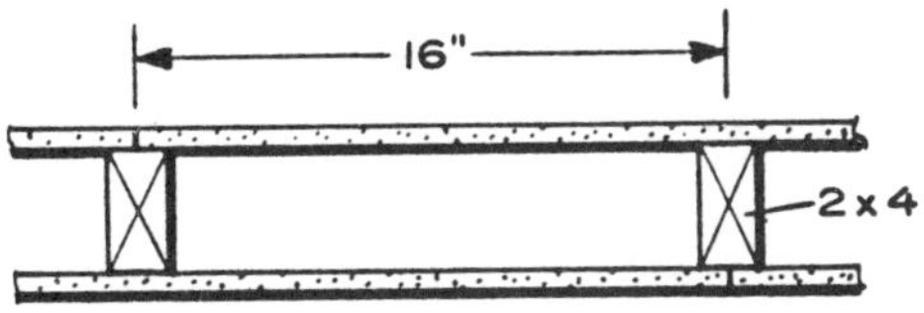 A	½" GYPSUM WALLBOARD ⅝" GYPSUM WALLBOARD	32 37
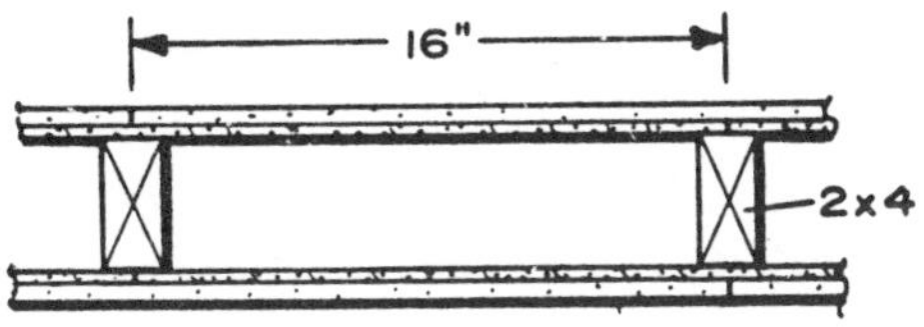 B	⅜" GYPSUM LATH (NAILED) PLUS ½" GYPSUM PLASTER WITH WHITECOAT FINISH (EACH SIDE)	39
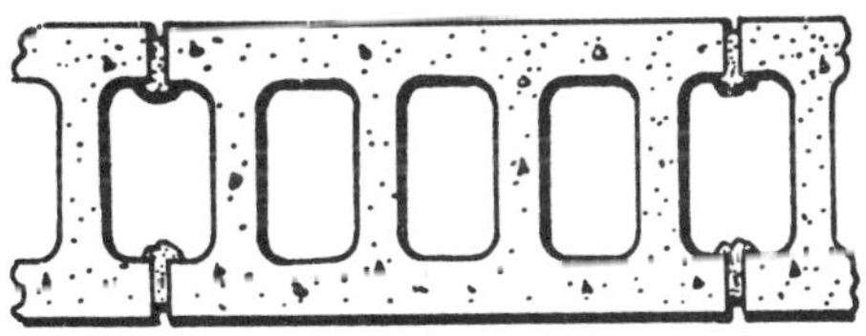 C	8" CONCRETE BLOCK	45
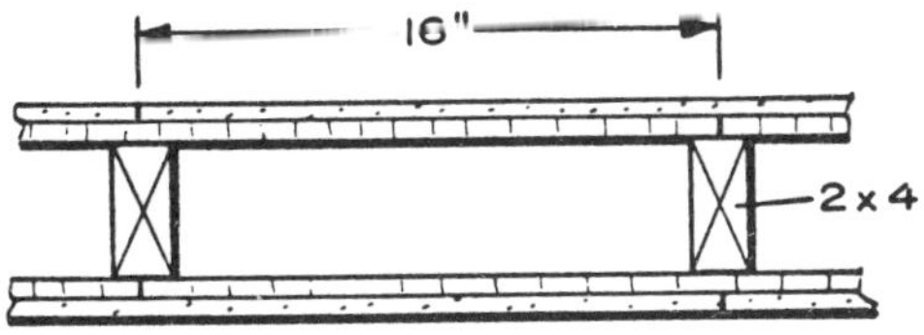 D	½" SOUND DEADENING BOARD (NAILED) ½" GYPSUM WALLBOARD (LAMINATED) (EACH SIDE)	46
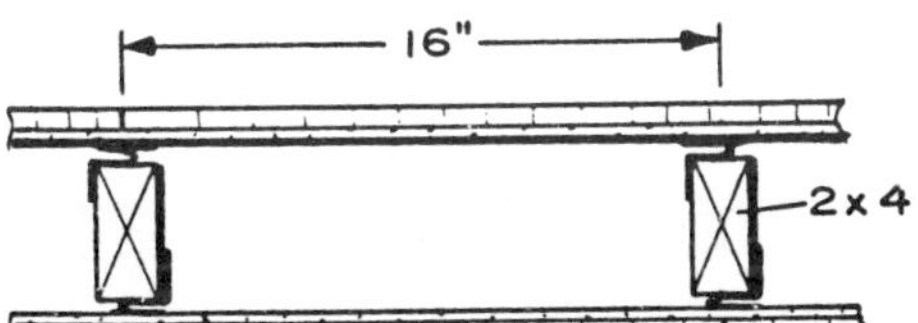 E	RESILIENT CLIPS TO ⅜" GYPSUM BACKER BOARD ½" FIBERBOARD (LAMINATED) (EACH SIDE)	52

Figure 105.—Sound insulation of single walls.

of ½-inch gypsum wallboard, will provide an STC value of 46 at a relatively low cost. A slightly better rating can be obtained by using ⅝-inch gypsum wallboard rather than ½-inch. A very satisfactory STC rating of 52 can be obtained by using resilient clips to fasten gypsum backer boards to the studs, followed by adhesive-laminated ½-inch fiberboard (fig. 105,*E*). This method further isolates the wall covering from the framing.

A similar isolation system consists of resilient channels nailed horizontally to 2- by 4-inch studs spaced 16 inches on center. Channels are spaced 24 inches apart vertically and ⅝-inch gypsum wallboard is screwed to the channels. An STC rating of 47 is thus obtained at a moderately low cost.

The use of a double wall, which may consist of a 2 by 6 or wider plate and staggered 2- by 4-inch studs, is sometimes desirable. One-half-inch gypsum wallboard on each side of this wall (fig. 106,*A*) results in an STC value of 45. However, two layers of ⅝-inch gypsum wallboard add little, if any, additional sound-transfer resistance (fig. 106,*B*). When 1½-inch blanket insulation is added to this construction (fig. 106,*C*), the STC rating increases to 49. This insulation may be installed as shown or placed between studs on one wall. A single wall with 3½ inches of insulation will show a marked improvement over an open stud space and is low in cost.

The use of ½-inch sound-deadening board and a lamination of gypsum wallboard in the double wall will result in an STC rating of 50 (fig. 106,*D*). The addition of blanket insulation to this combination will likely provide an even higher value, perhaps 53 or 54.

Floor-Ceiling Construction

Sound insulation between an upper floor and the

WALL DETAIL	DESCRIPTION	STC RATING
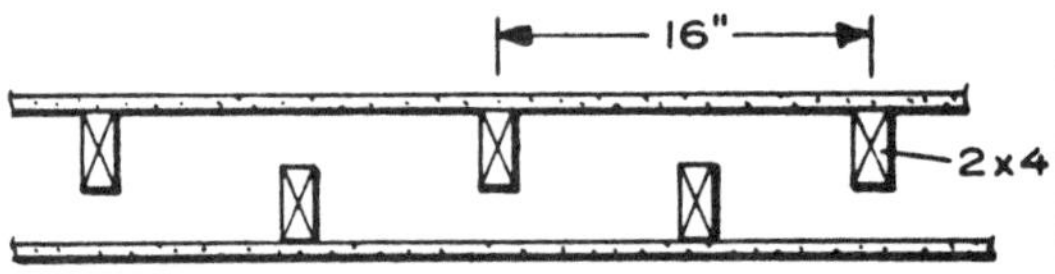 A	½" GYPSUM WALLBOARD	45
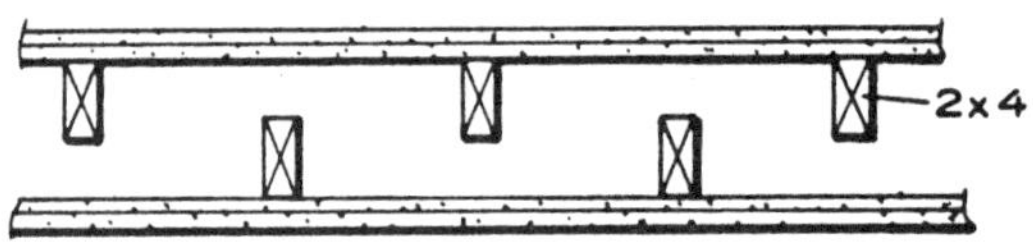 B	⅝" GYPSUM WALLBOARD (DOUBLE LAYER EACH SIDE)	45
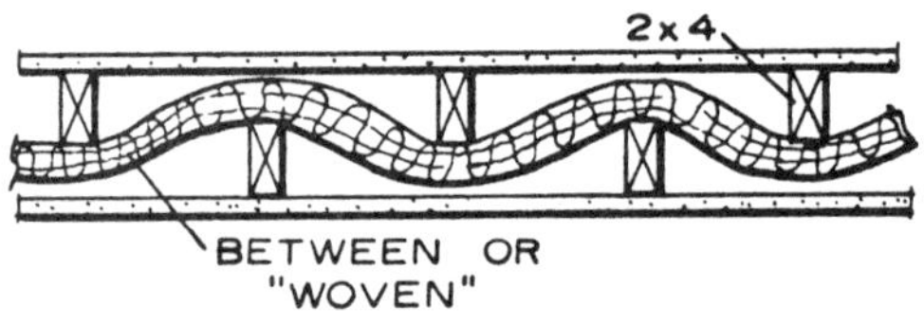 C	½" GYPSUM WALLBOARD 1½" FIBROUS INSULATION	49
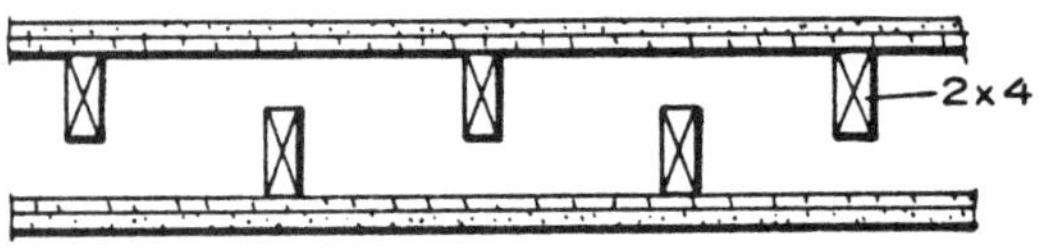 D	½" SOUND DEADENING BOARD (NAILED) ½" GYPSUM WALLBOARD (LAMINATED)	50

Figure 106.—Sound insulation of double walls.

ceiling of a lower floor not only involves resistance of airborne sounds but also that of impact noises. Thus, impact noise control must be considered as well as the STC value. Impact noise is caused by an object striking or sliding along a wall or floor surface, such as by dropped objects, footsteps, or moving furniture. It may also be caused by the vibration of a dishwasher, bathtub, food-disposal appa-atus, or other equipment. In all instances, the floor is set into vibration by the impact or contact and sound is radiated from both sides of the floor.

A method of measuring impact noise has been developed and is commonly expressed as the *Impact Noise Ratings* (*INR*).[6] The greater the positive value of the INR, the more resistant is the floor to impact noise transfer. For example, an INR of −2 is better than one of −17, and one of +5 INR is a further improvement in resistance to impact noise transfer.

Figure 107 shows STC and approximate INR(db)

[6] INR ratings in some publications are being abandoned in favor of IIC (Impact Insulation Class) ratings. See Glossary.

<table>
<tr><th rowspan="2">DETAIL</th><th rowspan="2">DESCRIPTION</th><th colspan="2">ESTIMATED VALUES</th></tr>
<tr><th>STC RATING</th><th>APPROX. INR</th></tr>
<tr><td>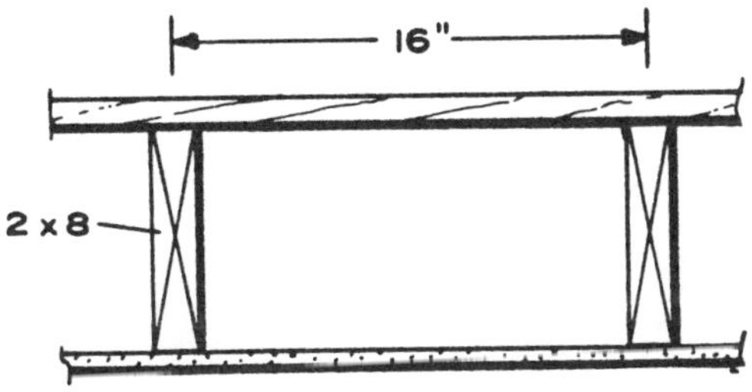

A</td><td>FLOOR
7/8" T. & G. FLOORING
CEILING
3/8" GYPSUM BOARD</td><td>30</td><td>-18</td></tr>
<tr><td>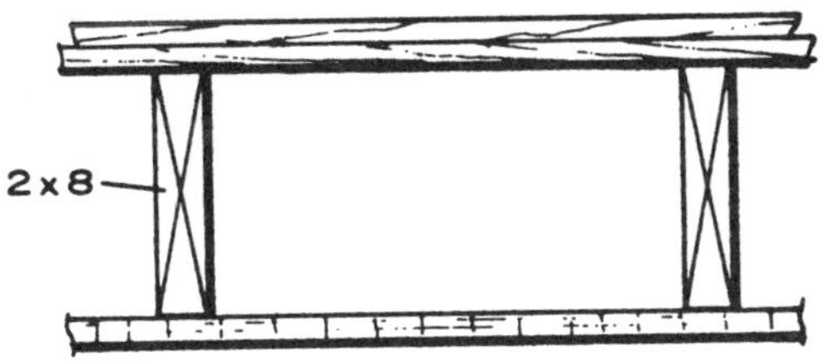

B</td><td>FLOOR
3/4" SUBFLOOR
3/4" FINISH FLOOR
CEILING
3/4" FIBERBOARD</td><td>42</td><td>-12</td></tr>
<tr><td>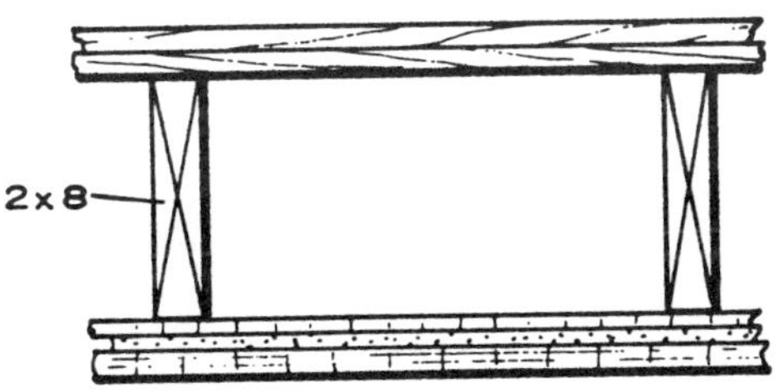

C</td><td>FLOOR
3/4" SUBFLOOR
3/4" FINISH FLOOR
CEILING
1/2" FIBERBOARD LATH
1/2" GYPSUM PLASTER
3/4" FIBERBOARD</td><td>45</td><td>-4</td></tr>
</table>

Figure 107.—Relative impact and sound transfer in floor-ceiling combinations (2- by 8-in. joists).

values for several types of floor constructions. Figure 107,*A*, perhaps a minimum floor assembly with tongued-and-grooved floor and ⅜-inch gypsum board ceiling, has an STC value of 30 and an approximate INR value of −18. This is improved somewhat by the construction shown in figure 107,*B*, and still further by the combination of materials in figure 107,*C*.

The value of isolating the ceiling joists from a gypsum lath and plaster ceiling by means of spring clips is illustrated in figure 108,*A*. An STC value of 52 and an approximate INR value of −2 result.

DETAIL	DESCRIPTION	ESTIMATED VALUES: STC RATING	ESTIMATED VALUES: APPROX. INR
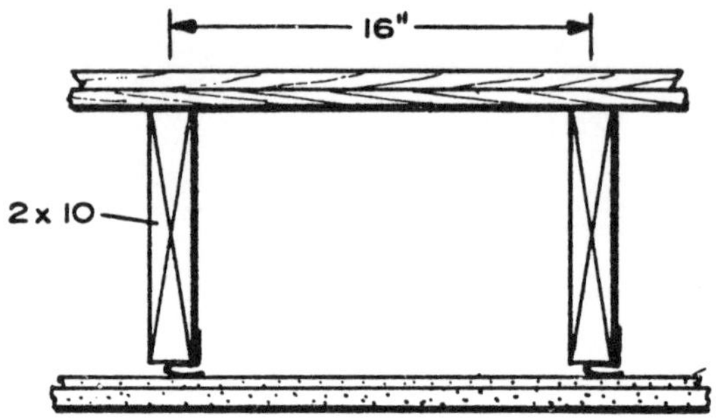 A	FLOOR ¾" SUBFLOOR (BUILDING PAPER) ¾" FINISH FLOOR CEILING GYPSUM LATH AND SPRING CLIPS ½" GYPSUM PLASTER	52	-2
 B	FLOOR ⅝" PLYWOOD SUBFLOOR ½" PLYWOOD UNDERLAYMENT ⅛" VINYL-ASBESTOS TILE CEILING ½" GYPSUM WALLBOARD	31	-17
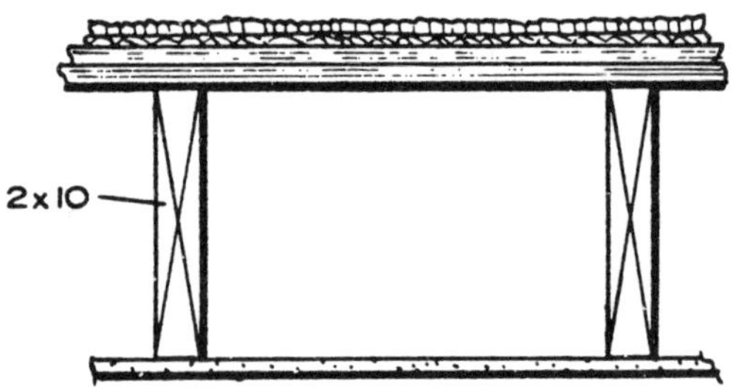 C	FLOOR ⅝" PLYWOOD SUBFLOOR ½" PLYWOOD UNDERLAYMENT FOAM RUBBER PAD ⅜" NYLON CARPET CEILING ½" GYPSUM WALLBOARD	45	+5

Figure 108.—Relative impact and sound transfer in floor-ceiling combinations (2- by 10-in. joists).

Foam-rubber padding and carpeting improve both the STC and the INR values. The STC value increases from 31 to 45 and the approximate INR from −17 to +5 (fig. 108,*B* and *C*). This can likely be further improved by using an isolated ceiling finish with spring clips. The use of sound-deadening board and a lamination of gypsum board for the ceiling would also improve resistance to sound transfer.

An economical construction similar to (but an improvement over) figure 108*C*, with a STC value of 48 and an approximate INR of +18, consists of the following: (a) A pad and carpet over ⅝-inch tongued-and-grooved plywood underlayment, (b) 3-inch fiberglass insulating batts between joists, (c) resilient channels spaced 24 inches apart, across the bottom of the joists, and (d) ⅝-inch gypsum board screwed to the bottom of the channels and finished with taped joints.

The use of separate floor joists with staggered ceiling joists below provides reasonable values but adds a good deal to construction costs. Separate joists with insulation between and a soundboard between subfloor and finish provide an STC rating of 53 and an approximate INR value of −3.

Sound Absorption

Design of the "quiet" house can incorporate another system of sound insulation, namely, sound absorption. Sound-absorbing materials can minimize the amount of noise by stopping the reflection of sound back into a room. Sound-absorbing materials do not necessarily have resistance to airborne sounds. Perhaps the most commonly used sound-absorbing material is acoustic tile. Wood fiber or similar materials are used in the manufacture of the tile, which is usually processed to provide some fire resistance and designed with numerous tiny sound traps on the tile surfaces. These may consist of tiny drilled or punched holes, fissured surfaces, or a combination of both.

Acoustic tile is most often used in the ceiling and areas where it is not subjected to excessive mechancial damage, such as above a wall wainscoating. It is normally manufactured in sizes from 12 by 12 to 12 by 48 inches. Thicknesses vary from ½ to ¾ inch, and the tile is usually factory finished ready for application. Paint or other finishes which fill or cover the tiny holes or fissures for trapping sound will greatly reduce its efficiency.

Acoustic tile may be applied by a number of methods—to existing ceilings or any smooth surface with a mastic adhesive designed specifically for this purpose, or to furring strips nailed to the underside of the ceiling joists. Nailing or stapling tile is the normal application method in this system. It is also used with a mechanical suspension system involving small "H," "Z," or "T" members. Manufacturers' recommendations should be followed in application and finishing.

CHAPTER 18

BASEMENT ROOMS

Many houses are now designed so that one or more of the rooms in lower floors are constructed on a concrete slab. In multilevel houses, this area may include a family room, a spare bedroom, or a study. Furthermore, it is sometimes necessary to provide a room in the basement of an existing house. Thus, in a new house or in remodeling the basement of an existing one, several factors should be considered, including insulation, waterproofing, and vapor resistance.

Floors

In the construction of a new building having basement rooms, provision should be made for reduction of heat loss and for prevention of ground moisture movement. As previously described in Chapter 4, "Concrete Floor Slabs on Ground," perimeter insulation reduces heat loss and a vapor barrier under the slab will prevent problems caused by a concrete floor damp from ground moisture (fig. 109). Providing these essential details, however, is somewhat more difficult in existing construction than in new construction.

The installation of a vapor barrier over an existing unprotected concrete slab is normally required when the floor is at or below the outside ground level and some type of finish floor is used. Flooring manufacturers often recommend that preparation of the slab for wood strip flooring consist of the following steps:

1. Mop or spread a coating of tar or asphalt mastic followed by an asphalt felt paper.
2. Lay short lengths of 2- by 4-inch screeds in a coating of tar or asphalt, spacing the rows about 12 inches apart, starting at one wall and ending at the opposite wall.
3. Place insulation around the perimeter, between screeds, where the outside ground level is near the basement floor elevation.

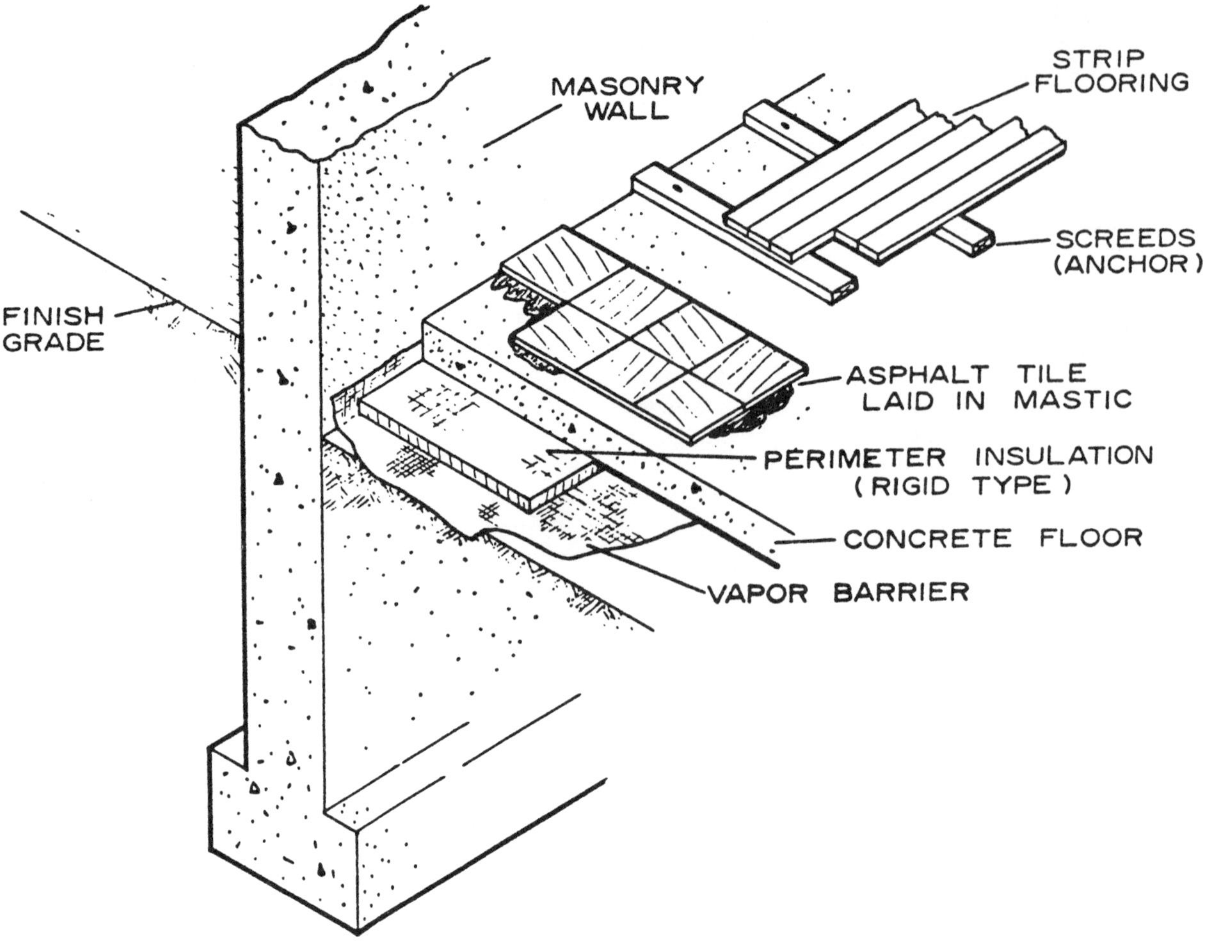

Figure 109.—Basement floor details for new construction.

4. Install wood strip flooring across the wood screeds.

This system can be varied somewhat by placing a conventional vapor barrier of good quality directly over the slab. Two- by four-inch furring strips spaced 12 to 16 inches apart are then anchored to the slab with concrete nails or with other types of commercial anchors. Some leveling of the 2 by 4's might be required. Strip flooring is then nailed to the furring strips after perimeter insulation is placed (fig. 110). If a wood block flooring is desired under these conditions, a plywood subfloor may be used over the furring strips. Plywood, ½ or ⅝ inch thick, is normally used if the edges are unblocked and furring strips are spaced 16 inches or more apart.

When insulation is not required around the perimeter because of the height of the outside grade above the basement floor, a much simpler method can be used for wood block or other type of tile finish. An asphalt mastic coating, followed by a good vapor barrier, serves as a base for the tile. An adhesive recommended by the flooring manufacturer is then used over the vapor barrier, after which the wood tile is applied. It is important that a smooth vapor-tight base be provided for the tile.

It is likely that such floor construction should be used only under favorable conditions where draintile is placed at the outside footings and soil conditions are favorable. When the slab or walls of an existing house are inclined to be damp, it is often difficult to insure a dry basement. Under such conditions, it is often advisable to use resilient tile or similar finish over some type of stable base such as plywood. This construction is to be preceded by installation of vapor barriers and protective coatings.

Walls

The use of an interior finish over masonry basement walls is usually desirable for habitable rooms. Furthermore, if the outside wall is partially exposed, it is advisable to use insulation between the wall and the

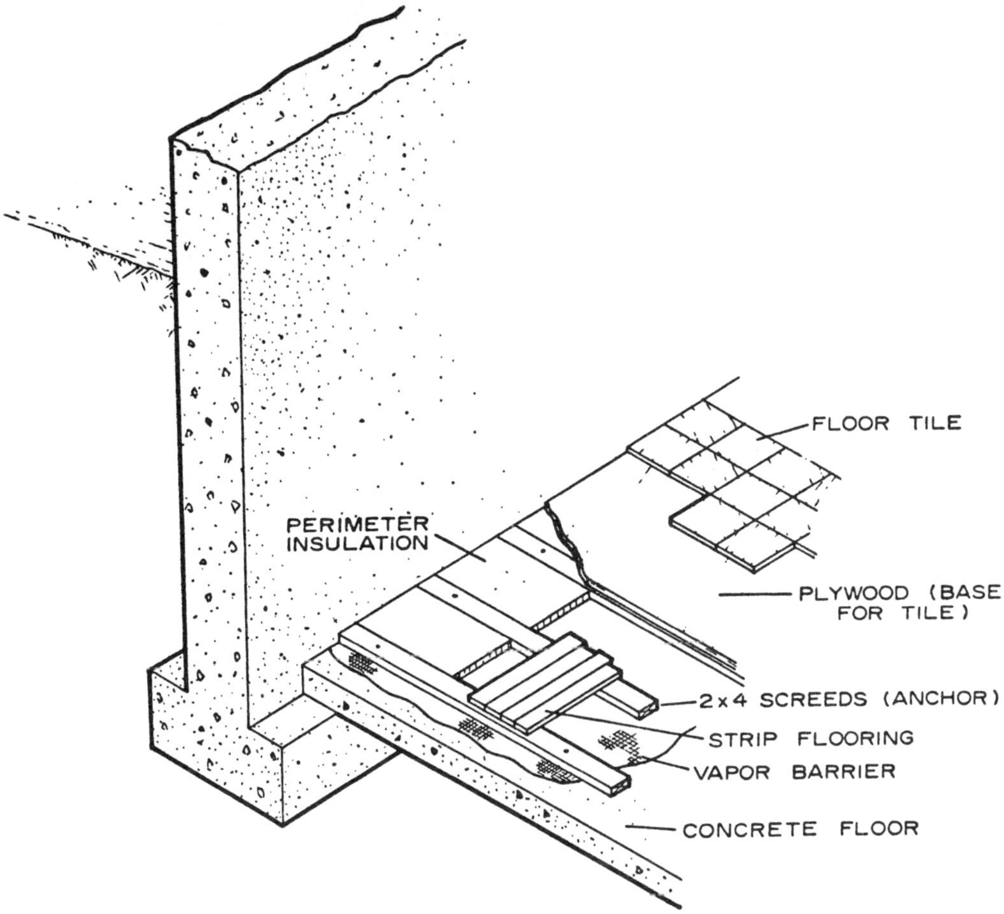

Figure 110.—Basement floor details for existing construction.

inside finish. Waterproofing the wall is important if there is any possibility of moisture entry. It can be done by applying one of the many waterproof coatings available to the inner surface of the masonry.

After the wall has been waterproofed, furring strips are commonly used to prepare the wall for interior finish. A 2- by 2-inch bottom plate is anchored to the floor at the junction of the wall and floor. A 2- by 2-inch or larger top plate is fastened to the bottom of the joists, to joist blocks, or anchored to the wall (fig. 111). Studs or furring strips, 2 by 2 inches or larger in size are then placed between top and bottom plates, anchoring them at the center when necessary with concrete nails or similar fasteners (fig. 111). Electrical outlets and conduit should be installed and insulation with vapor barrier placed between the furring strips. The interior finish of gypsum board, fiberboard, plywood, or other material is then installed. Furring strips are commonly spaced 16 inches on center, but this, of course, depends on the type and thickness of the interior finish.

Foamed plastic insulation is sometimes used on masonry walls without furring. It is important that the inner face of the wall be smooth and level without protrusions when this method is used. After the wall has been waterproofed, ribbons of adhesive are applied to the wall and sheets of foam insulation installed (fig. 112). Dry-wall adhesive is then applied and the gypsum board, plywood, or other finish pressed in place. Manufacturers' recommendations on adhesives and methods of installation should be followed. Most foam-plastic insulations have some vapor resistance in themselves, so the need for a separate vapor barrier is not as great as when blanket type insulation is used.

Ceilings

Some type of finish is usually desirable for the ceiling of the basement room. Gypsum board, plywood, or fiberboard sheets may be used and nailed directly to the joists. Acoustic ceiling tile and similar materials normally require additional nailing areas. This may be supplied by 1- by 2-inch or 1- by 3-inch strips nailed across the joists, and spaced to conform to

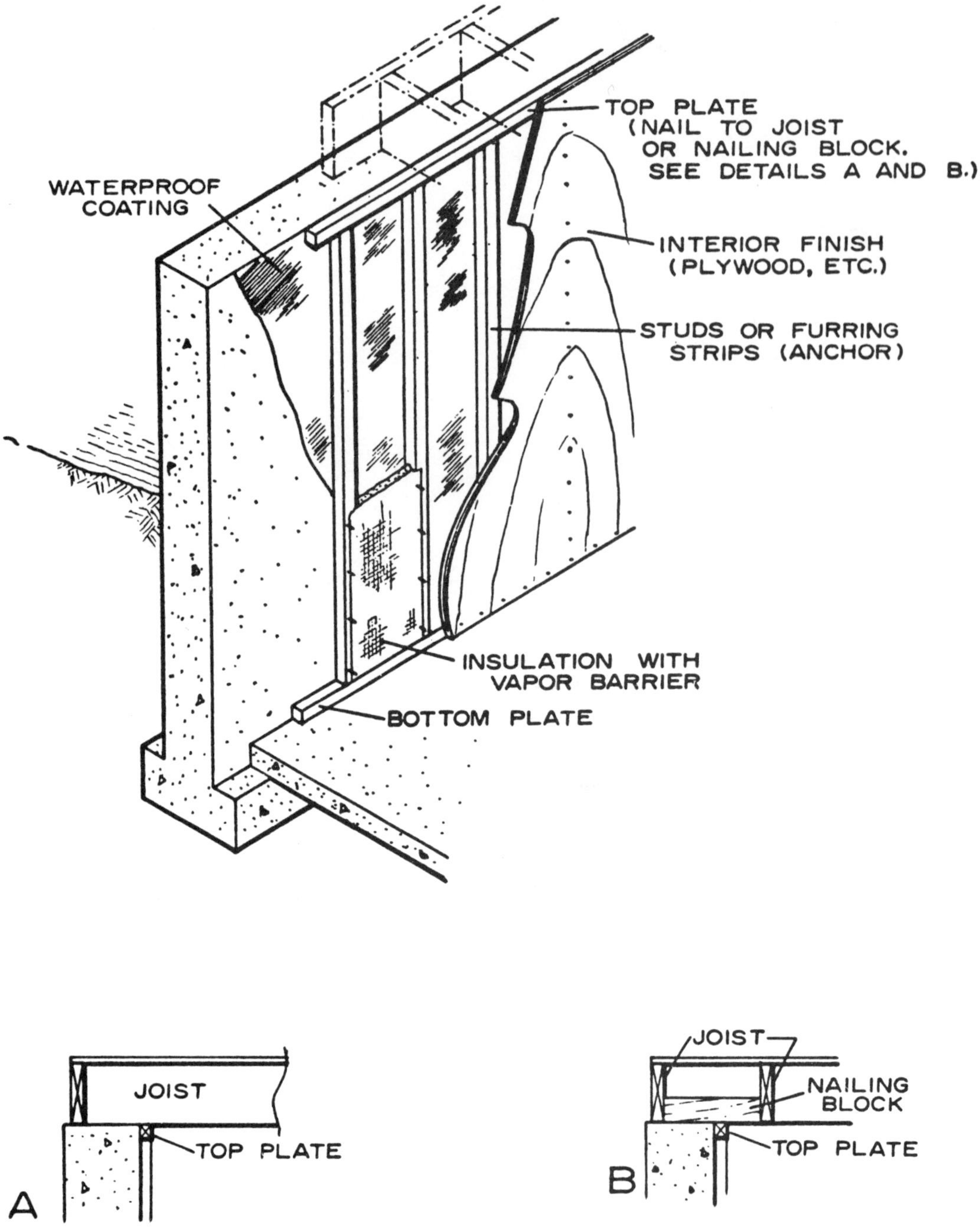

Figure 111.—Basement wall finish, with furring strips.

the size of the ceiling tile (fig. 113).

A *suspended ceiling* may also be desirable. This can consist of a system of light metal angles hung from the ceiling joists. Tiles are then dropped in place. This will also aid in decreasing sound transfer from the rooms above. Remember to install ceiling lights, heat supply and return ducts, or other utilities before finish is applied.

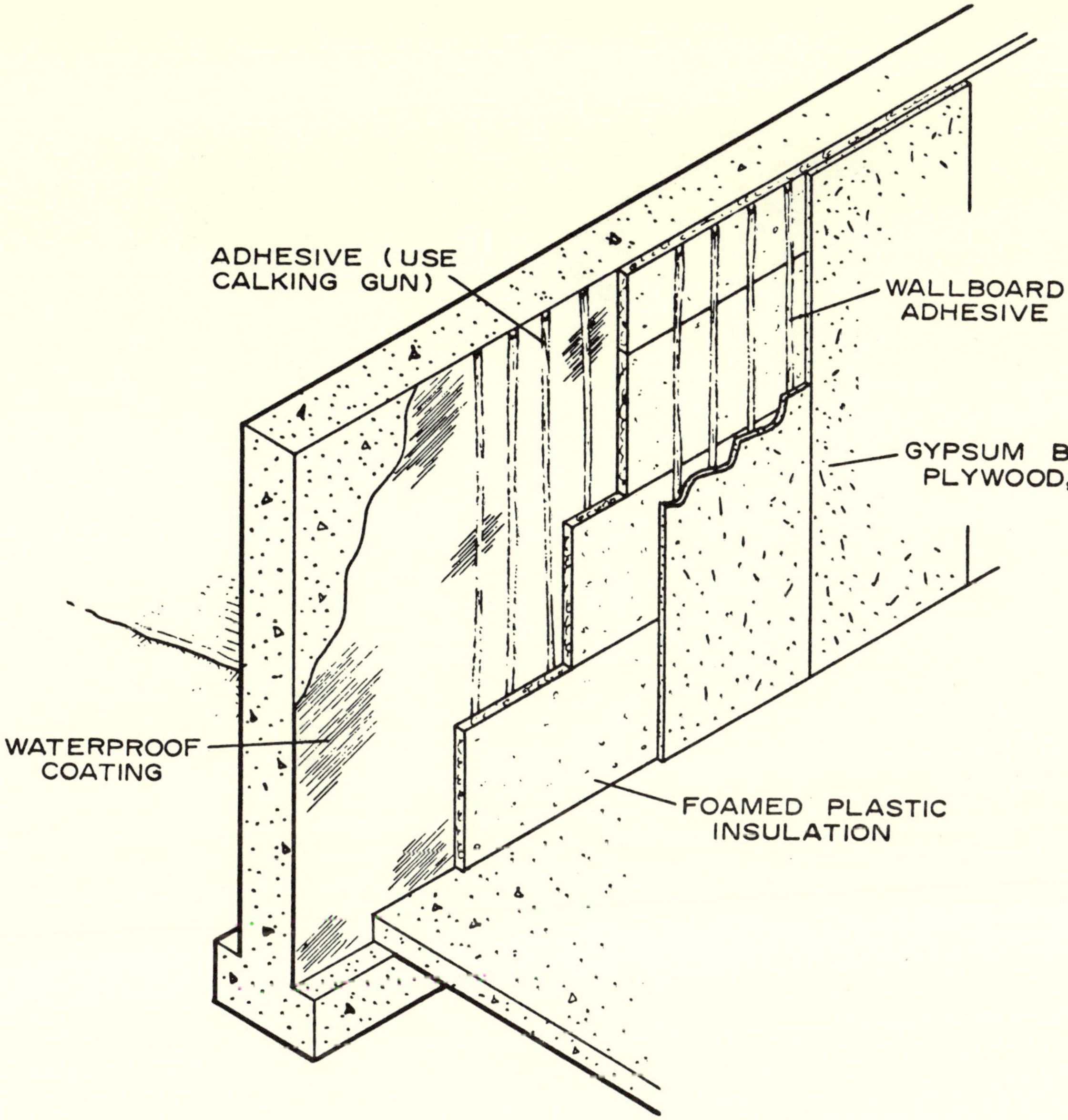

Figure 112.—Basement wall finish, without furring strips.

CHAPTER 19

INTERIOR WALL AND CEILING FINISH

Interior finish is the material used to cover the interior framed areas or structures of walls and ceilings. It should be prefinished or serve as a base for paint or other finishes including wallpaper. Depending on whether it is wood, gypsum wallboard, or plaster, size and thickness should generally comply with recommendations in this handbook. Finishes in bath and kitchen areas should have more rigid requirements because of moisture conditions. Several types of interior finishes are used in the modern home, mainly:

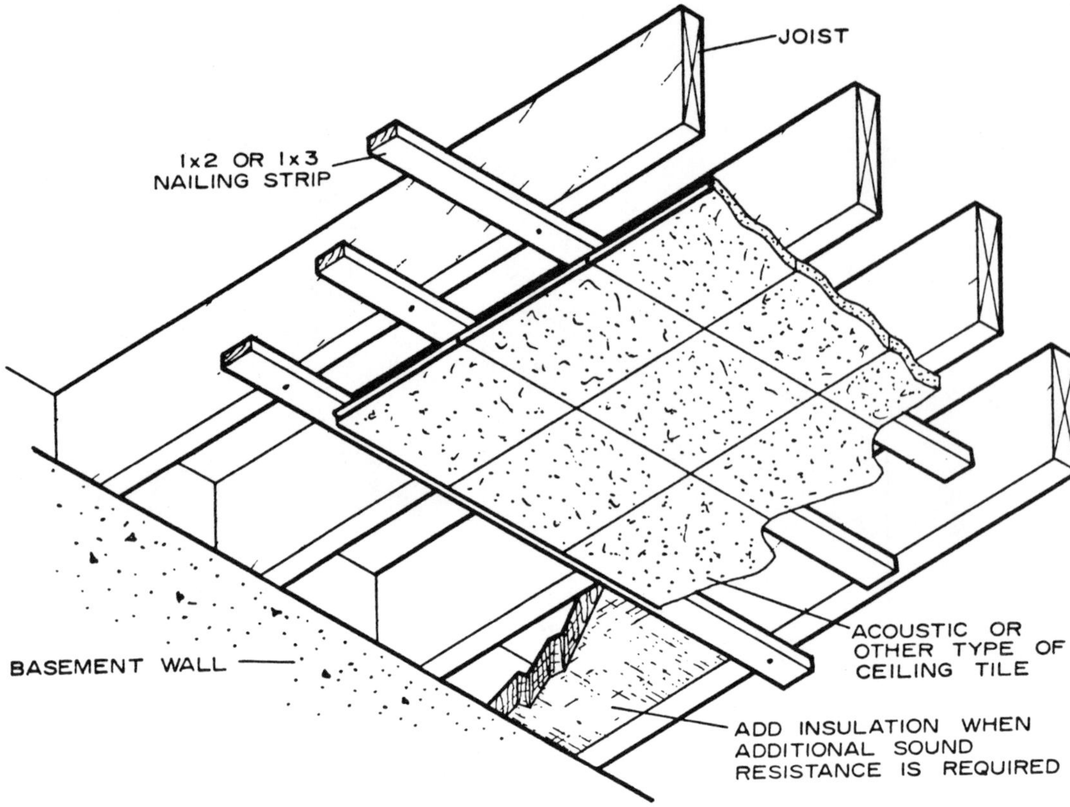

Figure 113.—Installation of ceiling tile.

(a) Lath and plaster, (b) wood paneling, fiberboard, or plywood, and (c) gypsum wallboard.

Types of Finishes

Though lath and plaster finish is widely used in home construction, use of dry-wall materials has been increasing. Dry wall is often selected by builders because there is usually a time saving in construction. A plaster finish, being a wet material, requires drying time before other interior work can be started—dry-wall finish does not. However, a gypsum dry wall demands a moderately low moisture content of the framing members to prevent "nail-pops." These result when frame members dry out to moisture equilibrium, causing the nailhead to form small "humps" on the surface of the board. Furthermore, stud alinement is more important for single-layer gypsum finish to prevent a wavy, uneven appearance. Thus, there are advantages to both plaster and gypsum dry-wall finishes and each should be considered along with the initial cost and future maintenance involved.

A plaster finish requires some type of base upon which to apply the plaster. *Rock lath* is perhaps the most common. *Fiberboard lath* is also used, and wood lath, quite common many years ago, is permitted in some areas. *Metal lath* or similar mesh forms are normally used only in bathrooms and as reinforcement, but provide a rigid base for plaster finish. They usually cost more, however, than other materials. Some of the rigid foam insulations cemented to masonry walls also serve as plaster bases.

There are many types of dry-wall finishes, but one of the most widely used is gypsum board in 4- by 8-foot sheets and in lengths up to 16 feet which are used for horizontal application. Plywood, hardboard, fiberboard, particleboard, wood paneling, and similar types, many in prefinished form, are also used.

Lath and Plaster

Plaster Base

A plaster finish requires some type of base upon which the plaster is applied. The base must have bonding qualities so that plaster adheres, or is keyed to the base which has been fastened to the framing members.

One of the most common types of plaster base, that may be used on sidewalls or ceilings, is *gypsum lath*, which is 16 by 48 inches and is applied horizontally across the framing members. It has paper faces with a gypsum filler. For stud or joist spacing of 16 inches on center, ⅜-inch thickness is used. For 24-inch on-center spacing, ½-inch thickness is required. This material can be obtained with a foil back that serves as a vapor barrier. If the foil faces an air space, it also has reflective insulating value. Gypsum lath may be obtained with perforations, which, by improving the bond, would lengthen the time the plaster would remain intact when exposed to fire. The building codes in some cities require such perforation.

Insulating fiberboard lath in ½-inch thickness and 16 by 48 inches in size is also used as a plaster base. It has greater insulating value than the gypsum lath, but horizontal joints must usually be reinforced with metal clips.

Metal lath in various forms such as diamond mesh, flat rib, and wire lath is another type of plaster base. It is usually 27 by 96 inches in size and is galvanized or painted to resist rusting.

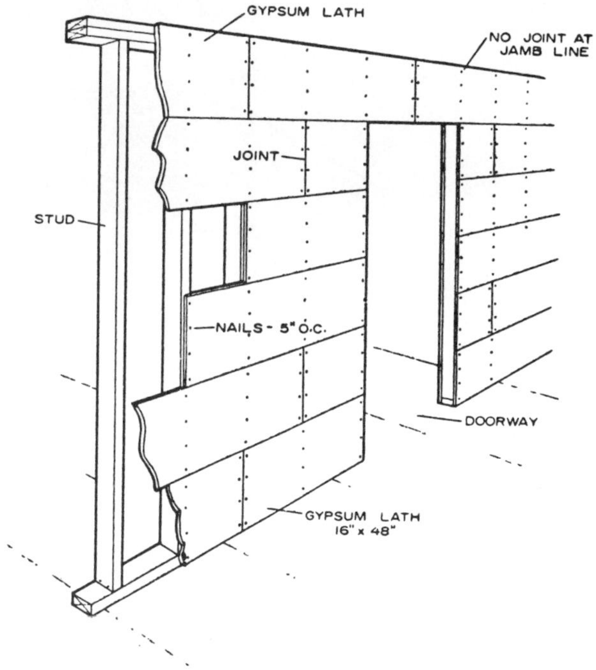

Figure 114.—Application of gypsum lath.

Installation of Plaster Base

Gypsum lath should be applied horizontally with joints broken (fig. 114). Vertical joints should be made over the center of studs or joists and nailed with 12- or 13-gage gypsum-lathing nails 1½ inches long and with a ⅜-inch flat head. Nails should be spaced 5 inches on center, or four nails for the 16-inch height, and used at each stud or joist crossing. Some manufacturers specify the ring-shank nails with a slightly greater spacing. Lath joints over heads of openings should not occur at the jamb lines (fig. 114).

Insulating lath should be installed much the same as gypsum lath, except that slightly longer blued nails should be used. A special waterproof facing is provided on one type of gypsum board for use as a ceramic tile base when the tile is applied with an adhesive.

Metal lath is often used as a plaster base around tub recesses and other bath and kitchen areas (fig. 115). It is also used when a ceramic tile is applied over a plastic base. It must be backed with water-resistant sheathing paper over the framing. The metal lath is applied horizontally over the waterproof backing with side and end joints lapped. It is nailed with No. 11 and No. 12 roofing nails long enough to provide about 1½-inch penetration into the framing member or blocking.

Plaster Reinforcing

Because some drying usually takes place in wood framing members after a house is completed, some shrinkage can be expected; in turn, this may cause plaster cracks to develop around openings and in corners. To minimize, if not eliminate, this cracking, expanded metal lath is used in certain key positions over the plaster-base material as reinforcement. Strips of expanded metal lath may be used over window and door openings (fig. 116,*A*). A strip about 10 by 20 inches is placed diagonally across each upper corner of the opening and tacked in place.

Metal lath should also be used under flush ceiling beams to prevent plaster cracks (fig. 116,*B*). On wood drop beams extending below the ceiling line, the metal

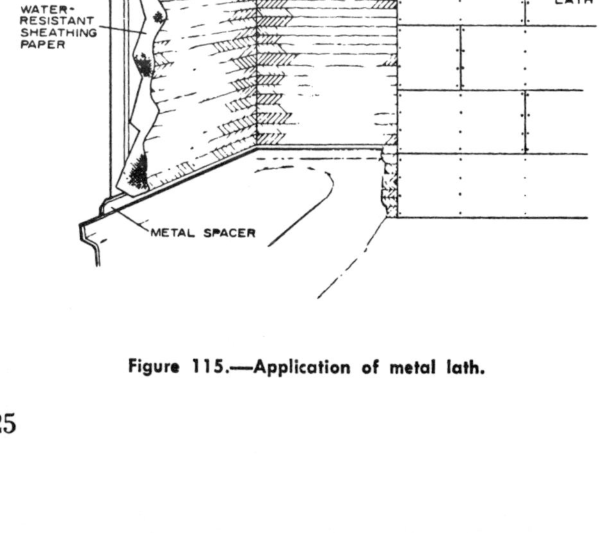

Figure 115.—Application of metal lath.

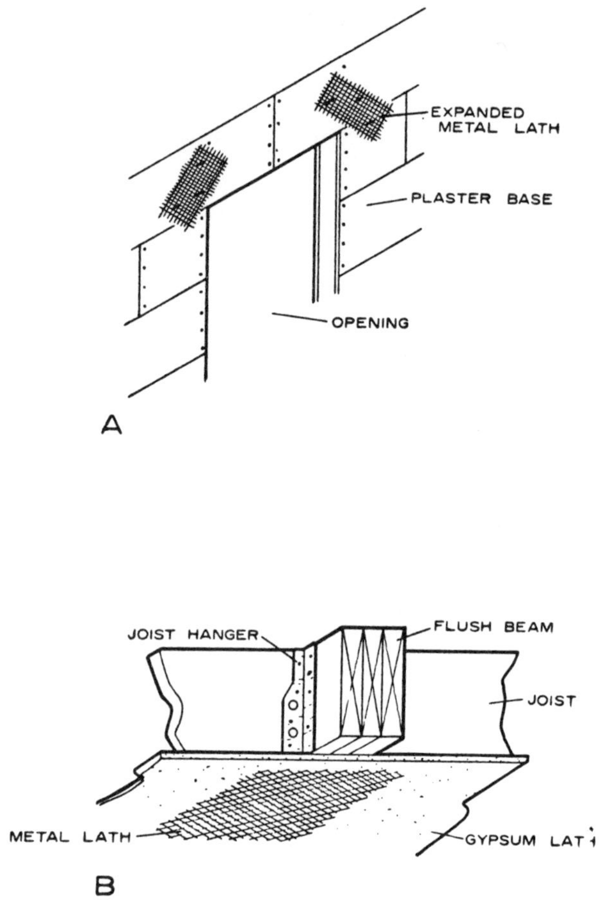

Figure 116.—Metal lath used to minimize cracking: *A*, At door and window openings; *B*, under flush beams.

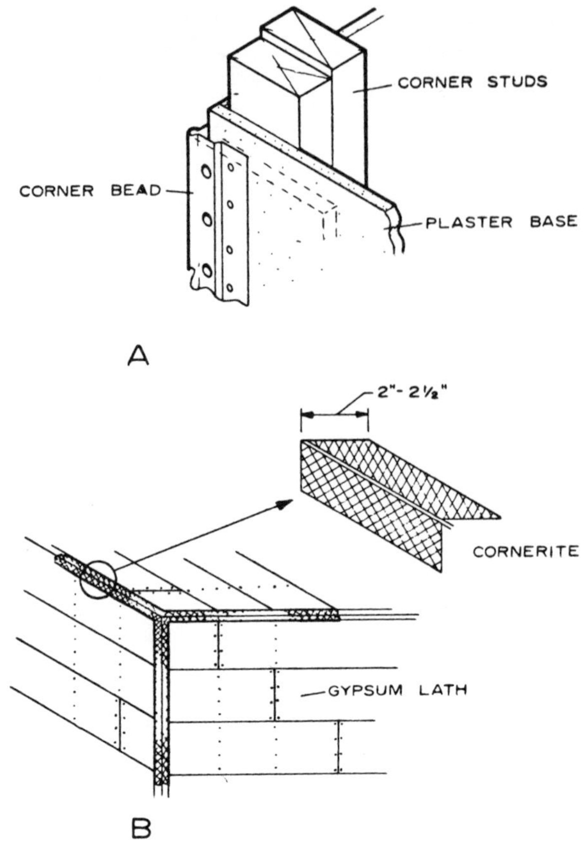

Figure 117.—Reinforcing of plaster at corners: *A*, Outside; *B*, inside.

lath is applied with self-furring nails to provide space for keying of the plaster.

Corner beads of expanded metal lath or of perforated metal should be installed on all exterior corners (fig. 117). They should be applied plumb and level. The bead acts as a leveling edge when walls are plastered and reinforces the corner against mechanical damage. To minimize plaster cracks, inside corners at the juncture of walls and of ceilings should also be reinforced. Metal lath or wire fabric (*cornerites*) are tacked lightly in place in these areas. Cornerites provide a key width of 2 to 2½ inches at each side for plaster.

Plaster Grounds

Plaster grounds are strips of wood used as guides or strike-off edges when plastering and are located around window and door openings and at the base of the walls. Grounds around interior door openings are often full-width pieces nailed to the sides over the studs and to the underside of the header (fig. 118,*A*). They are 5¼ inches in width, which coincides with standard jamb widths for interior walls with a plaster finish. They are removed after plaster has dried. Narrow strip grounds might also be used around these interior openings (fig. 118,*B*).

In window and exterior door openings, the frames are normally in place before plaster is applied. Thus, the inside edges of the side and head jamb can, and often do, serve as grounds. The edge of the window sill might also be used as a ground, or a narrow ⅞-inch-thick ground strip is nailed to the edge of the 2- by 4-inch sill. Narrow ⅞- by 1-inch grounds might also be used around window and door openings (fig. 118,*C*). These are normally left in place and are covered by the casing.

A similiar narrow ground or screed is used at the bottom of the wall in controlling thickness of the gypsum plaster and providing an even surface for the baseboard and molding (fig. 118,*A*). These strips are also left in place after plaster has been applied.

Plaster Materials and Method of Application

Plaster for interior finishing is made from combinations of sand, lime, or prepared plaster and water. Waterproof-finish wall materials (*Keene's cement*)

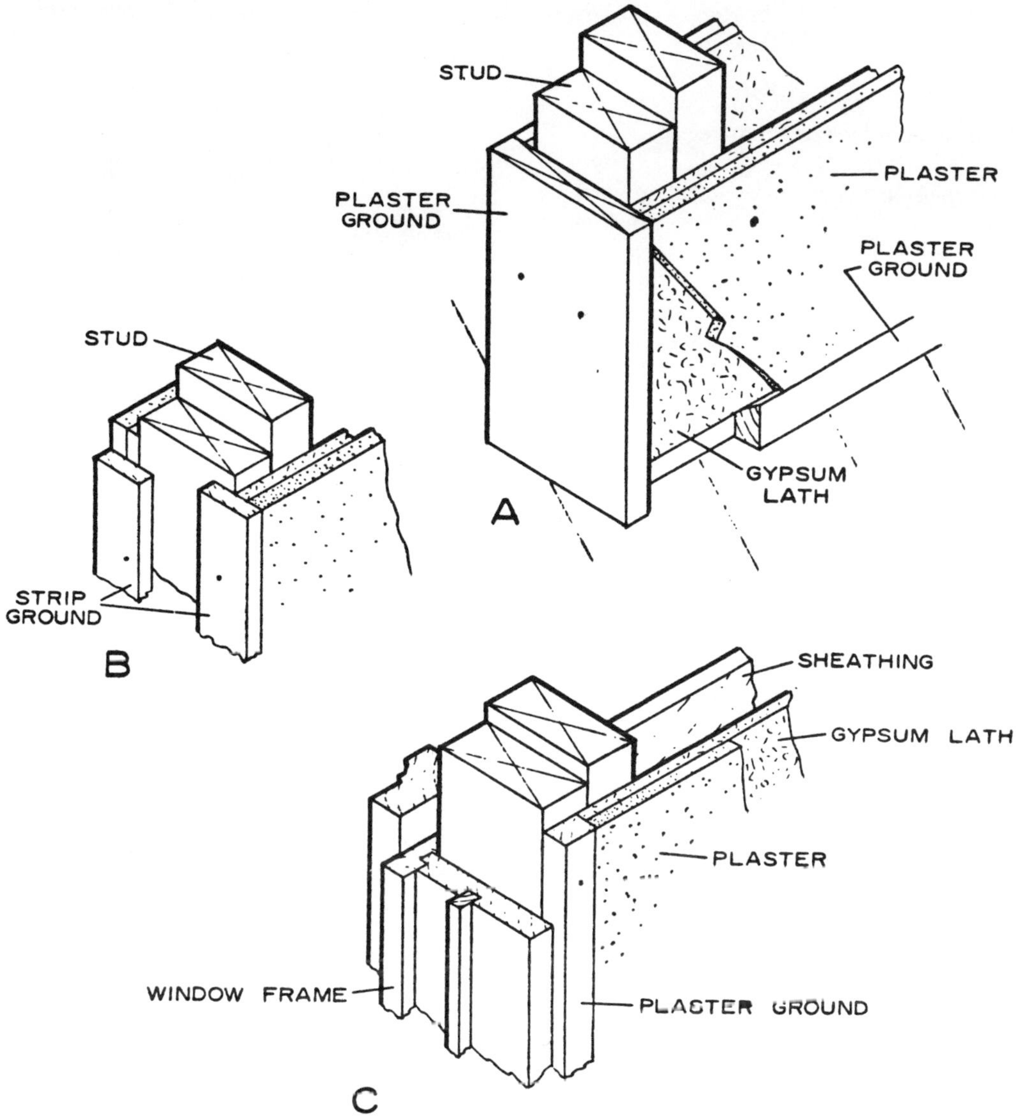

Figure 118.—Plaster grounds: *A*, At doorway and floor; *B*, strip ground at doorway; *C*, ground at window.

are available and should be used in bathrooms, especially in showers or tub recesses when tile is not used, and sometimes in the kitchen wainscot.

Plaster should be applied in three-coat or two-coat double-up work. The minimum thickness over 3/8-inch gypsum lath should be about 1/2 inch. The first plaster coat over metal lath is called the scratch coat and is scratched, after a slight set has occurred, to insure a good bond for the second coat. The second coat is called the brown or leveling coat, and leveling is done during the application of this coat.

The double-up work, combining the scratch and brown coat, is used on gypsum or insulating lath, and leveling and plumbing of walls and ceilings are done during application.

The final or finish coat consists of two general types —the *sand-float* and the *putty* finish. In the sand-float finish, lime is mixed with sand and results in a textured finish, the texture depending on the coarseness of the sand used. Putty finish is used without sand and has a smooth finish. This is common in kitchens and bathrooms where a gloss paint or enamel finish is used, and in other rooms where a smooth finish is desired. Keene's cement is often used as a finish plaster in bathrooms because of its durability.

The plastering operation should not be done in freezing weather without constant heat for protection from freezing. In normal construction, the heating unit is in place before plastering is started.

Insulating plaster, consisting of a vermiculite, perlite, or other aggregate with the plaster mix, may also be used for wall and ceiling finishes.

Dry-wall Finish

Dry-wall finish is a material that requires little, if any, water for application. More specifically, dry-wall finish includes gypsum board, plywood, fiberboard, or similar sheet material, as well as wood paneling in various thicknesses and forms.

The use of thin sheet materials such as gypsum board or plywood requires that studs and ceiling joists have good alinement to provide a smooth, even surface. Wood sheathing will often correct misalined studs on exterior walls. A "strong back" provides for alining of ceiling joists of unfinished attics (fig. 119,*A*) and can be used at the center of the span when ceiling joists are uneven.

Table 7 lists thicknesses of wood materials commonly used for interior covering.

TABLE 7.—*Minimum thicknesses for plywood, fiberboard, and wood paneling.*

Framing spaced (inches)	Thickness		
	Plywood	Fiberboard	Paneling
	In.	*In.*	*In.*
16	1/4	1/2	3/8
20	3/8	3/4	1/2
24	3/8	3/4	5/6

Gypsum Board

Gypsum board is a sheet material composed of a gypsum filler faced with paper. Sheets are normally 4 feet wide and 8 feet in length, but can be obtained in lengths up to 16 feet. The edges along the length are usually tapered, although some types are tapered on all edges. This allows for a filled and taped joint. This material may also be obtained with a foil back which serves as a vapor barrier on exterior walls. It is also available with vinyl or other prefinished surfaces. In new construction, 1/2-inch thickness is recommended for single-layer application. In laminated two-ply applications, two 3/8-inch-thick sheets are used. The 3/8-inch thickness, while considered minimum for 16-inch stud spacing in single-layer applications, is normally specified for repair and remodeling work.

Table 8 lists maximum member spacing for the various thicknesses of gypsum board.

When the single-layer system is used, the 4-foot-wide gypsum sheets are applied vertically or horizontally on the walls after the ceiling has been covered. Vertical application covers three stud spaces when studs are spaced 16 inches on center, and two when spacing is 24 inches. Edges should be centered on studs, and

TABLE 8.—*Gypsum board thickness (single layer)*

Installed long direction of sheet	Minimum thickness	Maximum spacing of supports (on center)	
		Walls	Ceilings
	In.	*In.*	*In.*
Parallel to framing members	3/8	16	
	1/2	24	16
	5/8	24	16
Right angles to framing members	3/8	16	16
	1/2	24	24
	5/8	24	24

only moderate contact should be made between edges of the sheet.

Fivepenny cooler-type nails (1 5/8 in. long) should be used with 1/2-inch gypsum, and fourpenny (1 3/8 in. long) with the 3/8-inch-thick material. Ring-shank nails, about 1/8 inch shorter, can also be used. Some manufacturers often recommend the use of special screws to reduce "bulging" of the surface ("nail-pops" caused by drying out of the frame members). If moisture content of the framing members is less than 15 percent when gypsum board is applied, "nail-pops" will be greatly reduced. It is good practice, when framing members have a high moisture content, to allow them to approach moisture equilibrium before application of the gypsum board. Nails should be spaced 6 to 8 inches for sidewalls and 5 to 7 inches for ceiling application (fig. 119,*B*). Minimum edge distance is 3/8 inch.

The horizontal method of application is best adapted to rooms in which full-length sheets can be used, as it minimizes the number of vertical joints. Where joints are necessary, they should be made at windows or doors. Nail spacing is the same as that used in vertical application. When studs are spaced 16 inches on center, horizontal nailing blocks between studs are normally not required when stud spacing is not greater than 16 inches on center and gypsum board is 3/8 inch or thicker. However, when spacing is greater, or an impact-resistant joint is required, nailing blocks may be used (fig. 119,*C*).

Another method of gypsum-board application (laminated two-ply) includes an undercourse of 3/8-inch material applied vertically and nailed in place. The finish 3/8-inch sheet is applied horizontally, usually in room-size lengths, with an adhesive. This adhesive is either applied in ribbons, or is spread with a notched trowel. The manufacturer's recommendations should be followed in all respects.

Nails in the finish gypsum wallboard should be driven with the heads slightly below the surface. The crowned head of the hammer will form a small dimple in the wallboard (fig. 120,*A*). A nail set should *not*

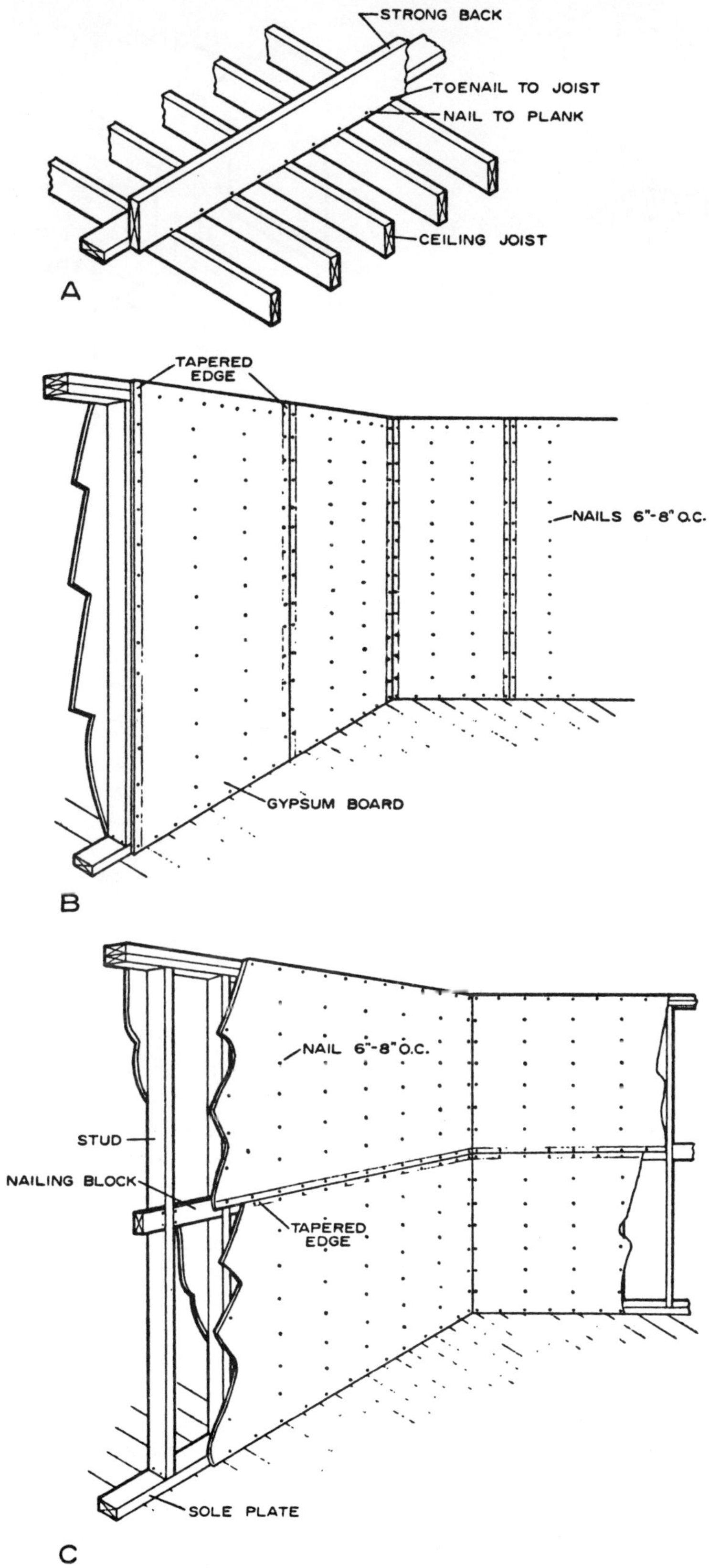

Figure 119.—Application of gypsum board finish: *A*, Strong back; *B*, vertical application; *C*, horizontal application.

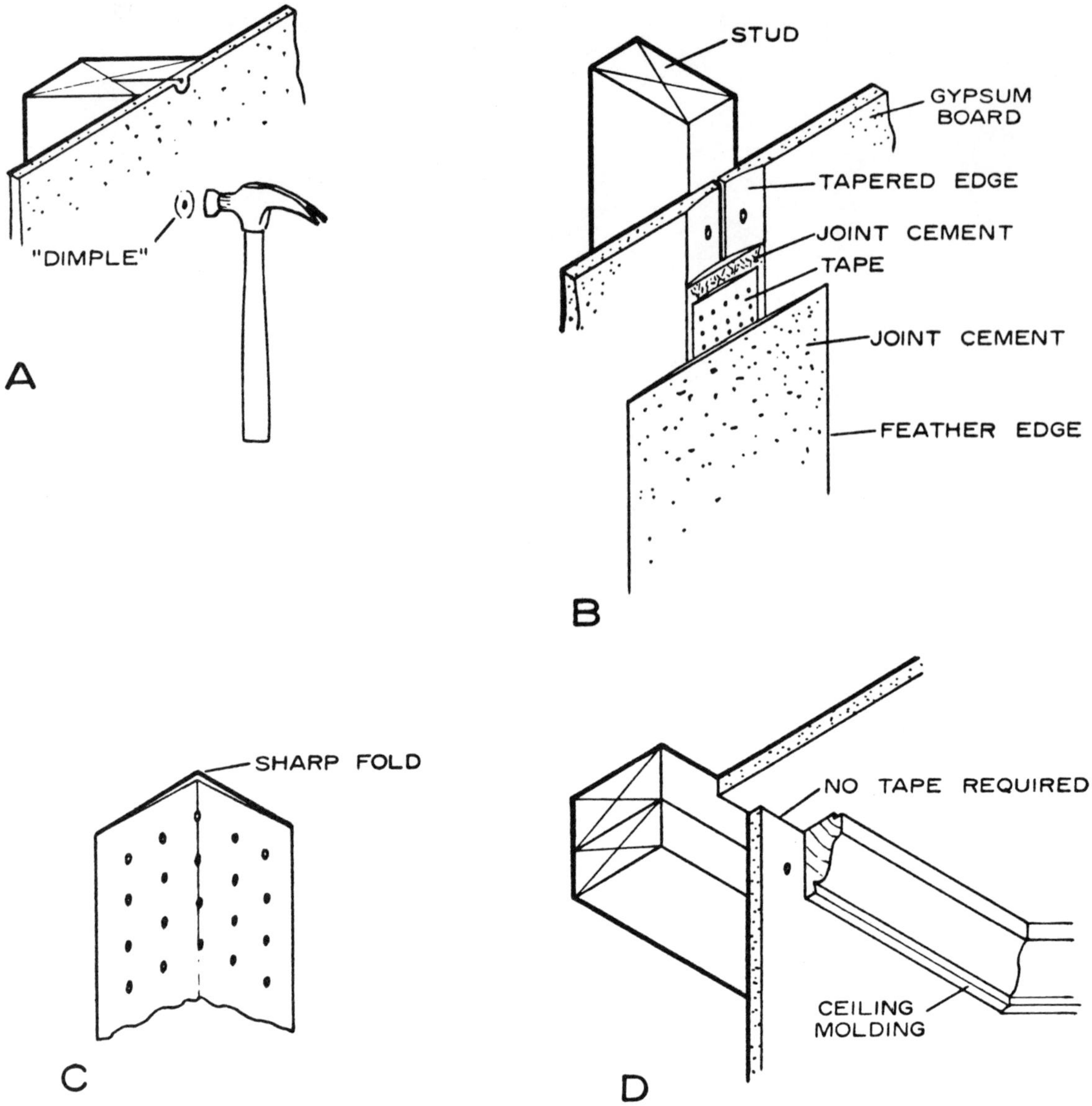

Figure 120.—Finishing gypsum dry wall: *A*, Nail set with crowned hammer; *B*, cementing and taping joint; *C*, taping at inside corners; *D*, alternate finish at ceiling.

be used, and care should be taken to avoid breaking the paper face.

Joint cement, "spackle," is used to apply the tape over the tapered edge joints and to smooth and level the surface. It comes in powder form and is mixed with water to a soft putty consistency so that it can be easily spread with a trowel or putty knife. It can also be obtained in premixed form. The general procedure for taping (fig. 120,*B*) is as follows:

1. Use a wide spackling knife (5 in.) and spread the cement in the tapered edges, starting at the top of the wall.

2. Press the tape into the recess with the putty knife until the joint cement is forced through the perforations.

3. Cover the tape with additional cement, feathering the outer edges.

4. Allow to dry, sand the joint lightly, and then apply the second coat, feathering the edges. A steel trowel is sometimes used in applying the second coat. For best results, a third coat may be applied, feathering beyond the second coat.

5. After the joint cement is dry, sand smooth (an electric hand vibrating sander works well).

6. For hiding hammer indentations, fill with joint cement and sand smooth when dry. Repeat with the second coat when necessary.

Interior corners may be treated with tape. Fold the tape down the center to a right angle (fig. 120,*C*) and (1) apply cement at the corner, (2) press the tape in place, and (3) finish the corner with joint cement. Sand smooth when dry and apply a second coat.

The interior corners between walls and ceilings may also be concealed with some type of molding (fig. 120,*D*). When moldings are used, taping this joint is not necessary. Wallboard corner beads at exterior corners will prevent damage to the gypsum board. They are fastened in place and covered with the joint cement.

Plywood

Prefinished plywood is available in a number of species, and its use should not be overlooked for accent walls or to cover entire room wall areas. Plywood for interior covering may be used in 4- by 8-foot and longer sheets. They may be applied vertically or horizontally, but with solid backing at all edges. For 16-inch frame-member spacing, 1/4-inch thickness is considered minimum. For 20- or 24-inch spacing, 3/8-inch plywood is the minimum thickness. Casing or finishing nails 1 1/4 to 1 1/2 inches long are used. Space them 8 inches apart on the walls and 6 inches apart on ceilings. Edge nailing distance should be not less than 3/8 inch. Allow 1/32-inch end and edge distance between sheets when installing. Most wood or wood-base panel materials should be exposed to the conditions of the room before installation. Place them around the heated room for at least 24 hours.

Adhesives may also be used to fasten prefinished plywood and other sheet materials to wall studs. These panel adhesives usually eliminate the need for more than two guide nails for each sheet. Application usually conforms to the following procedure: (a) Position the sheet and fasten it with two nails for guides at the top or side, (b) remove plywood and spread contact or similar adhesive on the framing members, (c) press the plywood in place for full contact using the nails for positioning, (d) pull the plywood away from the studs and allow adhesive to set, and (e) press plywood against the framing members and tap lightly with a rubber mallet for full contact. Manufacturers of adhesives supply full instructions for application of sheet materials.

Hardboard and Fiberboard

Hardboard and fiberboard are applied the same way as plywood. Hardboard must be at least 1/4 inch when used over open framing spaced 16 inches on center. Rigid backing of some type is required for 1/8-inch hardboard.

Fiberboard in tongued-and-grooved plank or sheet form must be 1/2 inch thick when frame members are spaced 16 inches on center and 3/4 inch when 24-inch spacing is used, as previously outlined. The casing or finishing nails must be slightly longer than those used for plywood or hardboard; spacing is about the same. Fiberboard is also used in the ceiling as acoustic tile and may be nailed to strips fastened to ceiling joists. It is also installed in 12- by 12-inch or larger tile forms on wood or metal hangers which are hung from the ceiling joists. This system is called a "suspended ceiling."

Wood Paneling

Various types and patterns of woods are available for application on walls to obtain desired decorative effects. For informal treatment, knotty pine, white-pocket Douglas-fir, sound wormy chestnut, and pecky cypress, finished natural or stained and varnished, may be used to cover one or more sides of a room. Wood paneling should be thoroughly seasoned to a moisture content near the average it reaches in service (fig. 121), in most areas about 8 percent. Allow the material to reach this condition by placing it around the wall of the heated room. Boards may be applied horizontally or vertically, but the same general methods of application should pertain to each. The following may be used as a guide in the application of matched wood paneling:

1. Apply over a vapor barrier and insulation when application is on the exterior wall framing or blocking (fig. 122).
2. Boards should not be wider than 8 inches except when a long tongue or matched edges are used.
3. Thickness should be at least 3/8 inch for 16-inch spacing of frame members, 1/2 inch for 20-inch spacing, and 5/8 inch for 24-inch spacing.
4. Maximum spacing of supports for nailing should be 24 inches on center (blocking for vertical applications).
5. Nails should be fivepenny or sixpenny casing or finishing nails.

Use two nails for boards 6 inches or less wide and three nails for 8-inch and wider boards. One nail can be blind-nailed in matched paneling.

Wood paneling in the form of small plywood squares can also be used for an interior wall covering (fig. 123). When used over framing and a vapor barrier, blocking should be so located that each edge has full bearing. Each edge should be fastened with casing or finish nails. When two sides are tongued and grooved, one edge (tongued side) may be blind-nailed. When paneling (16 by 48 in. or larger) crosses studs, it should also be nailed at each intermediate bearing. Matched (tongued-and-grooved) sides should be used when no horizontal blocking is provided or paneling is not used over a solid backing.

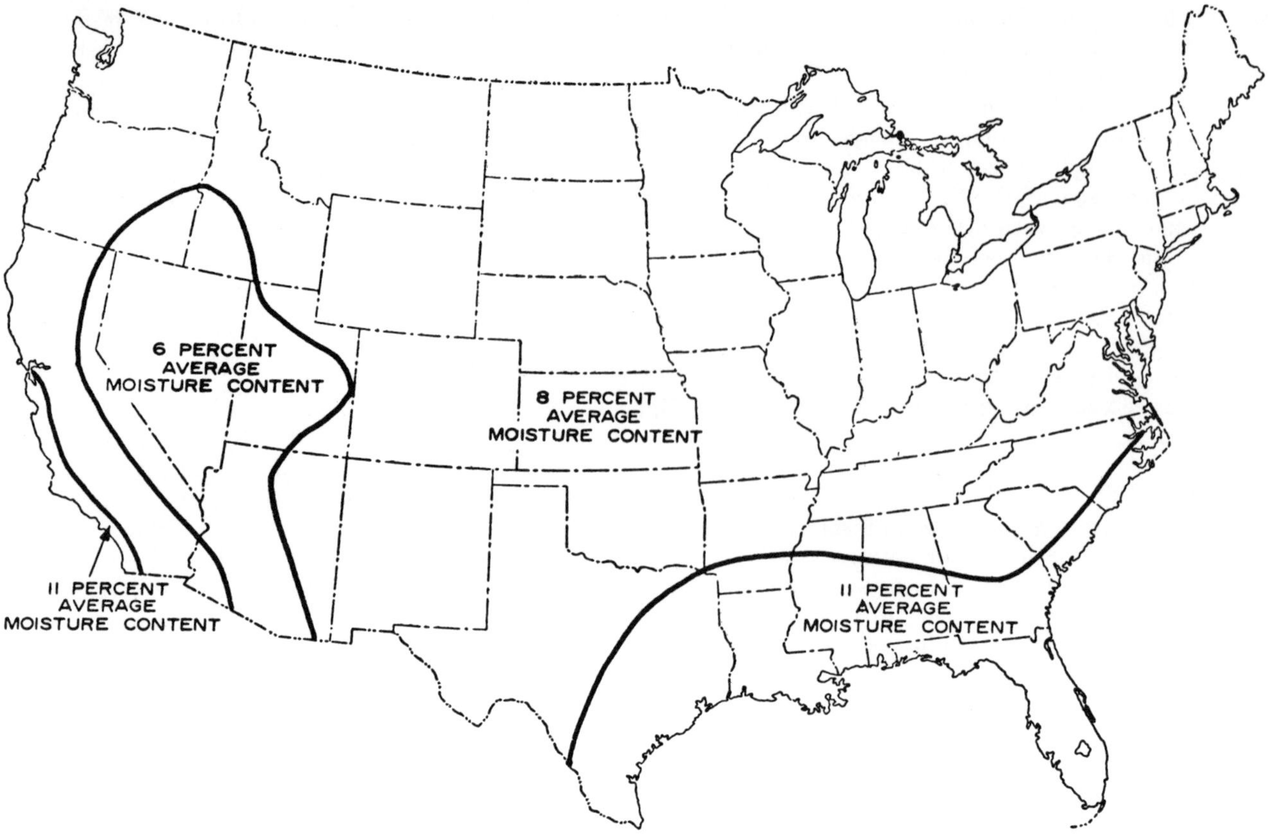

Figure 121.—Recommended average moisture content for interior finish woodwork in different parts of the United States.

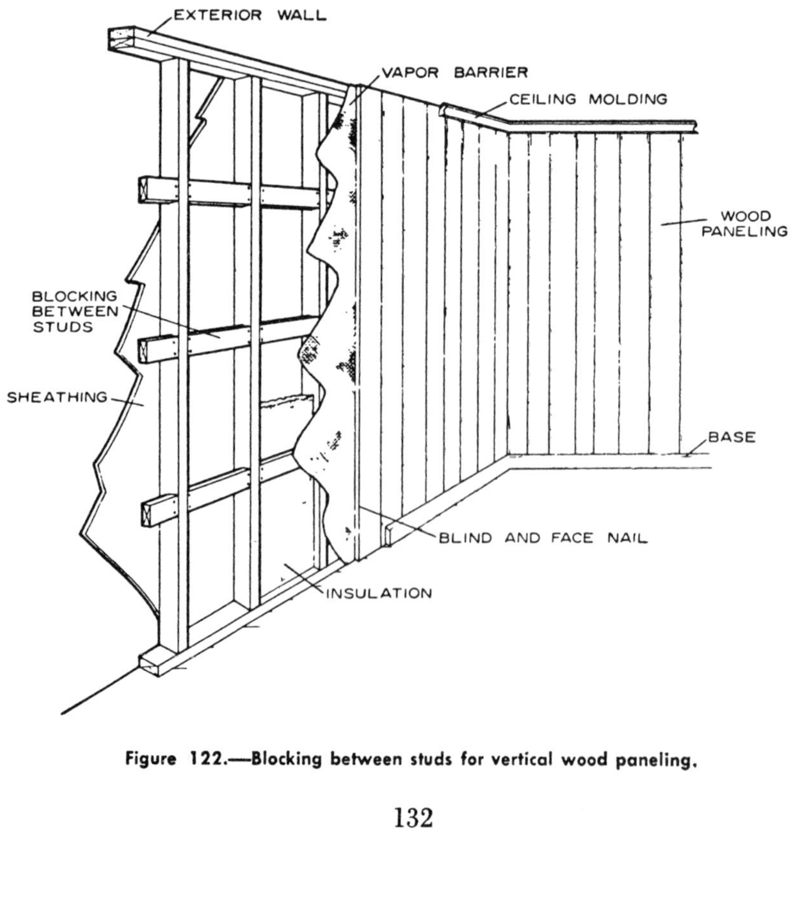

Figure 122.—Blocking between studs for vertical wood paneling.

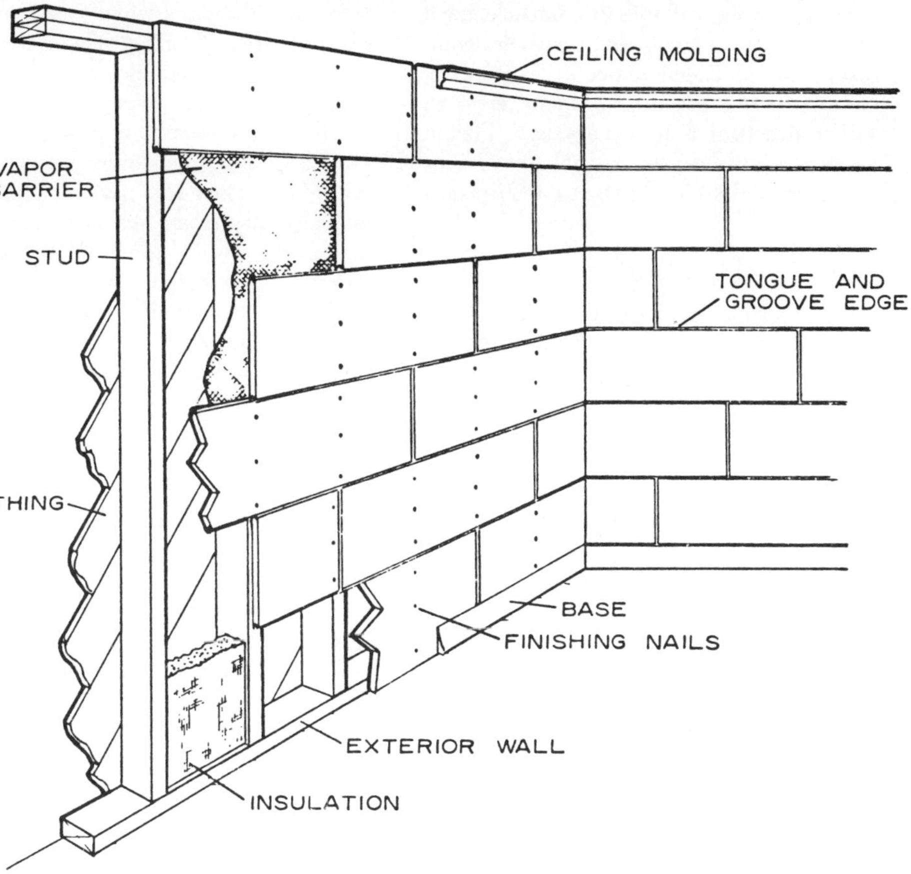

Figure 123.—Application of tongued-and-grooved paneling over studs.

CHAPTER 20

FLOOR COVERINGS

The term "finish flooring" refers to the material used as the final wearing surface that is applied to a floor. Perhaps in its simplest form it might be paint over a concrete floor slab. One of the many resilient tile floorings applied directly to the slab would likely be an improvement from the standpoint of maintenance, but not necessarily from the comfort standpoint.

Flooring Materials

Numerous flooring materials now available may be used over a variety of floor systems. Each has a property that adapts it to a particular usage. Of the practical properties, perhaps durability and maintenance ease are the most important. However, initial cost, comfort, and beauty or appearance must also be considered. Specific service requirements may call for special properties, such as resistance to hard wear in warehouses and on loading platforms, or comfort to users in offices and shops.

There is a wide selection of wood materials that may be used for flooring. Hardwoods and softwoods are available as strip flooring in a variety of widths and thicknesses and as random-width planks and block flooring. Other materials include linoleum, asphalt, rubber, cork, vinyl, and other materials in tile or sheet forms. Tile flooring is also available in a particleboard which is manufactured with small wood particles combined with resin and fabricated under high pressure. Ceramic tile and carpeting are used in many areas in ways not thought practical a few years ago. Plastic floor coverings used over concrete or stable wood subfloor are another variation in the types of finishes available.

Wood-strip Flooring

Softwood finish flooring costs less than most hardwood species and is often used to good advantage in bedroom and closet areas where traffic is light. It might also be selected to fit the interior decor. It is less dense than the hardwoods, less wear-resistant, and shows surface abrasions more readily. Softwoods most commonly used for flooring are southern pine, Douglas-fir, redwood, and western hemlock.

Table 9 lists the grades and description of softwood strip flooring. Softwood flooring has tongued-and-grooved edges and may be hollow-backed or grooved. Some types are also end-matched. Vertical-grain flooring generally has better wearing qualities than flat-grain flooring under hard usage.

Hardwoods most commonly used for flooring are red and white oak, beech, birch, maple, and pecan. Table 9 lists grades, types, and sizes. Manufacturers supply both prefinished and unfinished flooring.

Perhaps the most widely used pattern is a 25/32- by 2¼-inch *strip flooring.* These strips are laid lengthwise in a room and normally at right angles to the floor joists. Some type of a subfloor of diagonal boards or plywood is normally used under the finish floor. Strip flooring of this type is tongued-and-grooved and end-matched (fig. 124). Strips are random length and may vary from 2 to 16 feet or more. End-matched strip flooring in 25/32-inch thickness is generally hollow backed (fig. 124*A*,). The face is slightly wider than the bottom so that tight joints result when flooring is laid. The tongue fits tightly into the groove to prevent movement and floor "squeaks," All of these details are designed to provide beautiful finished floors that require a minimum of maintenance.

Another matched pattern may be obtained in ⅜- by 2-inch size (fig. 124,*B*). This is commonly used for remodeling work or when subfloor is edge-blocked or thick enough to provide very little deflection under loads.

Square-edged strip flooring (fig. 124,*C*) might also be used occasionally. It is usually ⅜ by 2 inches in size and is laid up over a substantial subfloor. Face-nailing is required for this type.

Wood-block flooring (fig. 125) is made in a number of patterns. Blocks may vary in size from 4 by 4 inches to 9 by 9 inches and larger. Thickness varies by type from 25/32 inch for laminated blocking or plywood

TABLE 9.—*Grade and description of strip flooring of several species and grain orientation*

Species	Grain orientation	Size: Thickness	Size: Width	First grade	Second grade	Third grade
		In.	*In.*			
			SOFTWOODS			
Douglas-fir and hemlock	Edge grain	25/32	2⅜–5 3/16	B and Better	C	D
	Flat grain	25/32	2⅜–5 3/16	C and Better	D	----------------
Southern pine	Edge grain and Flat grain	5/16–1 5/16	1¾–5 7/16	B and Better	C and Better	D (and No. 2)
			HARDWOODS			
Oak	Edge grain	25/32	1½–3¼	Clear	Select	----------------
	Flat grain	⅜	1½, 2	Clear	Select	No. 1 Common
		½	1½, 2			
Beech, birch, maple, and pecan[1]		25/32	1½–3¼	First grade	Second grade	----------------
		⅜	1½, 2			
		½	1½, 2			

[1] Special grades are available in which uniformity of color is a requirement.

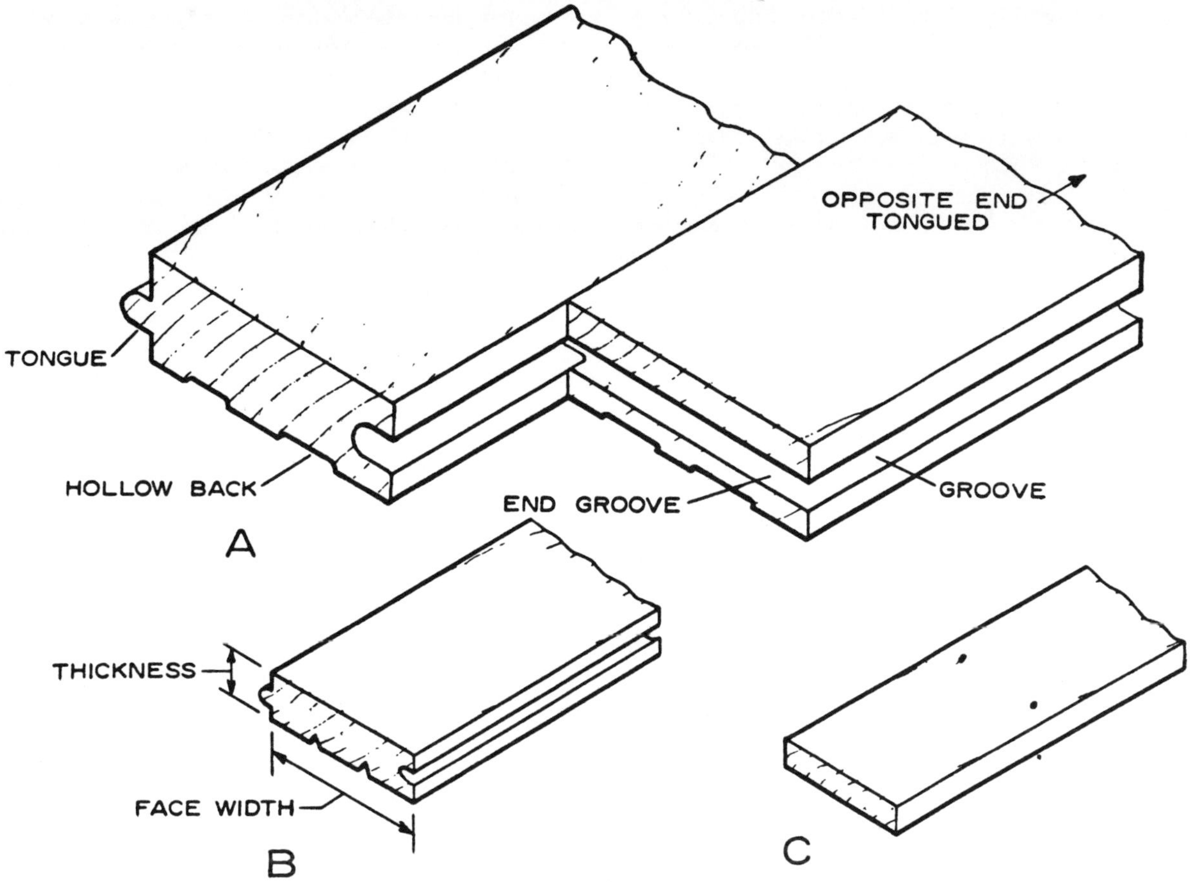

Figure 124.—Types of strip flooring: *A*, Side- and end-matched—25/32-inch; *B*, thin flooring strips—matched; *C*, thin flooring strips—square-edged.

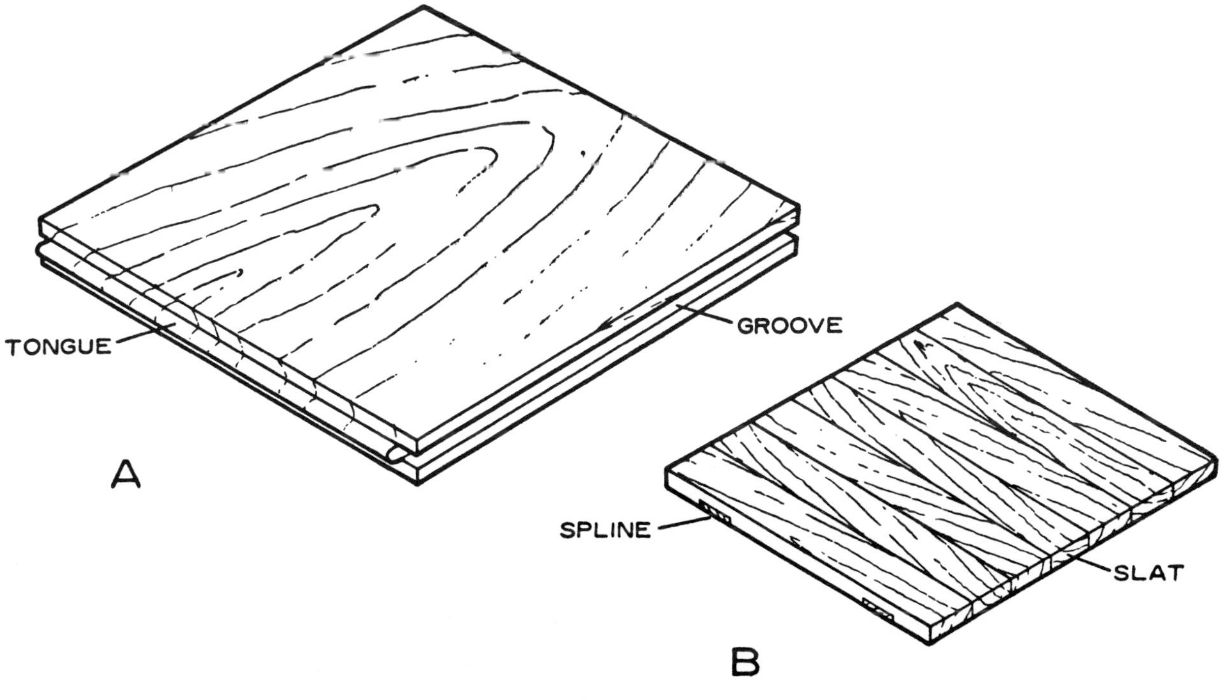

Figure 125.—Wood block flooring: *A*, Tongued-and-grooved; *B*, square-edged—splined.

block tile (fig. 125,*A*) to ⅛-inch stabilized veneer. Solid wood tile is often made up of narrow strips of wood splined or keyed together in a number of ways. Edges of the thicker tile are tongued and grooved, but thinner sections of wood are usually square-edged (fig. 125,*B*). Plywood blocks may be ⅜ inch and thicker and are usually tongued-and-grooved. Many block floors are factory-finished and require only waxing after installation. While stabilized veneer squares are still in the development stage, it is likely that research will produce a low-cost wood tile which can even compete with some of the cheaper nonwood resilient tile now available.

Installation of Wood Strip Flooring

Flooring should be laid after plastering or other interior wall and ceiling finish is completed and dried out, windows and exterior doors are in place, and most of the interior trim, except base, casing, and jambs, are applied, so that it may not be damaged by wetting or by construction activity.

Board subfloors should be clean and level and covered with a deadening felt or heavy building paper. This felt or paper will stop a certain amount of dust, will somewhat deaden sound, and, where a crawl space is used, will increase the warmth of the floor by preventing air infiltration. To provide nailing into the joists wherever possible, location of the joists should be chalklined on the paper as a guide. Plywood subfloor does not normally require building paper.

Strip flooring should normally be laid crosswise to the floor joists (fig. 126,*A*). In conventionally designed houses, the floor joists span the width of the building over a center supporting beam or wall. Thus, the finish flooring of the entire floor area of a rectangular house will be laid in the same direction. Flooring with "L" or "T" shaped plans will usually have a direction change at the wings, depending on joist direction. As joists usually span the short way in a living room, the flooring will be laid lengthwise to the room. This is desirable appearance-wise and also will reduce shrinkage and swelling effects on the flooring during seasonal changes.

Flooring should be delivered only during dry weather and stored in the warmest and driest place available in the house. The recommended average moisture content for flooring at time of installation varies somewhat in different sections of the United States. The moisture content map (fig. 121) outlines these recommendations. Moisture absorbed after delivery to the house site is one of the most common causes of open joints between flooring strips that appear after several months of the heating season.

Floor squeaks are usually caused by movement of one board against another. Such movement may occur because: (a) Floor joists are too light, causing excessive deflection, (b) sleepers over concrete slabs are not held down tightly, (c) tongues are loose fitting, or (d) nailing is poor. Adequate nailing is an important means of minimizing squeaks, and another is to apply the finish floors only after the joists have dried to 12 percent moisture content or less. A much better job results when it is possible to nail the finish floor through the subfloor into the joists than if the finish floor is nailed *only* to the subfloor.

Various types of nails are used in nailing different thicknesses of flooring. For $^{25}/_{32}$-inch flooring, it is best to use eightpenny flooring nails; for ½-inch, sixpenny; and for ⅜-inch fourpenny casing nails. (All the foregoing are blind-nailed.) For thinner square-edge flooring, it is best to use a 1½-inch flooring brad and face-nail every 7 inches with two nails, one near each edge of the strip, into the subfloor.

Other types of nails, such as the ring-shank and screw-shank type, have been developed in recent years for nailing of flooring. In using them, it is well to check with the floor manufacturer's recommendations as to size and diameter for specific uses. Flooring brads are also available with blunted points to prevent splitting of the tongue.

Figure 126,*B* shows the method of nailing the first strip of flooring placed ½ to ⅝ inch away from the wall. The space is to allow for expansion of the flooring when moisture content increases. The nail is driven straight down through the board at the groove edge. The nails should be driven into the joist and near enough to the edge so that they will be covered by the base or shoe molding. The first strip of flooring can also be nailed through the tongue. Figure 127,*A* shows in detail how nails should be driven into the tongue of the flooring at an angle of 45° to 50°. The nail should not be driven quite flush so as to prevent damaging the edge by the hammerhead (fig. 127,*B*). The nail can be set with the end of a large-size nail set or by laying the nail set flatwise against the flooring (fig. 127,*B*). Nailing devices using standard flooring or special nails are often used by flooring contractors. One blow of the hammer on the plunger drives and sets the nail.

To prevent splitting the flooring, it is sometimes desirable to predrill through the tongue, especially at the ends of the strip. For the second course of flooring from the wall, select pieces so that the butt joints will be well separated from those in the first course. Under normal conditions, each board should be driven up tightly. Crooked pieces may require wedging to force them into alinement or may be cut and used at the ends of the course or in closets. In completing the flooring, a ½- to ⅝-inch space is provided between the wall and the last flooring strip. Because of the closeness of the wall, this strip is usually face-nailed so that the base or shoe covers the set nailheads.

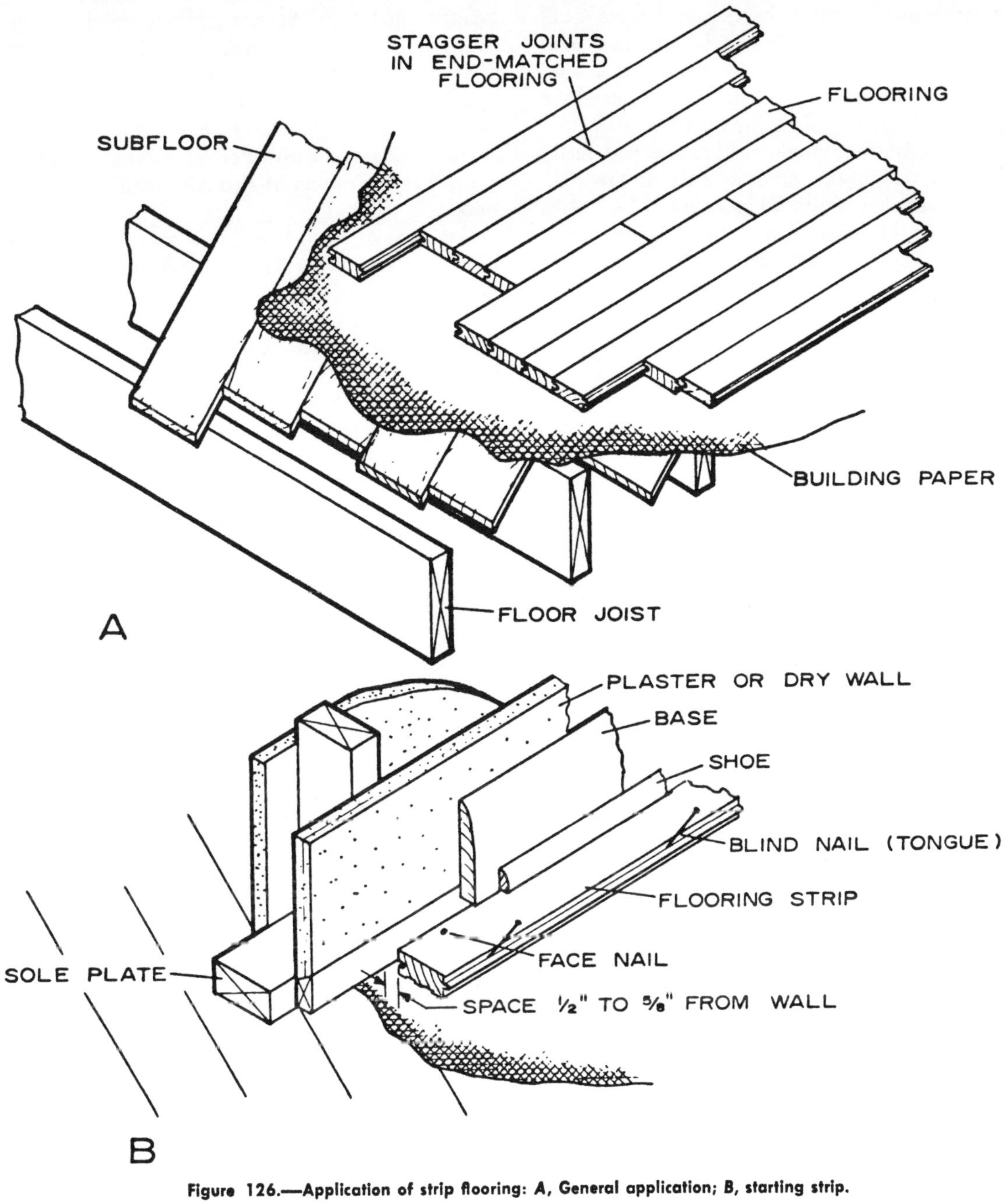

Figure 126.—Application of strip flooring: *A*, General application; *B*, starting strip.

Installation of Wood Flooring Over Concrete Slabs

Installation of wood floor over concrete slabs was described briefly in the chapter "Concrete Floor Slabs on Ground" and illustrated in figure 15. As outlined, one of the important factors in satisfactory performance is the use of a good vapor barrier under the slab to resist the movement of ground moisture and vapor. The vapor barrier is placed under the slab during construction. However, an alternate method must be used when the concrete is already in place (fig. 110.)

Another system of preparing a base for wood flooring when there is no vapor barrier under the slab is shown in figure 128. To resist decay, treated 1- by 4-inch furring strips are anchored to the existing slab, shimming when necessary to provide a level base. Strips should be spaced no more than 16 inches on center. A good waterproof or water-vapor resistant coating on the concrete before the treated strips are applied is usually recommended to aid in reducing moisture movement. A vapor barrier, such as a 4-mil polyethylene or similar membrane, is then laid over the anchored 1-by 4-inch wood strips and a second set of 1 by 4's nailed to the first. Use 1½-inch-long nails spaced 12 to 16 inches apart in a staggered pattern. The moisture content of these second members should be about the same as that of the strip flooring to be applied (6 to 11 pct., fig. 121). Strip flooring can then be installed as previously described.

When other types of finish floor, such as a resilient tile, are used, plywood is placed over the 1 by 4's as a base.

Wood and Particleboard Tile Flooring

Wood and particleboard tile are, for the most part, applied with adhesive on a plywood or similar base. The exception is 25/32-inch wood block floor, which has tongues on two edges and grooves on the other two edges. If the base is wood, these tiles are commonly nailed through the tongue into the subfloor. However, wood block may be applied on concrete slabs with an adhesive. Wood block flooring is installed by changing the grain direction of alternate blocks. This minimizes the effects of shrinking and swelling of the wood.

One type of wood floor tile is made up of a number of narrow slats to form 4- by 4-inch and larger squares. Four or more of these squares, with alternating grain direction, form a block. Slats, squares, and blocks are held together with an easily removed membrane. Adhesive is spread on the concrete slab or underlayment with a notched trowel and the blocks installed immediately. The membrane is then removed and the blocks tamped in place for full adhesive contact. Manufac-

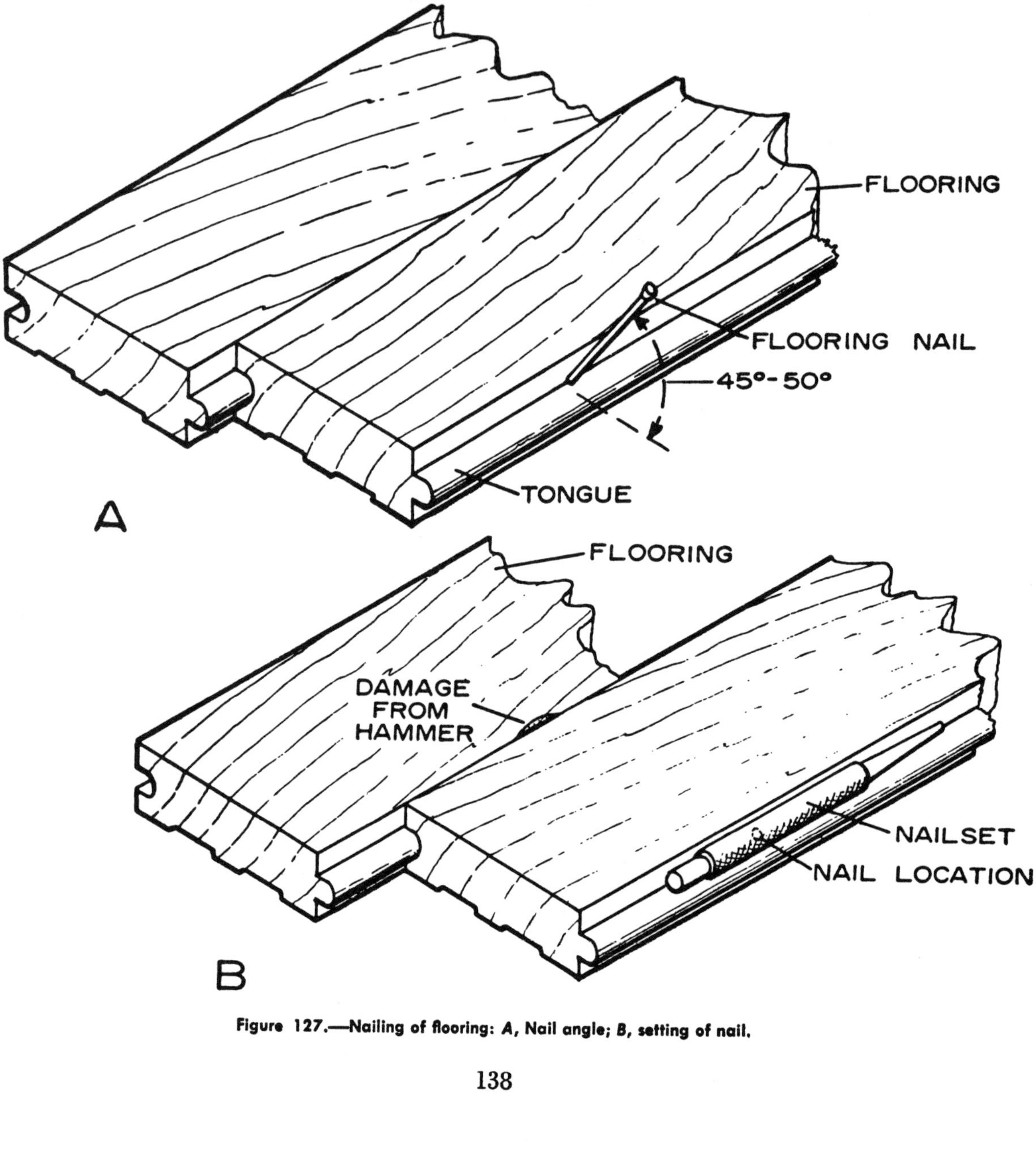

Figure 127.—Nailing of flooring: *A*, Nail angle; *B*, setting of nail.

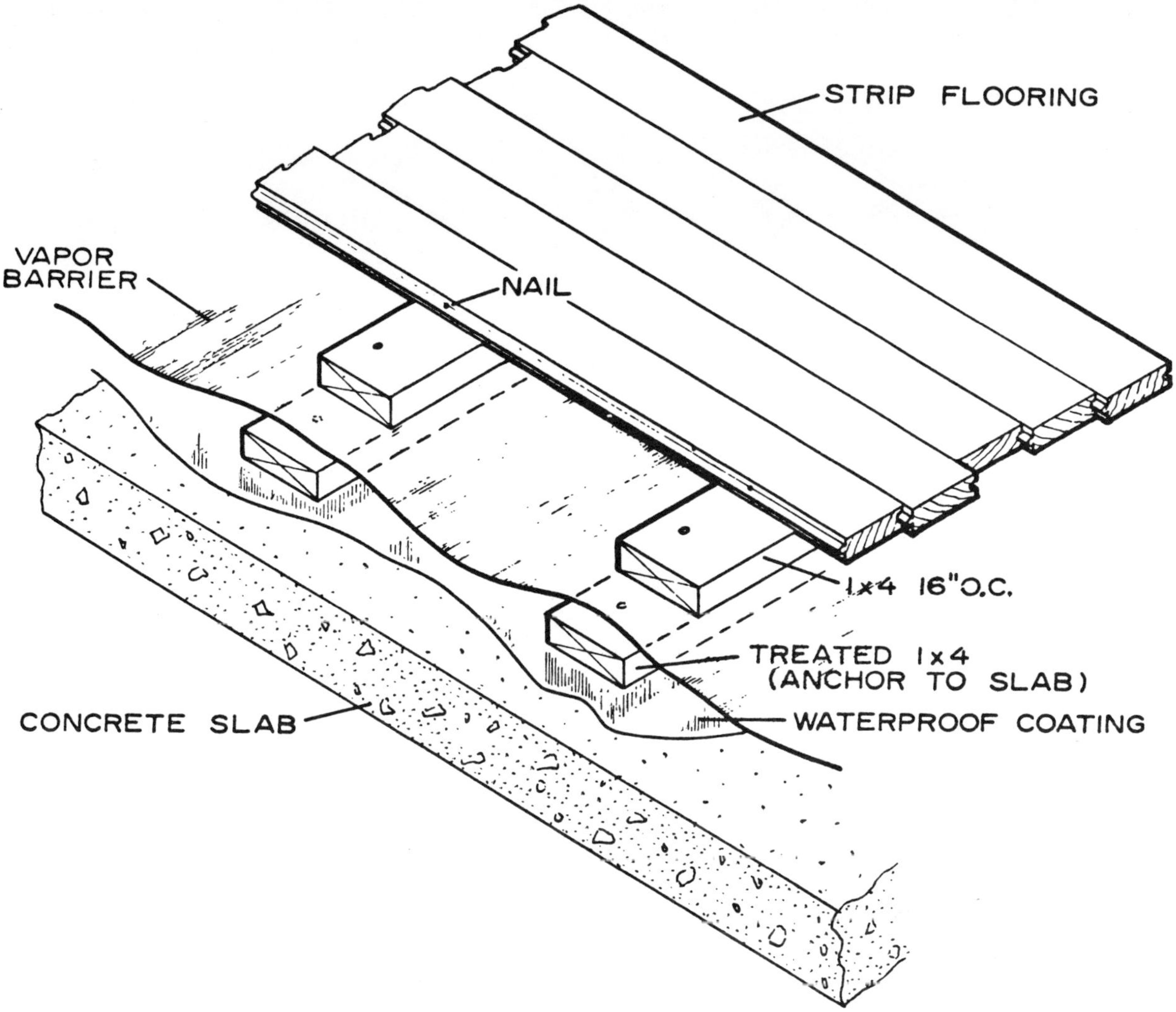

Figure 128.—Base for wood flooring on concrete slab (without an underlying vapor barrier).

turer's recommendations for adhesive and method of application should always be followed. Similar tile made up of narrow strips of wood are fastened together with small rabbeted cleats, tape or similar fastening methods. They too are normally applied with adhesive in accordance with manufacturer's directions.

Plywood squares with tongued-and-grooved edges are another popular form of wood tile. Installation is much the same as for the wood tile previously described. Usually, tile of this type is factory-finished.

A wood-base product used for finish floors is particleboard tile. It is commonly 9 by 9 by 3/8 inches in size with tongued-and-grooved edges. The back face is often marked with small saw kerfs to stabilize the tile and provide a better key for the adhesive. Manufacturer's directions as to the type of adhesive and method of installation are usually very complete; some even include instructions on preparation of the base upon which the tile is to be laid. This tile should not be used over concrete.

Base for Resilient Floors

Resilient floors should *not* be installed directly over a board or plank subfloor. Underlayment grade of wood-based panels such as plywood, particleboard, and hardboard is widely used for suspended floor applications (fig. 129*A*).

Four- by 8-foot plywood or particleboard panels, in a range of thickness from 3/8 to 3/4 inch, are generally selected for use in new construction. Four- by 4-foot or larger sheets of untempered hardboard, plywood, or particleboard of 1/4- or 3/8-inch thickness is used in remodeling work because of the floor thicknesses involved. The underlayment grade of particleboard is

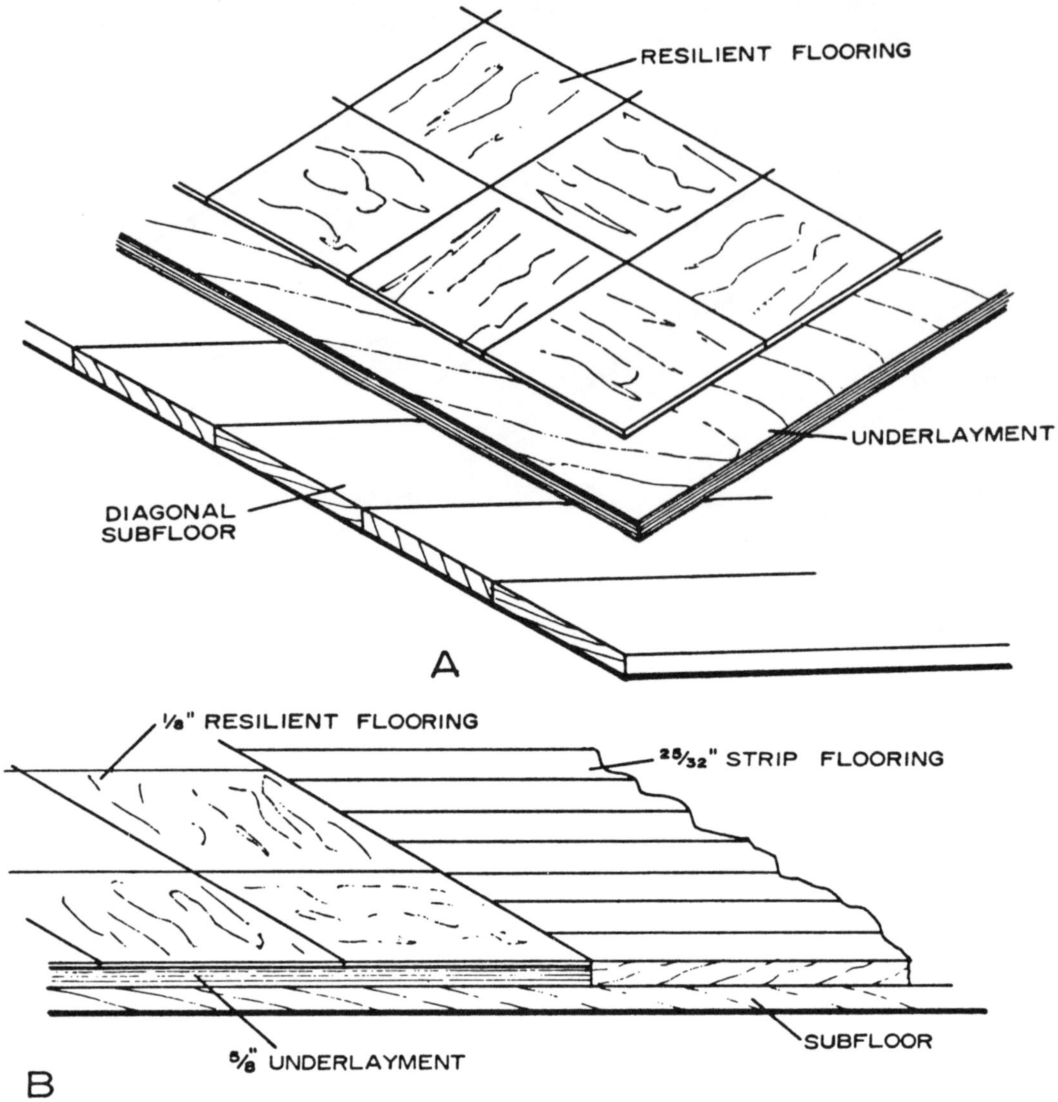

Figure 128.—Base for wood flooring and concrete slab (without an underlying vapor barrier).

a standard product and is available from many producers. Manufacturer's instructions should be followed in the care and use of the product. Plywood underlayment is also a standard product and is available in interior types, exterior types, and interior types with an exterior glueline. The underlayment grade provides for a sanded panel with a C-plugged or better face ply and a C-ply or better immediately under the face. This construction resists damage to the floor surface from concentrated loads such as chair legs, etc.

Generally, underlayment panels are separate and installed over structurally adequate subfloors. Combination subfloor-underlayment panels of plywood construction find increasing usage. Panels for this dual purpose use generally have tongued-and-grooved or blocked edges and C-plugged or better faces to provide a smooth, even surface for the resilient floor covering.

The method of installing plywood combination subfloor and underlayment has been covered in the section on Plywood Subfloor. Underlayment should be laid up as outlined in that section with 1/32-inch edge and end spacing. Sand smooth to provide a level base for the resilient flooring. To prevent nails from showing on the surface of the tile, joists and subfloor should have a moisture content near the average value they reach in service.

The thickness of the underlayment will vary somewhat, depending on the floors in adjoining rooms. The installation of tile in a kitchen area, for example, is usually made over a 5/8-inch underlayment when finish floors in the adjoining living or dining areas are $^{25}/_{32}$-inch strip flooring (fig. 129,*B*). When thinner wood floors are used in adjoining rooms, adjustments are made in the thickness of the underlayment.

Concrete for resilient floors should be prepared as shown in figures 14, 15, or 16, with a good vapor barrier installed somewhere between the soil and the finish floor, preferably just under the slab. Concrete should be leveled carefully when a resilient floor is to be used directly on the slab to minimize dips and waves.

Tile should not be laid on a concrete slab until it has completely dried. One method which may be used to determine this is to place a small square of polyethylene or other low-perm material on the slab overnight. If the underside is dry in the morning, the slab is usually considered dry enough for the installation of the tile.

Types of Resilient Floors

Linoleum

Linoleum may be obtained in various thicknesses and grades, usually in 6-foot-wide rolls. It should not be laid on concrete slabs on the ground. Manufacturer's directions should be followed. After the linoleum is laid, it is usually rolled to insure complete adhesion to the floor.

Asphalt Tile

Asphalt tile is one of the lower cost resilient coverings and may be laid on a concrete slab which is in contact with the ground. However, the vapor barrier under the slab is still necessary. Asphalt tile is about ⅛ inch thick and usually 9 by 9 or 12 by 12 inches in size. Because most types are damaged by grease and oil, it is not used in kitchens.

Asphalt tile is ordinarily installed with an adhesive spread with a notched trowel. Both the type of adhesive and size of notches are usually recommended by the manufacturer.

Other Tile Forms

Vinyl, vinyl asbestos, rubber, cork, and similar coverings are manufactured in tile form, and several types are available for installation in 6-foot-wide rolls. These materials are usually laid over some type of underlayment and not directly on a concrete slab. Standard tile size is 9 by 9 inches but it may also be obtained in 12- by 12-inch size and larger. Decorative strips may be used to outline or to accent the room's perimeter.

In installing all types of square or rectangular tile, it is important that the joints do not coincide with the joints of the underlayment. For this reason, it is recommended that a layout be made before tile is laid. Normally, the manufacturer's directions include laying out a base line at or near the center of the room and parallel to its length. The center or near center, depending on how the tile will finish at the edges, is used as a starting point. This might also be used as a point in quartering the room with a second guideline at exact right angles to the first. The tile is then laid in quarter-room sections after the adhesive is spread.

Seamless

A liquid-applied seamless flooring, consisting of resin chips combined with a urethane binder, is a relatively new development in floor coverings. It is applied in a 2-day cycle and can be used over a concrete base or a plywood subfloor. Plywood in new construction should be at least a C-C plugged exterior grade in ⅝-inch thickness, or ⅜-inch plywood over existing floors. This type of floor covering can be easily renewed.

Carpeting

Carpeting many areas of a home from living room to kitchen and bath is becoming more popular as new carpeting materials are developed. The cost, however, may be considerably higher than a finished wood floor, and the life of the carpeting before replacement would be much less than that of the wood floor. Many wise home builders will specify oak floors even though they expect to carpet some areas. The resale value of the home is then retained even if the carpeting is removed. However, the advantage of carpeting in sound absorption and resistance to impact should be considered. This is particularly important in multi-floor apartments where impact noise reduction is an extremely important phase of construction. If carpeting is to be used, subfloor can consist of ⅝-inch (minimum) tongued and grooved plywood (over 16-inch joist spacing). Top face of the plywood should be C plugged grade or better. Mastic adhesives are also being used to advantage in applying plywood to floor joists. Plywood, particleboard, or other underlayments are also used for a carpet base when installed over a subfloor.

Ceramic Tile

Ceramic tile and similar floor coverings in many sizes and patterns for bath, lavatory, and entry areas may be installed by the cement-plaster method or by the use of adhesives. The cement-plaster method requires a concrete-cement setting bed of 1¼ inches minimum thickness (fig. 130). Joists are chamfered (beveled) and cleats used to support waterproof plywood subfloor or forms cut between the joists. The cement base is reinforced with woven wire fabric or expanded metal lath.

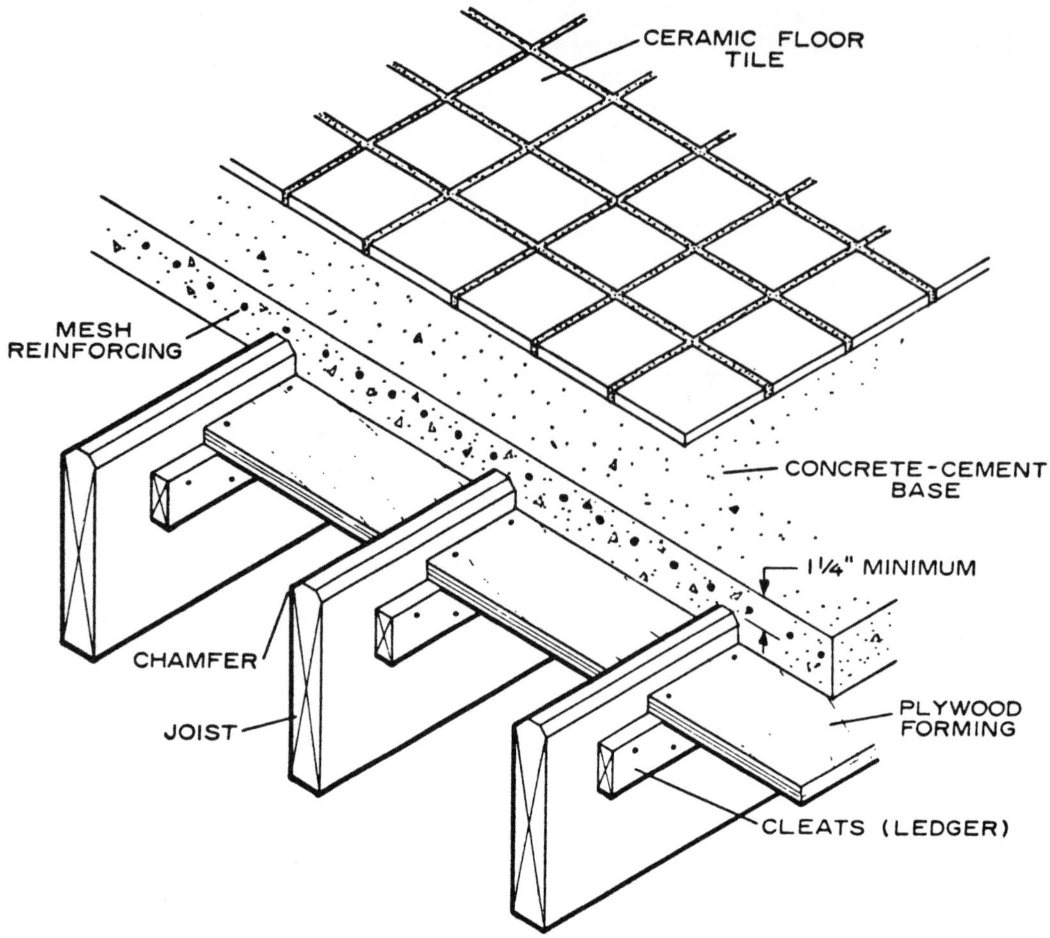

Figure 130.—Cement base for ceramic floor tile.

Tile should be soaked before it is installed. It is pressed firmly in place in the still plastic setting bed, mortar is compressed in the joints, and the joints tooled the same day tile is laid. Laying tile in this manner normally requires a workman skilled in this system. It should then be covered with waterproof paper for damp curing.

Adhesive used for ceramic floor tile should be the type recommended by the manufacturer. When installed over wood joists, a waterproof plywood ¾-inch thick with perimeter and intermediate nailing provides a good base. Before installing tile, a waterproof sealer or a thin coat of tile adhesive is applied to the plywood. Tile should be set over a full covering of adhesive using the "floating method" with a slight twisting movement for full embedment. "Buttering" or using small pats of adhesive on each tile is not acceptable. Tile should not be grouted until volatiles from the adhesive have evaporated. After grouting, joints should be fully tooled.

CHAPTER 21

INTERIOR DOORS, FRAMES, AND TRIM

Interior trim, doorframes, and doors are normally installed after the finish floor is in place. Cabinets, built-in bookcases and fireplace mantels, and other millwork units are also placed and secured at this time. Some contractors may install the interior doorframes *before* the finish floor is in place, allowing for the flooring at the bottom of the jambs. This is usually done when the jambs act as plaster grounds. However, because excessive moisture is present and edges of the jambs are often marred, this practice is usually undesirable.

Decorative Treatment

The decorative treatment for interior doors, trim, and other millwork may be paint or a natural finish with stain, varnish, or other non-pigmented material. The paint or natural finish desired for the woodwork in various rooms often determines the type of species of wood to be used. Interior finish that is to be painted should be smooth, close-grained, and free from pitch streaks. Some species having these requirements in a high degree include ponderosa pine, northern white pine, redwood, and spruce. When hardness and resistance to hard usage are additional requirements, species such as birch, gum, and yellow-poplar are desirable.

For natural finish treatment, a pleasing figure, hardness, and uniform color are usually desirable. Species with these requirements include ash, birch, cherry, maple, oak, and walnut. Some require staining for best appearance.

The recommended moisture content for interior finish varies from 6 to 11 percent, depending on the climatic conditions. The areas of varying moisture content in the United States are shown in figure 121.

Trim Parts for Doors and Frames

Doorframes

Rough openings in the stud walls for interior doors are usually framed out to be 3 inches more than the door height and 2½ inches more than the door width. This provides for the frame and its plumbing and leveling in the opening. Interior doorframes are made up of two side *jambs* and a head jamb and include stop moldings upon which the door closes. The most common of these jambs is the one-piece type (fig. 131,*A*). Jambs may be obtained in standard 5¼-inch widths for plaster walls and 4⅝-inch widths for walls with ½-inch dry-wall finish. The two-and three-piece adjustable jambs are also standard types (fig. 131,*B* and *C*). Their principal advantage is in being adaptable to a variety of wall thicknesses.

Some manufacturers produce interior doorframes with the door fitted and prehung, ready for installing. Application of the casing completes the job. When used with two- or three-piece jambs, casings can even be installed at the factory.

Common minimum widths for single interior doors are: (a) Bedroom and other habitable rooms, 2 feet 6 inches; (b) bathrooms, 2 feet 4 inches; (c) small closet and linen closets, 2 feet. These sizes vary a great deal, and sliding doors, folding door units, and similar types are often used for wardrobes and may be 6 feet or more in width. However, in most cases, the jamb, stop, and casing parts are used in some manner to frame and finish the opening.

Standard interior and exterior door heights are 6 feet 8 inches for first floors, but 6-foot 6-inch doors are sometimes used on the upper floors.

Casing

Casing is the edge trim around interior door openings and is also used to finish the room side of windows and exterior door frames. Casing usually varies in width from 2¼ to 3½ inches, depending on the style. Casing may be obtained in thicknesses from ½ to ¾ inch, although 11/16 inch is standard in many of the narrow-line patterns. Two common patterns are shown in figure 131,*D* and *E*.

Interior Doors

As in exterior door styles, the two general interior types are the flush and the panel door. Novelty doors, such as the folding door unit, might be flush or louvered. Most standard interior doors are 1⅜ inches thick.

The flush interior door is usually made up with a hollow core of light framework of some type with thin plywood or hardboard (fig. 132,*A*). Plywood-faced flush doors may be obtained in gum, birch, oak, mahogany, and woods of other species, most of which are suitable for natural finish. Nonselected grades are usually painted as are hardboard-faced doors.

The panel door consists of solid *stiles* (vertical side members), *rails* (cross pieces), and *panel filters* of various types. The five-cross panel and the Colonial-type panel doors are perhaps the most common of this style (fig. 132,*B* and *C*). The louvered door (fig. 132,*D*) is also popular and is commonly used for closets because it provides some ventilation. Large openings for wardrobes are finished with sliding or folding doors, or with flush or louvered doors (fig. 132,*E*). Such doors are usually 1⅛ inches thick.

Hinged doors should open or swing in the direction of natural entry, against a blank wall whenever possible, and should not be obstructed by other swinging doors. Doors should *never* be hinged to swing into a hallway.

Doorframe and Trim Installation

When the frame and doors are not assembled and prefitted, the side jambs should be fabricated by nailing through the notch into the head jamb with three sevenpenny or eightpenny coated nails (fig. 131,*A*). The assembled frames are then fastened in the rough openings by shingle wedges used between the side jamb and the stud (fig. 133,*A*). One jamb is plumbed and leveled using four or five sets of shingle wedges for the height of the frame. Two eightpenny finishing nails are used at each wedged area, one driven so that the doorstop will cover it (fig. 133,*A*). The opposite side jamb is now fastened in place with shingle wedges and

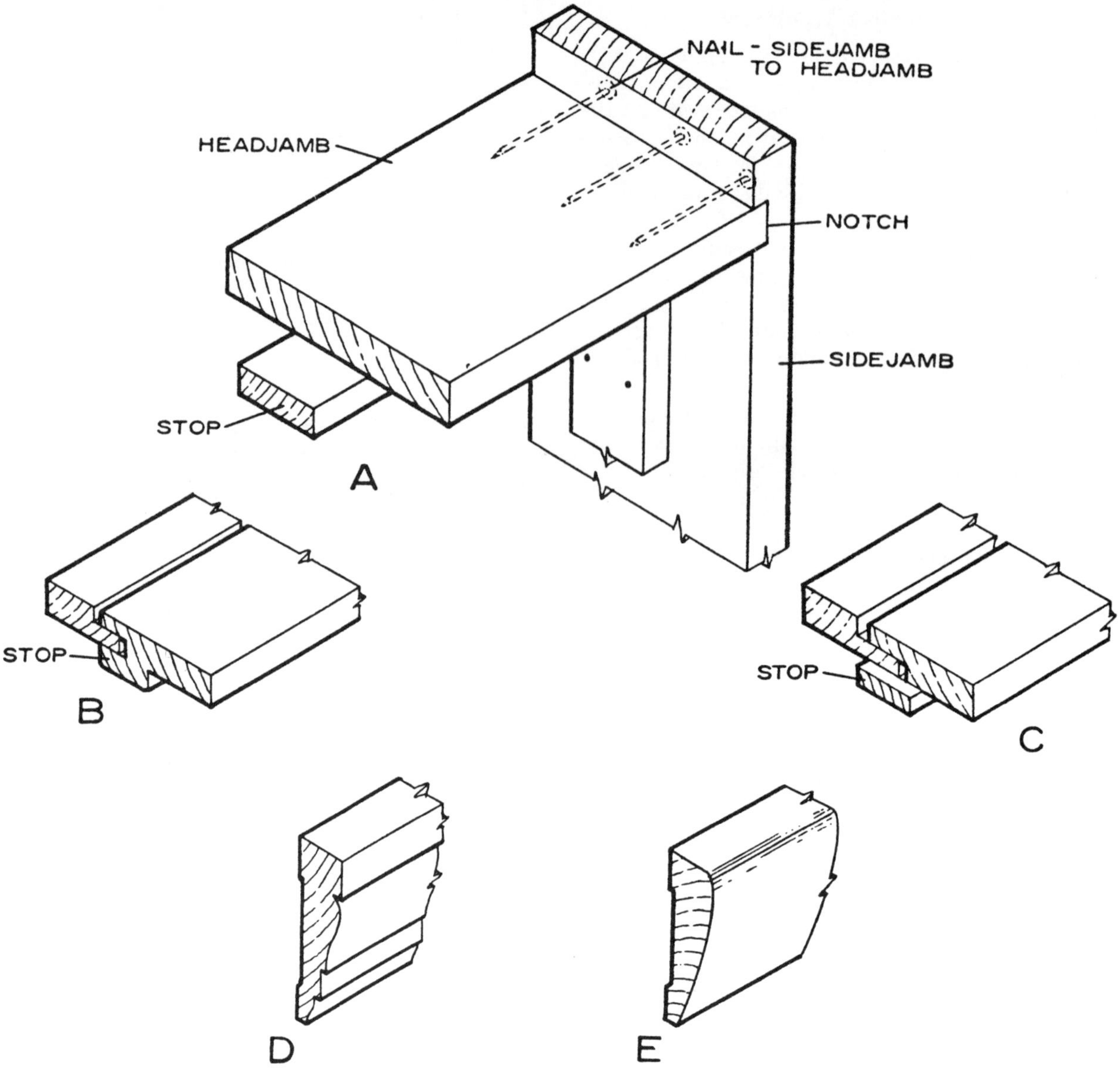

Figure 131.—Interior door parts: *A*, **Door jambs and stops;** *B*, **two-piece jamb;** *C*, **three-piece jamb;** *D*, **Colonial casing;** *E*, **ranch casing.**

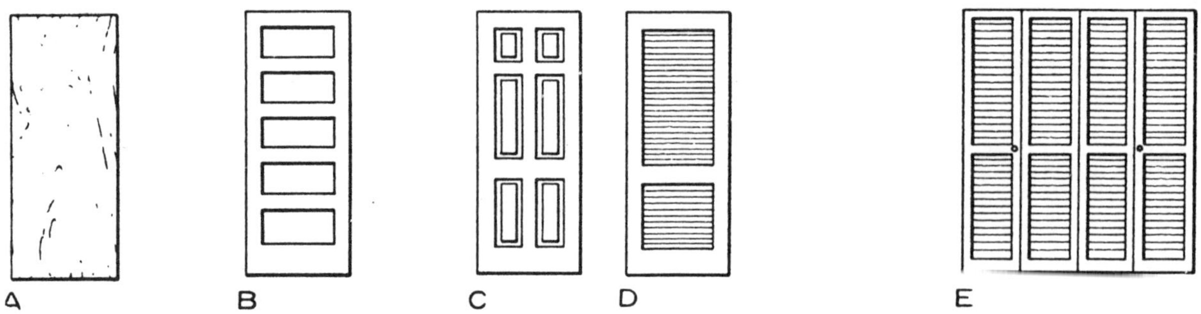

Figure 132.—Interior doors: *A*, **Flush;** *B*, **panel (five-cross);** *C*, **panel (Colonial);** *D*, **louvered;** *E*, **folding (louvered).**

finishing nails, using the first jamb as a guide in keeping a uniform width.

Casings are nailed to both the jamb and the framing studs or header, allowing about a $\frac{3}{16}$-inch edge distance from the face of the jamb (fig. 133,*A*) Finish or casing nails in sixpenny or sevenpenny sizes, depending on the thickness of the casing, are used to nail into the stud. Fourpenny or fivepenny finishing nails or $1\frac{1}{2}$-inch brads are used to fasten the thinner edge of the casing to the jamb. In hardwood, it is usually advisable to predrill to prevent splitting. Nails in the casing are located in pairs (fig. 133,*A*) and

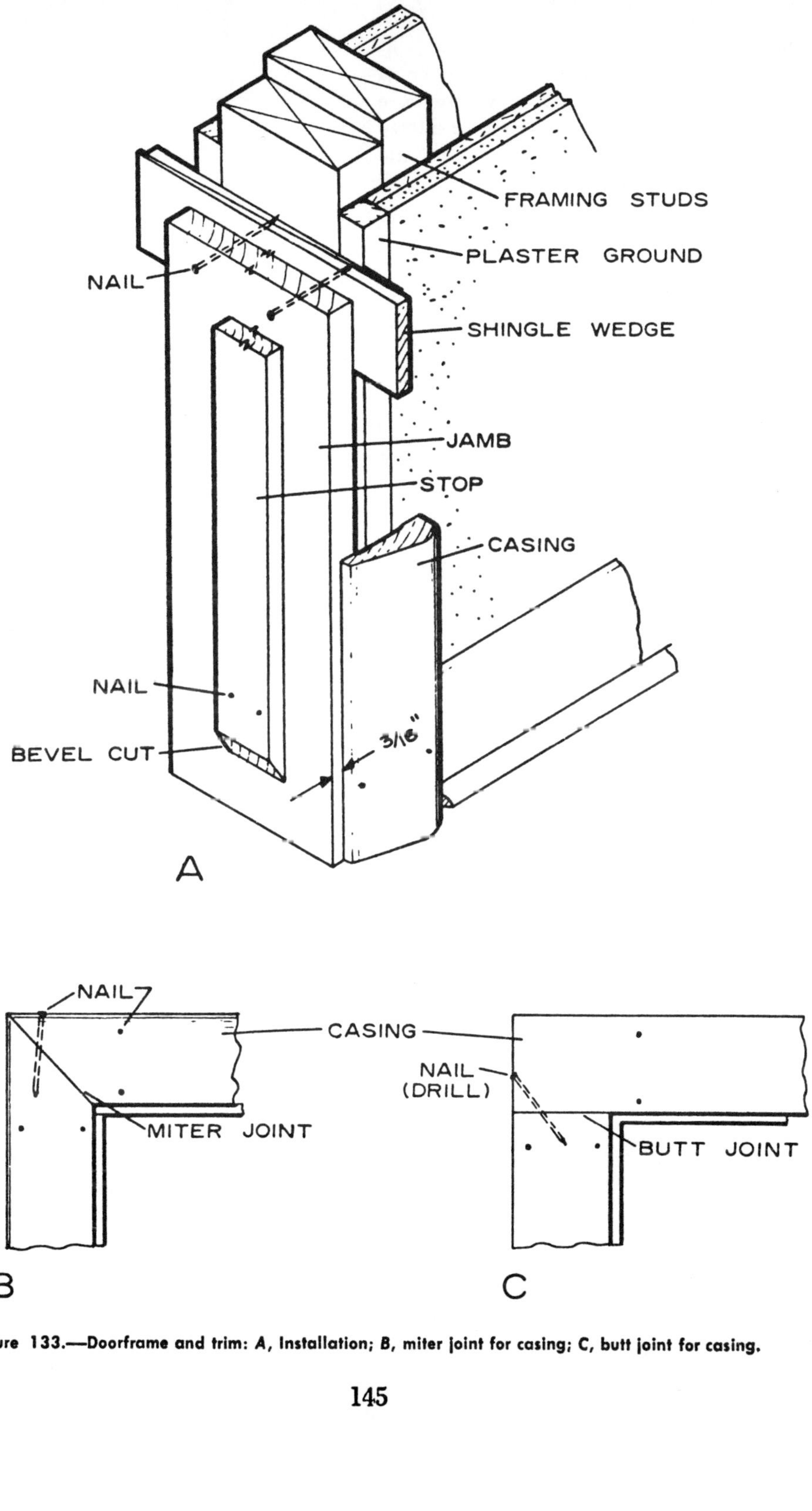

Figure 133.—Doorframe and trim: *A*, Installation; *B*, miter joint for casing; C, butt joint for casing.

spaced about 16 inches apart along the full height of the opening and at the head jamb.

Casing with any form of molded shape must have a mitered joint at the corners (fig. 133,*B*) When casing is square-edged, a butt joint may be made at the junction of the side and head casing (fig. 133,*C*). If the moisture content of the casing is well above that recommended in figure 121, a mitered joint may open slightly at the outer edge as the material dries. This can be minimized by using a small glued spline at the corner of the mitered joint. Actually, use of a spline joint under any moisture condition is considered good practice, and some prefitted jamb, door, and casing units are provided with splined joints. Nailing into the joint after drilling will aid in retaining a close fit (fig. 133,*B* and *C*).

The door opening is now complete except for fitting and securing the hardware and nailing the stops in proper position. Interior doors are normally hung with two 3½- by 3½-inch loose-pin butt hinges. The door is fitted into the opening with the clearances shown in figure 134. The clearance and location of hinges, lock set, and doorknob may vary somewhat, but they are generally accepted by craftsmen and conform to most millwork standards. The edge of the lock stile should be beveled slightly to permit the door to clear the jamb when swung open. If the door is to swing across heavy carpeting, the bottom clearance may be slightly more.

Thresholds are used under exterior doors to close the space allowed for clearance. Weather strips around exterior door openings are very effective in reducing air infiltration.

In fitting doors, the stops are usually temporarily nailed in place until the door has been hung. Stops for doors in single-piece jambs are generally 7/16 inch thick and may be ¾ to 2¼ inches wide. They are installed with a mitered joint at the junction of the side and head jambs. A 45° bevel cut at the bottom of the stop, about 1 to 1½ inches above the finish floor, will eliminate a dirt pocket and make cleaning or refinishing of the floor easier (fig. 133,*A*)

Some manufacturers supply prefitted door jambs and doors with the hinge slots routed and ready for installation. A similar door buck of sheet metal with formed stops and casing is also available.

Installation of Door Hardware

Hardware for doors may be obtained in a number of finishes, with brass, bronze, and nickel perhaps the most common. Door sets are usually classed as: (a) Entry lock for exterior doors, (b) bathroom set (inside lock control with safety slot for opening from the outside), (c) bedroom lock (keyed lock), and (d) passage set (without lock).

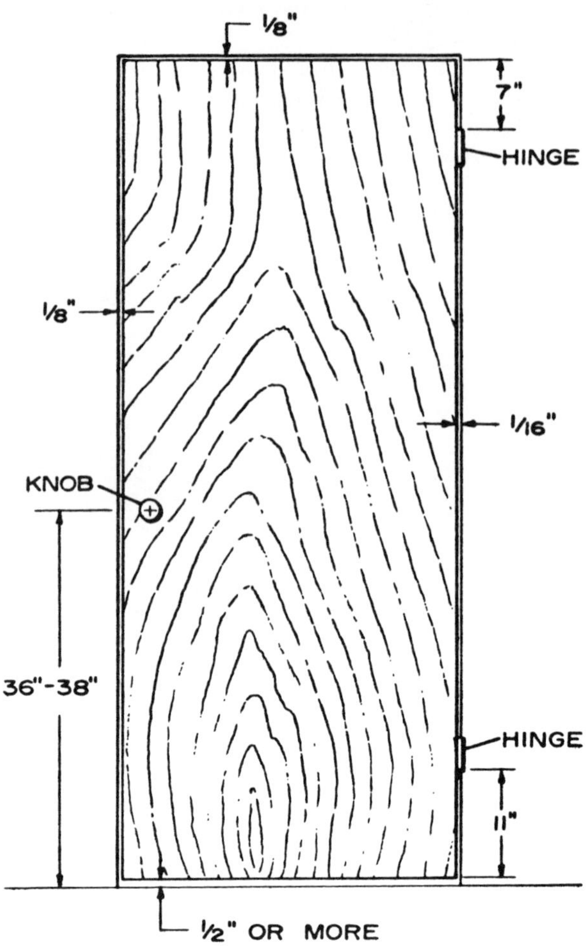

Figure 134.—Door clearances.

Hinges

Using three hinges for hanging 1¾-inch exterior doors and two hinges for the lighter interior doors is common practice. There is some tendency for exterior doors to warp during the winter because of the difference in exposure on the opposite sides. The three hinges reduce this tendency. Three hinges are also useful on doors that lead to unheated attics and for wider and heavier doors that may be used within the house.

Loose-pin butt hinges should be used and must be of the proper size for the door they support. For 1¾-inch-thick doors. use 4- by 4-inch butts; for 1⅜-inch doors, 3½- by 3½-inch butts. After the door is fitted to the framed opening, with the proper clearances, hinge halves are fitted to the door. They are routed into the door edge with about a 3/16-inch back distance (fig. 135,*A*). One hinge half should be set flush with the surface and must be fastened square with the edge of the door. Screws are included with each pair of hinges.

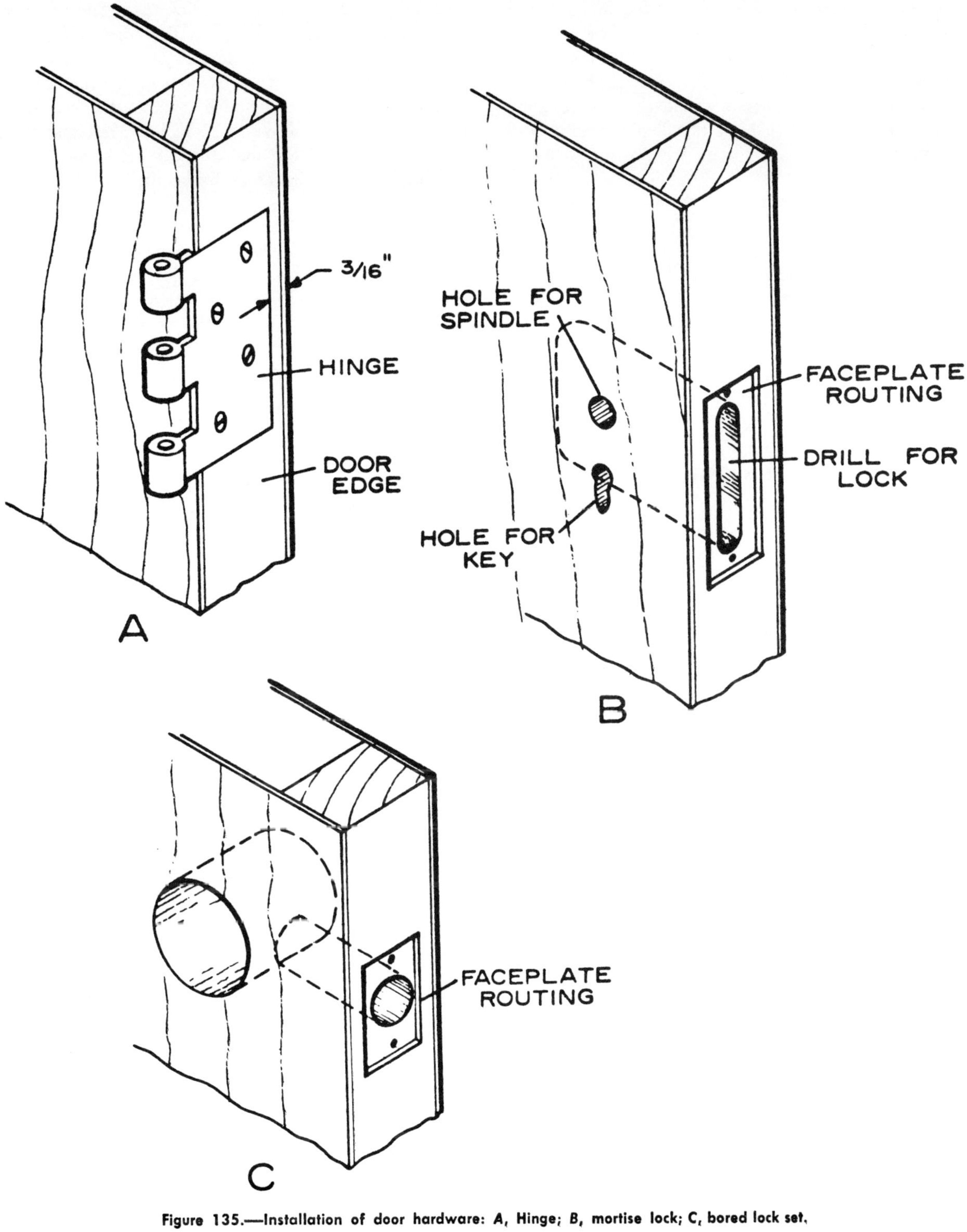

Figure 135.—Installation of door hardware: *A*, Hinge; *B*, mortise lock; C, bored lock set.

The door is now placed in the opening and blocked up at the bottom for proper clearance. The jamb is marked at the hinge locations, and the remaining hinge half is routed and fastened in place. The door is then positioned in the opening and the pins slipped in place. If hinges have been installed correctly and the jambs are plumb, the door will swing freely.

Locks

Types of door locks differ with regard to installation, first cost, and the amount of labor required to set them. Lock sets are supplied with instructions that should be followed for installation. Some types require drilling of the edge and face of the door and routing of the edge to accommodate the lock set and faceplate (fig. 135,*B*). A more common bored type (fig. 135,*C*) is much easier to install as it requires only one hole drilled in the edge and one in the face of the door. Boring jigs and faceplate markers are available to provide accurate installation. The lock should be installed so that the doorknob is 36 to 38 inches above the floorline. Most sets come with paper templates marking the location of the lock and size of the holes to be drilled.

Strike Plate

The strike plate, which is routed into the door jamb, holds the door in place by contact with the latch. To install, mark the location of the latch on the door jamb and locate the strike plate in this way. Rout out the marked outline with a chisel and also rout for the latch (fig. 136,*A*). The strike plate

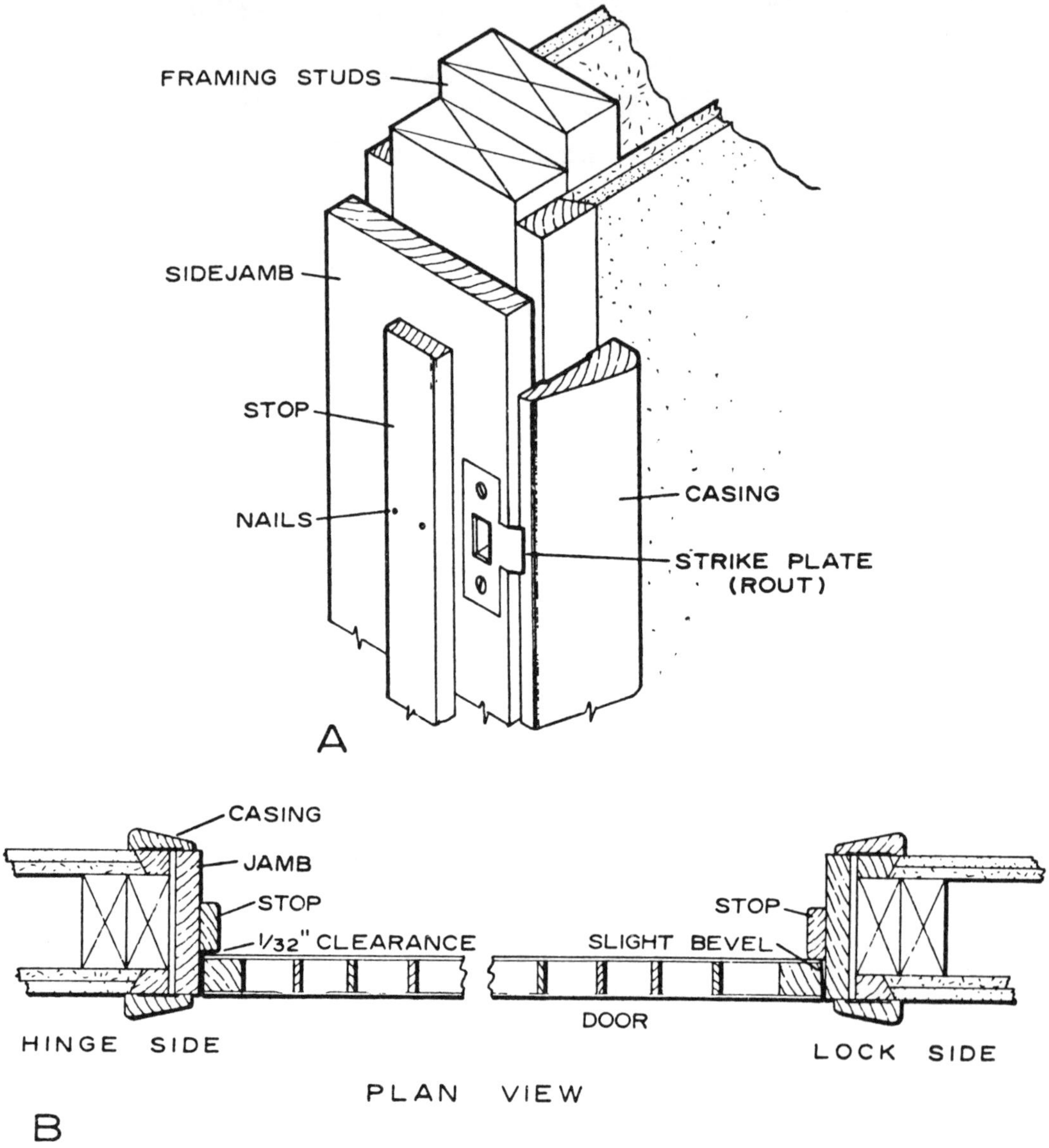

Figure 136.—Door details: *A*, Installation of strike plate; *B*, location of stops.

should be flush with or slightly below the face of the door jamb. When the door is latched, its face should be flush with the edge of the jamb.

Doorstops

The stops which have been set temporarily during fitting of the door and installation of the hardware may now be nailed in place permanently. Finish nails or brads, $1\frac{1}{2}$ inches long, should be used. The stop at the lock side should be nailed first, setting it tight against the door face when the door is latched. Space the nails 16 inches apart in pairs (fig. 136,*A*).

The stop behind the hinge side is nailed next, and a $\frac{1}{32}$-inch clearance from the door face should be allowed (fig. 133,*B*) to prevent scraping as the door is opened. The head-jamb stop is then nailed in place. Remember that when door and trim are painted, some of the clearances will be taken up.

Wood-trim Installation

The casing around the window frames on the interior of the house should be the same pattern as that used around the interior door frames. Other trim which is used for a double-hung window frame includes the sash stops, stool, and apron (fig. 137,*A*). Another method of using trim around windows has the entire opening enclosed with casing (fig. 137,*B*). The stool is then a filler member between the bottom sash rail and the bottom casing.

The *stool* is the horizontal trim member that laps the window sill and extends beyond the casing at the sides, with each end notched against the plastered

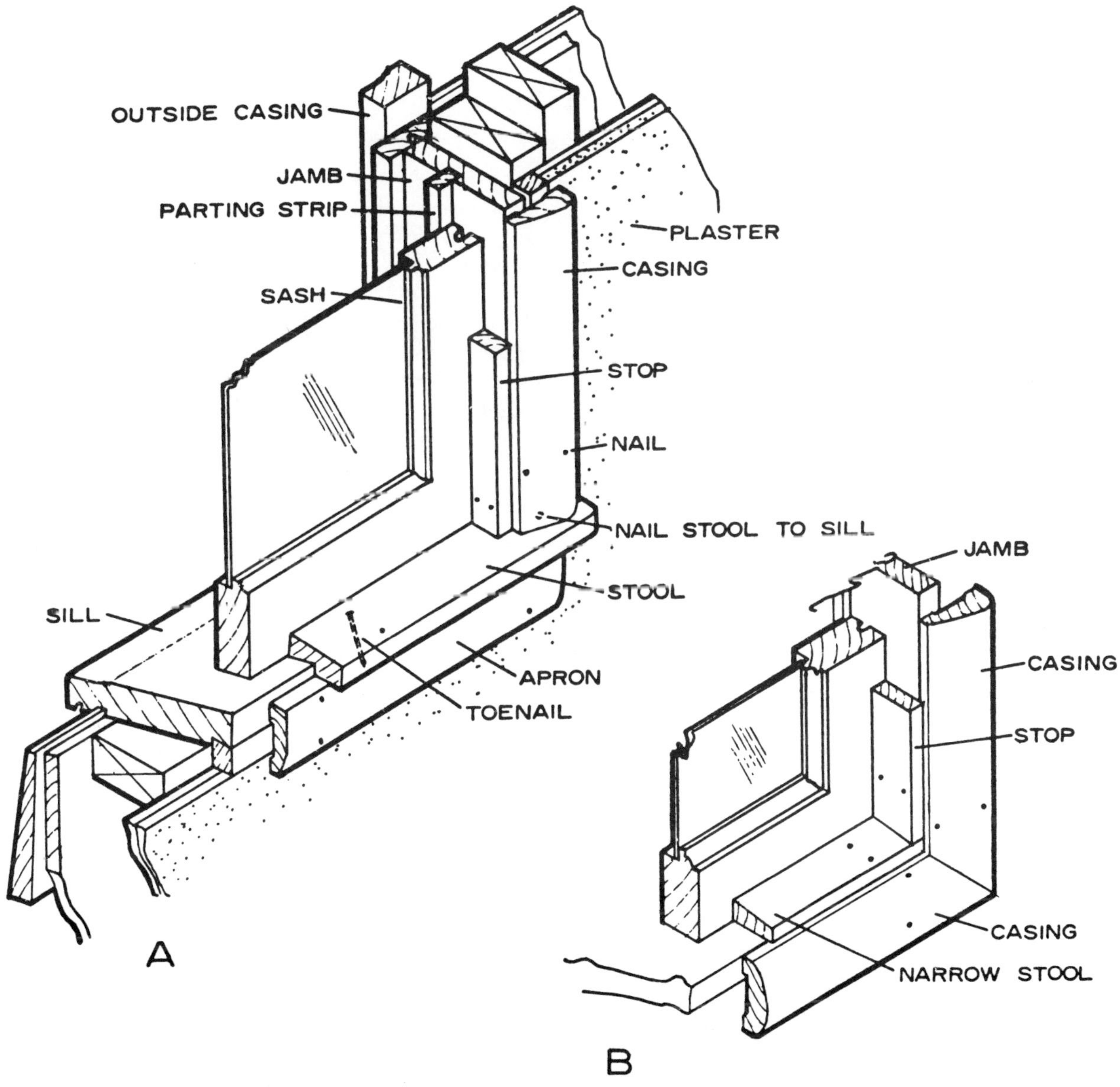

Figure 137.—Installation of window trim: *A*, With stool and apron; *B*, enclosed with casing.

wall. The *apron* serves as a finish member below the stool. The window stool is the first piece of window trim to be installed and is notched and fitted against the edge of the jamb and the plaster line, with the outside edge being flush against the bottom rail of the window sash (fig. 137,*A*). The stool is blind-nailed at the ends so that the casing and the stop will cover the nailheads. Predrilling is usually necessary to prevent splitting. The stool should also be nailed at midpoint to the sill and to the apron with finishing nails. Face-nailing to the sill is sometimes substituted or supplemented with toenailing of the outer edge to the sill (fig. 137,*A*).

The casing is applied and nailed as described for doorframes (fig. 133,*A*), except that the inner edge is flush with the inner face of the jambs so that the stop will cover the joint between the jamb and casing. The window stops are then nailed to the jambs so that the window sash slides smoothly. Channel-type weather stripping often includes full-width metal subjambs into which the upper and lower sash slide, replacing the parting strip. Stops are located against these instead of the sash to provide a small amount of pressure. The apron is cut to a length equal to the outer width of the casing line (fig. 137,*A*). It is nailed to the window sill and to the 2- by 4-inch framing sill below.

When casing is used to finish the bottom of the window frame as well as the sides and top, the narrow stool butts against the side window jamb. Casing is then mitered at the bottom corners (fig. 137,*B*) and nailed as previously described.

Base and Ceiling Moldings

Base Moldings

Base molding serves as a finish between the finished wall and floor. It is available in several widths and forms. Two-piece base consists of a baseboard topped with a small base cap (fig. 138,*A*). When plaster is not straight and true, the small base molding will

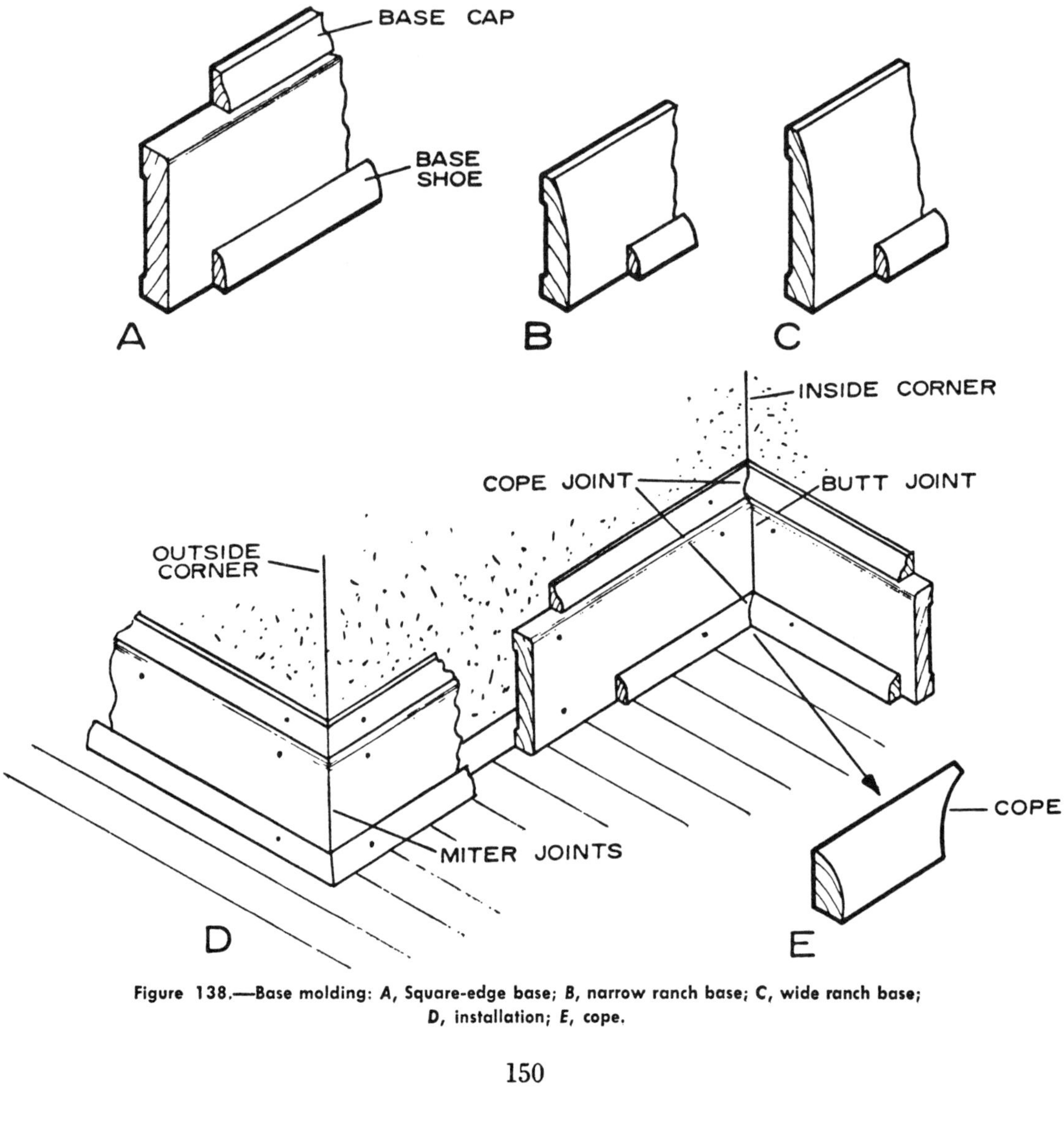

Figure 138.—Base molding: *A*, Square-edge base; *B*, narrow ranch base; *C*, wide ranch base; *D*, installation; *E*, cope.

conform more closely to the variations than will the wider base alone. A common size for this type of baseboard is 5/8 by 3¼ inches or wider. One-piece base varies in size from 7/16 by 2¼ inches to ½ by 3¼ inches and wider (fig. 138,*B* and *C*). Although a wood member is desirable at the junction of the wall and carpeting to serve as a protective "bumper", wood trim is sometimes eliminated entirely.

Most baseboards are finished with a base shoe, ½ by ¾ inch in size (fig. 138,*A*, *B*, and *C*). A single-base molding without the shoe is sometimes placed at the wall-floor junction, especially where carpeting might be used.

Installation of Base Molding

Square-edged baseboard should be installed with a butt joint at inside corners and a mitered joint at outside corners (fig. 138,*D*). It should be nailed to each stud with two eightpenny finishing nails. Molded single-piece base, base moldings, and base shoe should have a coped joint at inside corners and a mitered joint at outside corners. A coped joint is one in which the first piece is square-cut against the plaster or base and the second molding coped. This is accomplished by sawing a 45° miter cut and with a coping saw trimming the molding along the inner line of the miter (fig. 138,*E*). The base shoe should be nailed into the subfloor with long slender nails and not into the baseboard itself. Thus, if there is a small amount of shrinkage of the joists, no opening will occur under the shoe.

Ceiling Moldings

Ceiling moldings are sometimes used at the junction of wall and ceiling for an architectural effect or to terminate dry-wall paneling of gypsum board or wood (fig. 139,*A*). As in the base moldings, inside

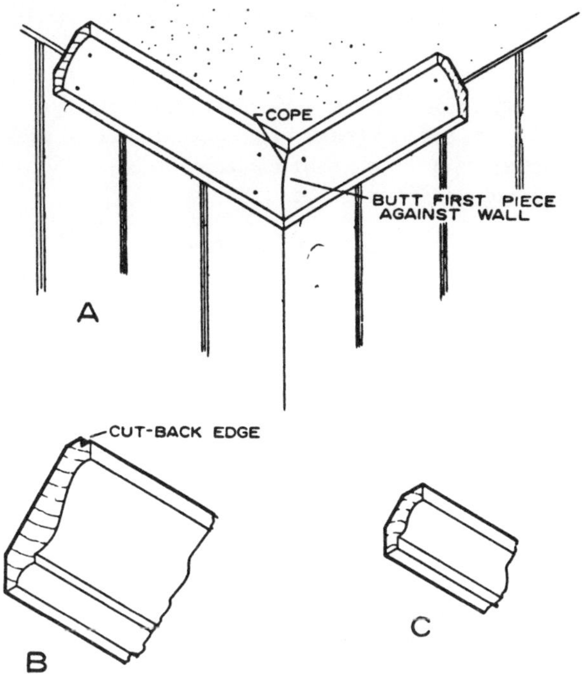

Figure 139.—Ceiling moldings; *A*, Installation (inside corner); *B*, crown molding; *C*, small crown molding.

corners should also be cope-jointed. This insures a tight joint and retains a good fit if there are minor moisture changes.

A cutback edge at the outside of the molding will partially conceal any unevenness of the plaster and make painting easier where there are color changes (fig. 139, *B*). For gypsum dry-wall construction, a small simple molding might be desirable (fig. 139,*C*). Finish nails should be driven into the upper wall-plates and also into the ceiling joists for large moldings when possible.

CHAPTER 22

CABINETS AND OTHER MILLWORK

Millwork, as a general term, usually includes most of those wood materials and house components which require manufacturing. This not only covers the interior trim, doors, and other items previously described, but also such items as kitchen cabinets, fireplace mantels, china cabinets, and similar units. Most of these units are produced in a millwork manufacturing plant and are ready to install in the house. They differ from some other items because they usually require only fastening to the wall or floor.

While many units are custom made, others can be ordered directly from stock. For example, kitchen cabinets are often stock items which may be obtained in 3-inch-width increments, usually beginning at widths of 12 or 15 inches and on up to 48 inch widths.

As in the case of interior trim, the cabinets, shelving, and similar items can be made of various wood species. If the millwork is to be painted, ponderosa pine, southern pine, Douglas-fir, gum, and similar species may be used. Birch, oak, redwood, and knotty pine, or other species with attractive surface variations, are some of the woods that are finished with varnish or sealers.

Recommended moisture content for book cases and other interior millwork may vary from 6 to 11 percent in different parts of the country. These areas, together with the moisture contents, are shown on the moisture-content map (fig. 121).

Kitchen Cabinets

The kitchen usually contains more millwork than the rest of the rooms combined. This is in the form of wall and base cabinets, broom closets, and other items. An efficient plan with properly arranged cabinets will not only reduce work and save steps for the housewife, but will often reduce costs because of the need for a smaller area. Location of the refrigerator, sink, dishwasher, and range, together with the cabinets, is also important from the standpoint of plumbing and electrical connections. Good lighting, both natural and artificial, is also important in designing a pleasant kitchen.

Kitchen cabinets, both base and wall units, should be constructed to a standard of height and depth. Figure 140 shows common base cabinet counter heights and depths as well as clearances for wall cabinets. While the counter height limits range from 30 to 38 inches, the standard height is usually 36 inches.

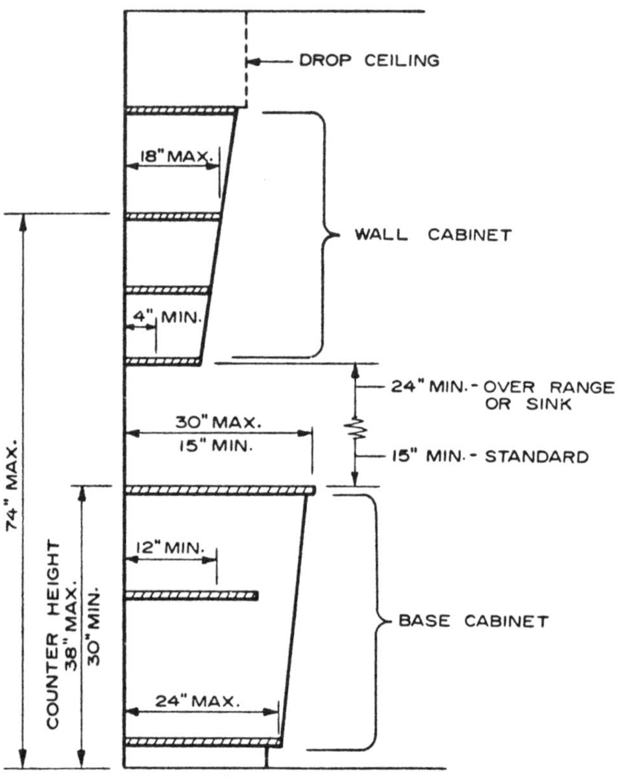

Figure 140.—Kitchen cabinet dimensions.

Wall cabinets vary in height depending on the type of installation at the counter. The tops of wall cabinets are located at the same height, either free or under a 12- to 14-inch drop ceiling or storage cabinet. Wall cabinets are normally 30 inches high, but not more than 21 inches when a range or sink is located under them. Wall cabinets can also be obtained in 12-, 15-, 18-, and 24-inch heights. The shorter wall cabinets are usually placed over refrigerators.

Narrow wall cabinets are furnished with single doors and the wider ones with double doors (fig. 141,*A*). Base cabinets may be obtained in full-door or full-drawer units or with both drawers and doors (fig. 141,*B*). Sink fronts or sink-base cabinets, corner cabinets, broom closets, and desks are some of the special units which may be used in planning the ideal kitchen. Cabinets are fastened to the wall through cleats located at the back of each cabinet. It is good practice to use long screws to penetrate into each wall stud.

Four basic layouts are commonly used in the design of a kitchen. The *U-type* with the sink at the bottom of the U and the range and refrigerator on opposite sides is very efficient (fig. 142,*A*).

The *L-type* (fig. 142,*B*), with the sink and range on one leg and the refrigerator on the other, is sometimes used with a dining space in the opposite corner.

The "parallel wall" or *pullman kitchen plan* (fig. 142,*C*) is often used in narrow kitchens and can be quite efficient with proper arrangement of the sink, range, and refrigerator.

The *sidewall type* (fig. 142,*D*) usually is preferred for small apartments. All cabinets, the sink, range, and refrigerator are located along one wall. Counter space is usually somewhat limited in this design when kitchens are small.

Closets and Wardrobes

The simple clothes closet is normally furnished with a shelf and a rod for hanging clothes. Others may have small low cabinets for the storage of shoes and similar items. Larger wardrobes with sliding or folding doors may be combined with space for hanging clothes as well as containing a dresser complete with drawers and mirror. Many built-in combinations are possible, all of which reduce the amount of bedroom furniture needed.

Linen closets may be simply a series of shelves behind a flush or panel door. Others may consist of an open cabinet with doors and drawers built directly into a notch or corner of the wall located near the bedrooms and bath.

Mantels

The type of *mantel* used for a fireplace depends on the style and design of the house and its interior finish. The contemporary fireplace may have no mantel

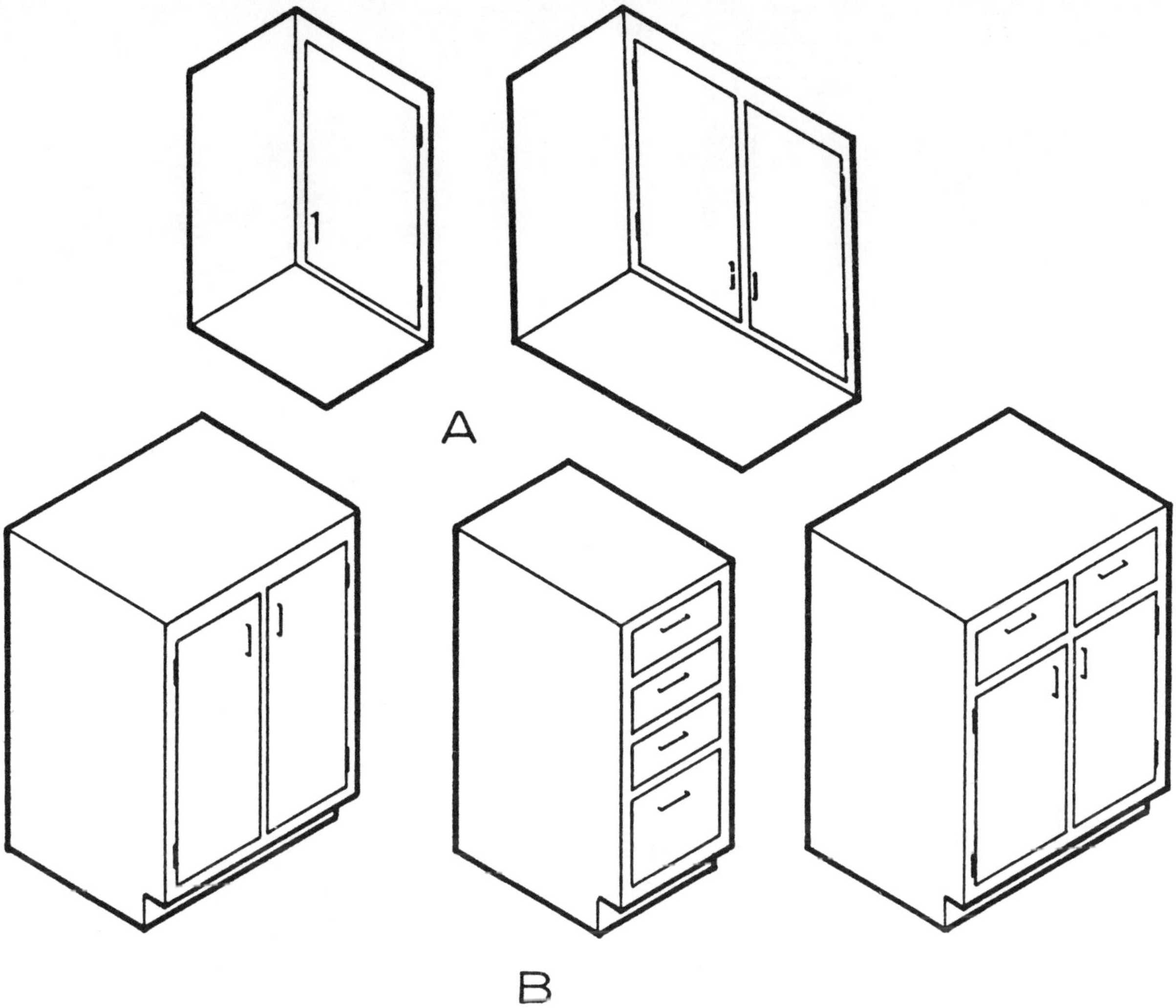

Figure 141.—Kitchen cabinets: *A*, Wall cabinets; *B*, base cabinets.

at all, or at best a simple wood molding used as a transition between the masonry and the wall finish. However, the colonial or formal interior usually has a well-designed mantel enclosing the fireplace opening. This may vary from a simple mantel (fig. 143), to a more elaborate unit combining paneling and built-in cabinets along the entire wall. In each design, however, it is important that no wood or other combustible material be placed within 3½ inches of the edges of the fireplace opening. Furthermore, any projection more than 1½ inches in front of the fireplace, such as the mantel shelf, should be at least 12½ inches above the opening. Mantels are fastened to the header and framing studs above and on each side of the fireplace.

China Cases

Another millwork item often incorporated in the dining room of a formal or traditional design is the china case. It is usually designed to fit into one or two corners of the room. This corner cabinet often has glazed doors above and single- or double-panel doors below (fig. 144). It may be 7 feet or more high with a drop ceiling above with a face width of about 3 feet. Shelves are supplied in both the upper and lower cabinets.

China cases or storage shelves in dining rooms of contemporary houses may be built in place by the contractor. A row of cabinets or shelves may act as a separator between dining room and kitchen and serve as a storage area for both rooms.

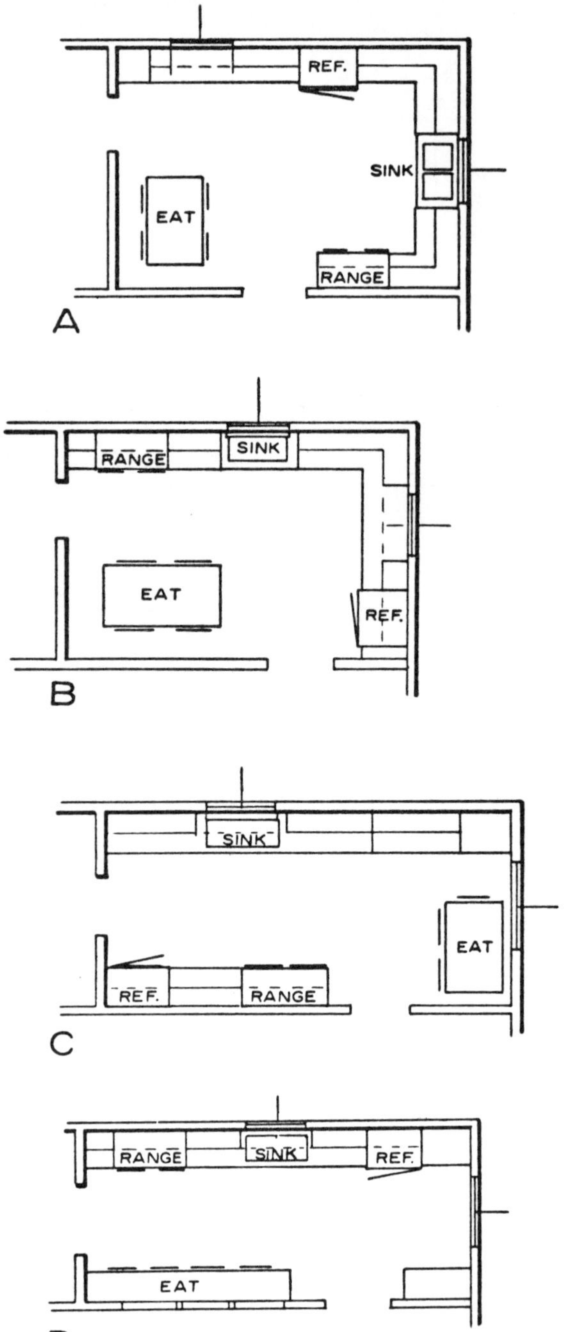

Figure 142.—Kitchen layouts *A*, U-type; *B*, L-type; C, "parallel wall" type; *D*, sidewall type.

Figure 143.—Fireplace mantel.

Figure 144.—Corner china case.

CHAPTER 23

STAIRS

Stairways in houses should be designed and constructed to afford safety and adequate headroom for the occupants as well as space for the passage of furniture (*11*). The two types of stairs commonly used in houses are (a) the finished main stairs leading to the second floor or split-level floors and (b) the basement or service stairs leading to the basement or garage area. The main stairs are designed to provide easy ascent and descent and may be made a feature of the interior design. The service stairs to basement areas are usually somewhat steeper and are constructed of less expensive materials, although safety and convenience are still prime factors in their design.

Construction

Most finish and service stairs are constructed in place. The main stairs are assembled with prefabricated parts, which include housed stringers, treads, and risers. Basement stairs may be made simply of 2- by 12-inch carriages and plank treads. In split-level design or a midfloor outside entry, stairways are often completely finished with plastered walls, handrails, and appropriate moldings.

Wood species appropriate for main stairway components include oak, birch, maple, and similar hardwoods. Treads and risers for the basement or service stairways may be of Douglas-fir, southern pine, and similar species. A hardwood tread with a softwood or lower grade hardwood riser may be combined to provide greater resistance to wear.

Types of Stairways

Three general types of stairway runs most commonly used in house construction are the straight run (fig. 145,*A*), the long "L" (fig. 145,*B*), and the narrow "U" (fig. 146,*A*). Another type is similar to the Long "L" except that "winders" or "pie-shaped" treads (fig. 146,*B*) are substituted for the landing. This type of stairs is not desirable and should be avoided whenever possible because it is obviously not as convenient or as safe as the long "L." It is used where the stair run is not sufficient for the more conventional stairway containing a landing. In such instances, the winders should be adjusted to replace the landings so that the width of the tread, 18 inches from the narrow end, will not be less than the tread width on the straight run (fig. 147,*A*). Thus if the standard tread is 10 inches wide, the winder tread should be at least 10 inches wide at the 18-inch line.

Another basic rule in stair layout concerns the landing at the top of a stairs when the door opens into the stairway, such as on a stair to the basement. This landing, as well as middle landings, should not be less than 2 feet 6 inches long (fig. 147,*B*).

Sufficient headroom in a stairway is a primary requisite. For main stairways, clear vertical distance should not be less than 6 feet 8 inches (fig. 148,*A*). Basement or service stairs should provide not less than a 6-foot 4-inch clearance.

The minimum tread width and riser height must also be considered. For closed stairs, a 9-inch tread width and an 8¼-inch riser height should be considered a minimum even for basement stairways (fig. 148,*B*). Risers with less height are always more desirable. The nosing projection should be at least 1⅛ inches; however, if the projection is too much greater, the stairs will be awkward and difficult to climb.

Ratio of Riser to Tread

There is a definite relation between the height of a riser and the width of a tread, and all stairs should be laid out to conform to well-established rules governing these relations. If the combination of run and rise is too great, there is undue strain on the leg muscles and on the heart of the climber; if the combination is too small, his foot may kick the riser at each step and an attempt to shorten stride may be tiring. Experience has proved that a riser 7½ to 7¾ inches high with appropriate tread width combines both safety and comfort.

A rule of thumb which sets forth a good relation between the height of the riser and the width of the tread is:

The tread width multiplied by the riser height in inches should equal to 72 to 75. The stairs shown in figure 148,*B* would conform to this rule—9 times 8¼ = 74¼. If the tread is 10 inches, however, the riser should be 7½ inches, which is more desirable for common stairways. Another rule sometimes used is: The tread width plus twice the riser height should equal about 25.

These desirable riser heights should, therefore, be used to determine the number of steps between floors. For example, 14 risers are commonly used for main stairs between the first and second floors. The 8-foot ceiling height of the first floor plus the upper-story floor joists, subfloor, and finish floor result in a floor-to-floor height of about 105 inches. Thus, 14 divided into 105 is exactly 7½ inches, the height of each riser. Fifteen risers used for this height would result in a 7-inch riser height.

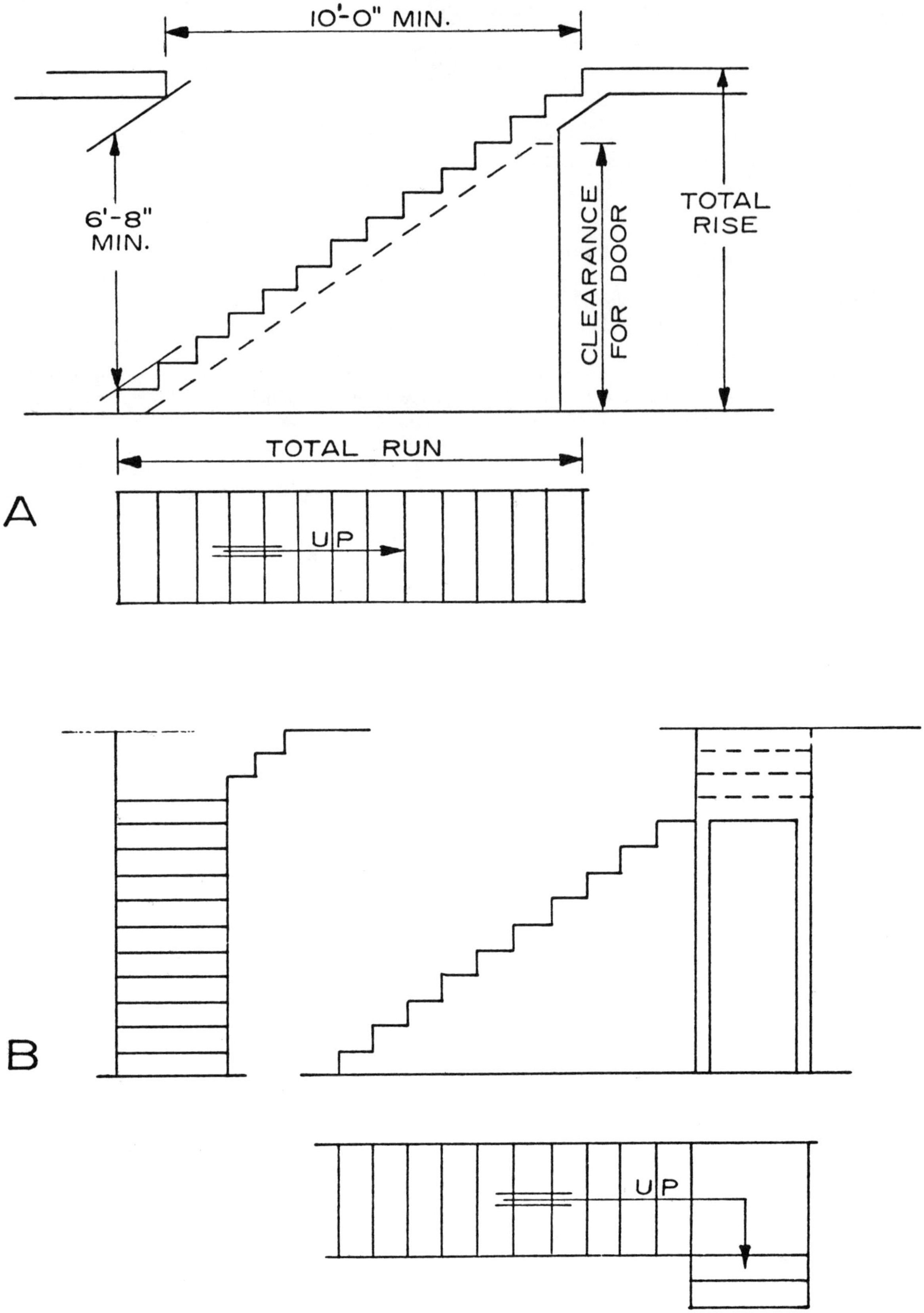

Figure 145.—Common types of stair runs: *A*, Straight; *B*, long "L."

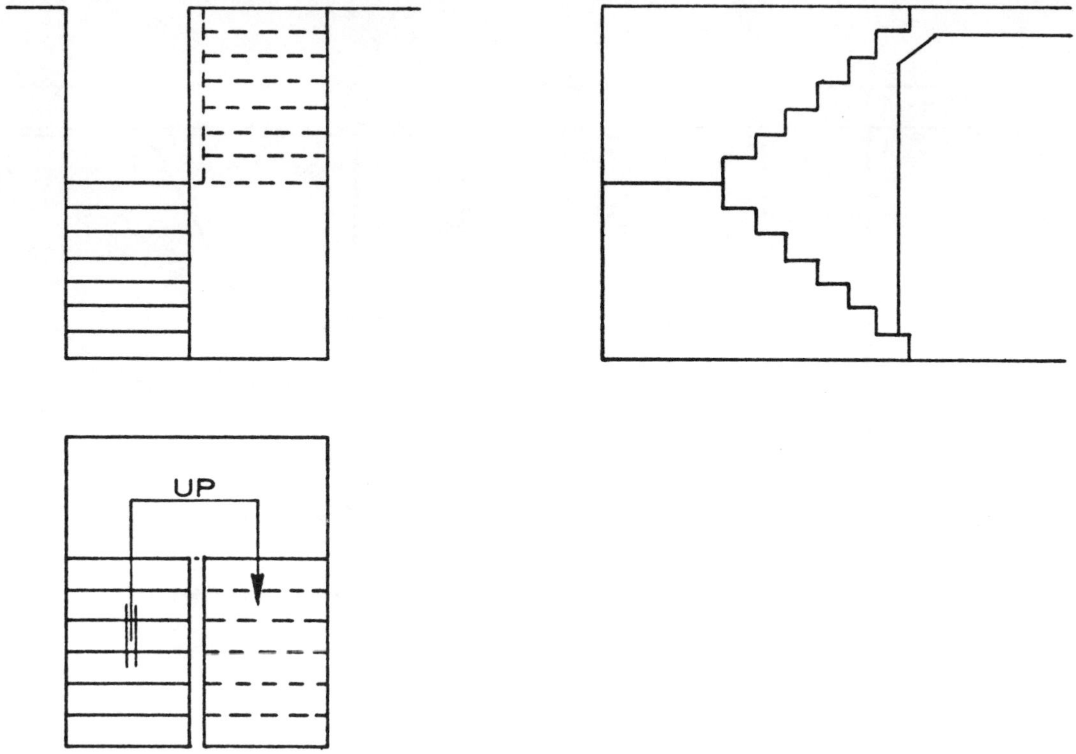

A

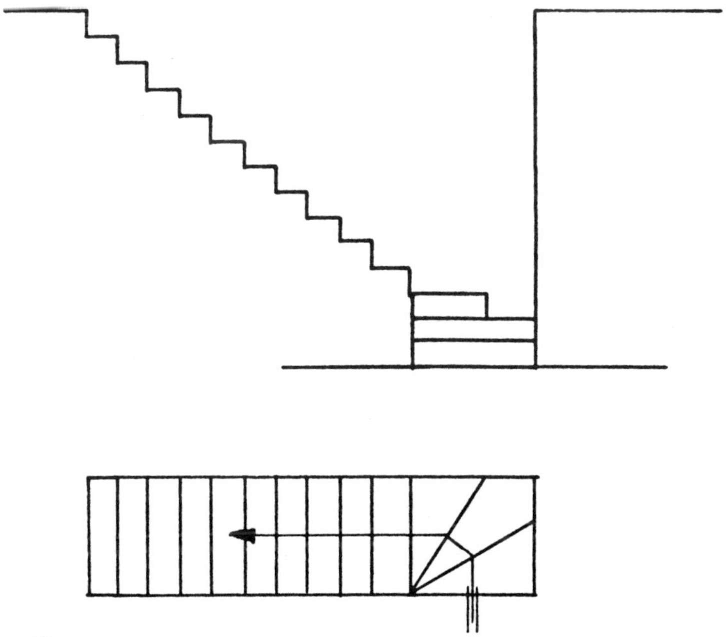

B

Figure 146.—Space-saving stairs: *A*, Narrow "U"; *B*, winder.

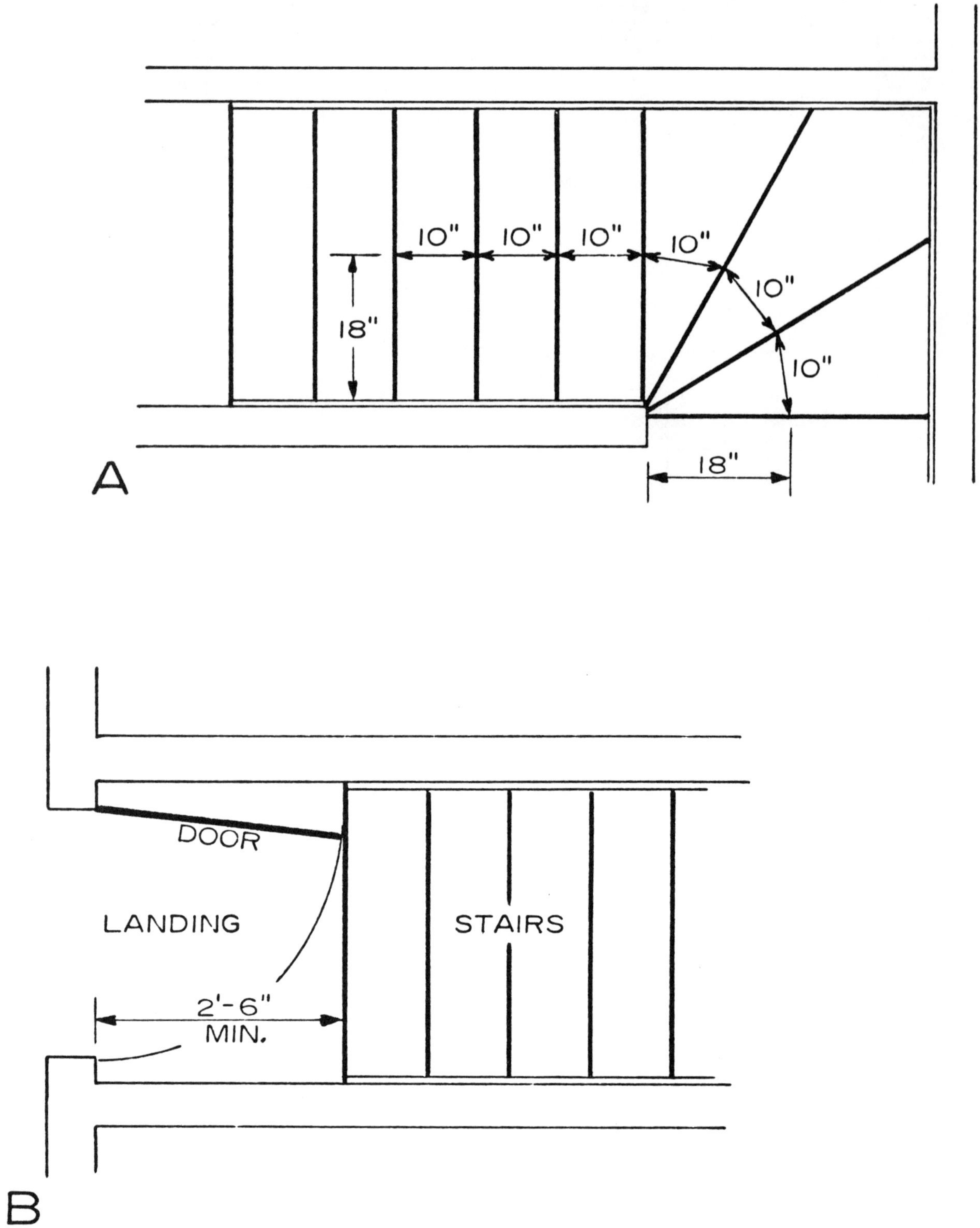

Figure 147.—Stair layout: *A*, Winder treads; *B*, landings.

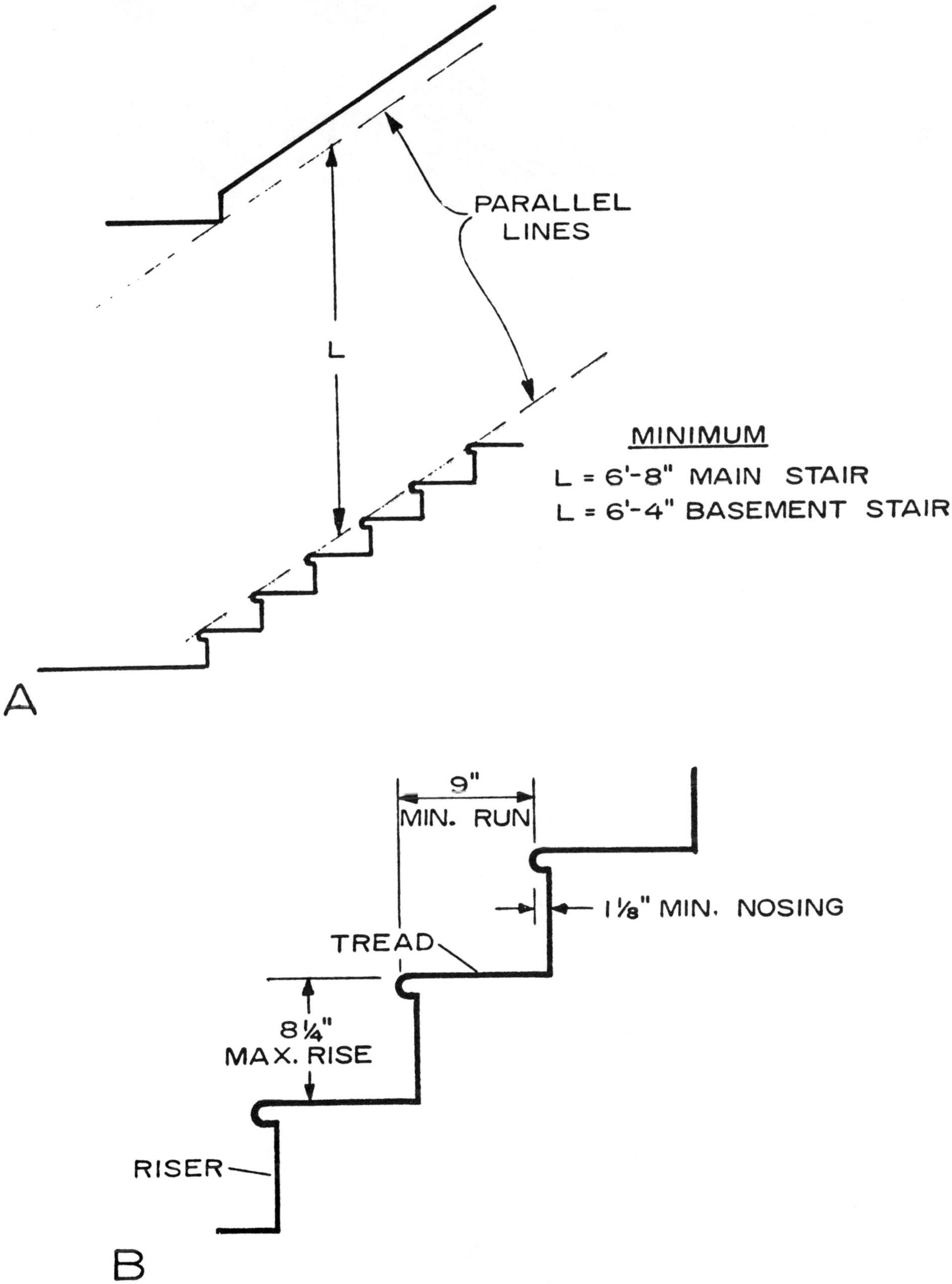

Figure 148.—Stairway dimensions: *A*, Minimum headroom; *B*, closed stair dimensions.

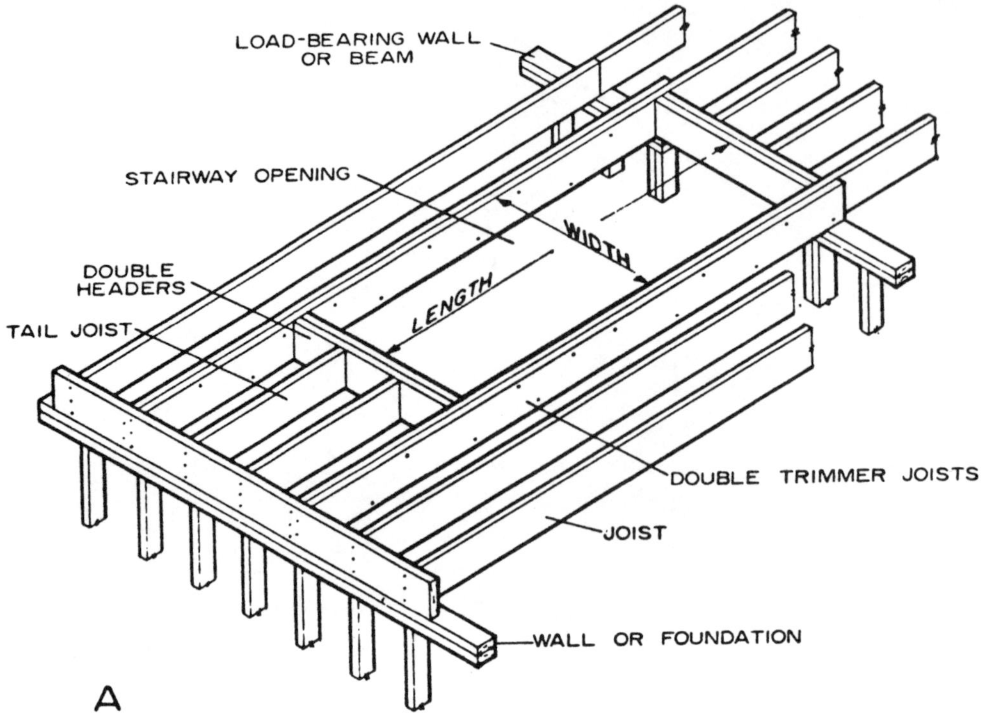

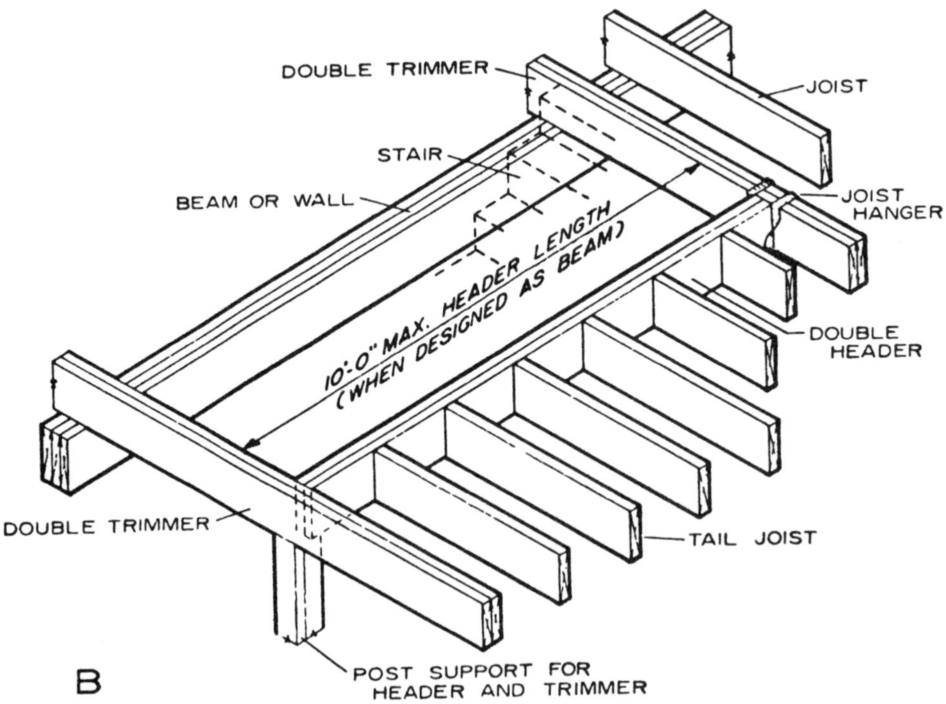

Figure 149.—Framing for stairs: *A*, Length of opening parallel to joists; *B*, length of opening perpendicular to joists.

Stair Widths and Handrails

The width of the main stairs should be not less than 2 feet 8 inches clear of the handrail. However, many main stairs are designed with a distance of 3 feet 6 inches between the centerline of the enclosing sidewalls. This will result in a stairway with a width of about 3 feet. Split-level entrance stairs are even wider. For basement stairs, the minimum clear width is 2 feet 6 inches.

A continuous handrail should be used on at least one side of the stairway when there are more than three risers. When stairs are open on two sides, there should be protective railings on each side.

Framing for Stairs

Openings in the floor for stairways, fireplaces, and chimneys are framed out during construction of the floor system (figs. 27 and 29). The long dimension of stairway openings may be either parallel or at right angles to the joists. However, it is much easier to frame a stairway opening when its length is parallel to the joists. For basement stairways, the rough openings may be about 9 feet 6 inches long by 32 inches wide (two joist spaces). Openings in the second floor for the main stair are usually a minimum of 10 feet long. Widths may be 3 feet or more. Depending on the short header required for one or both ends, the opening is usually framed as shown in figure 149,*A* when joists parallel the length of the opening. Nailing should conform to that shown in figures 27 and 29.

When the length of the stair opening is perpendicular to the length of the joists, a long doubled header is required (fig. 149,*B*). A header under these conditions without a supporting wall beneath is usually limited to a 10-foot length. A load-bearing wall under all or part of this opening simplifies the framing immensely, as the joists will then bear on the top plate of the wall rather than be supported at the header by joist hangers or other means. Nailing should conform to that shown in figures 27 and 29.

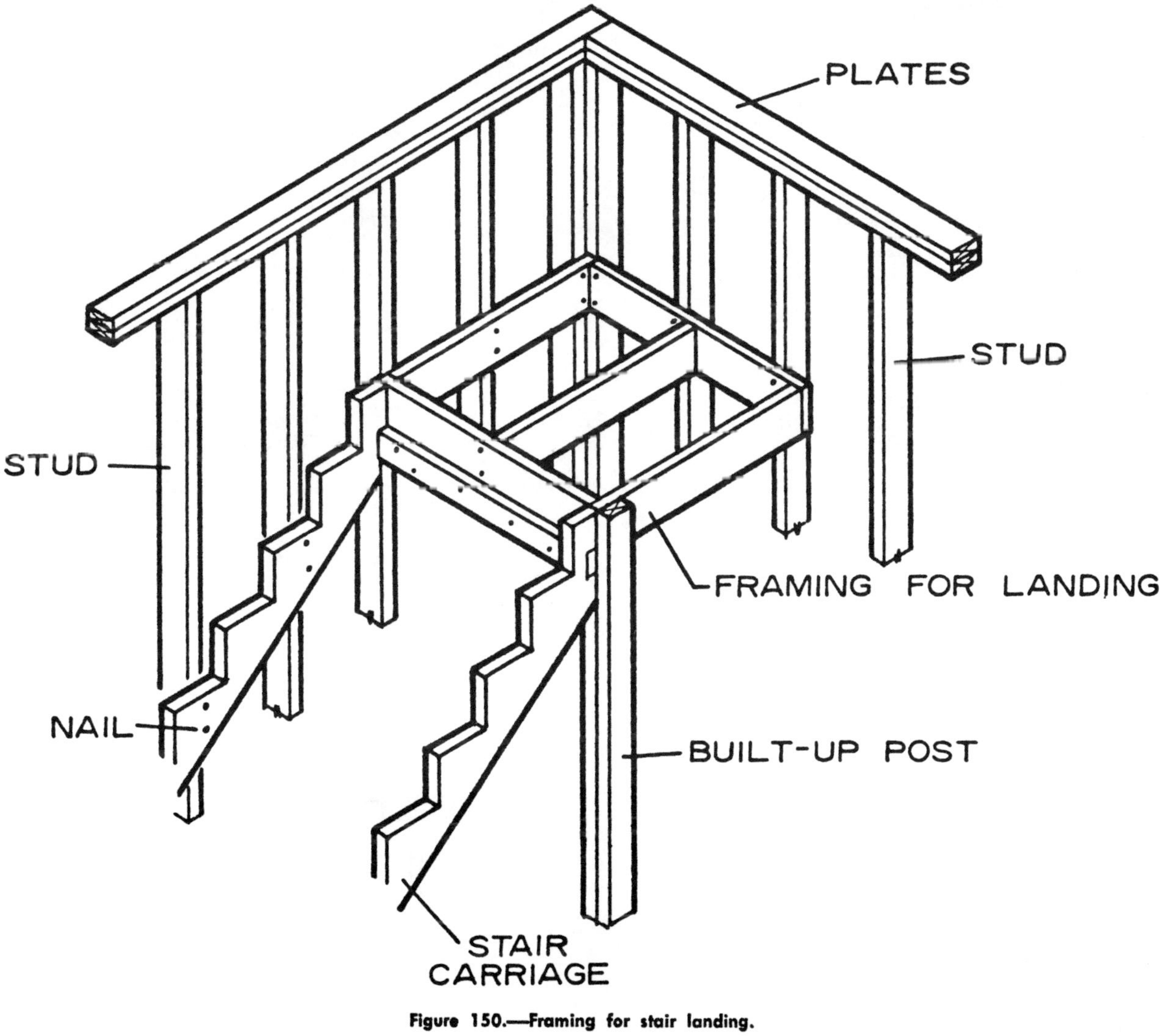

Figure 150.—Framing for stair landing.

The framing for an L-shaped stairway is usually supported in the basement by a post at the corner of the opening or by a load-bearing wall beneath. When a similar stair leads from the first to the second floor, the landing can be framed-out (fig. 150). The platform frame is nailed into the enclosing stud walls and provides a nailing area for the subfloor as well as a support for the stair carriages.

Stairway Details

Basement Stairs

Stair carriages which carry the treads and support the loads on the stair are made in two ways. Rough stair carriages commonly used for basement stairs are made from 2- by 12-inch planks. The effective depth below the tread and riser notches must be at least 3½ inches (fig. 151,*A*). Such carriages are usually placed only at each side of the stairs; however, an intermediate carriage is required at the center of the stairs when the treads are 1$\frac{1}{16}$ inches thick and the stairs wider than 2 feet 6 inches. Three carriages are also required when treads are 1⅝ inches thick and stairs are wider than 3 feet. The carriages are fastened to the joist header at the top of the stairway or rest on a supporting ledger nailed to the header (fig. 151,*B*).

Firestops should be used at the top and bottom of all stairs, as shown (fig. 151,*A*).

Perhaps the simplest system is one in which the carriages are not cut out for the treads and risers. Rather, cleats are nailed to the side of the unnotched carriage and the treads nailed to them. This design, however, is likely not as desirable as the notched carriage system when walls are present. Carriages can also be supported by walls located below them.

The bottom of the stair carriages may rest and be anchored to the basement floor. Perhaps a better method is to use an anchored 2- by 4- or 2- by 6-inch treated kicker plate (fig. 151,*C*).

Basement stair treads can consist of simple 1½-inch-thick plank treads without risers. However, from the standpoint of appearance and maintenance, the use of 1⅛-inch finished tread material and nominal 1-inch boards for risers is usually justified. Finishing nails fasten them to the plank carriages.

A somewhat more finished staircase for a fully enclosed stairway might be used from the main floor to the attic. It combines the rough notched carriage with a finish stringer along each side (fig. 152,*A*). The finish stringer is fastened to the wall, before carriages are fastened. Treads and risers are cut to fit snugly between the stringers and fastened to the rough carriage with finishing nails (fig. 152,*A*). This may be varied somewhat by nailing the rough carriage directly to the wall and notching the finished stringer to fit (fig. 152,*B*). The treads and risers are installed as previously described.

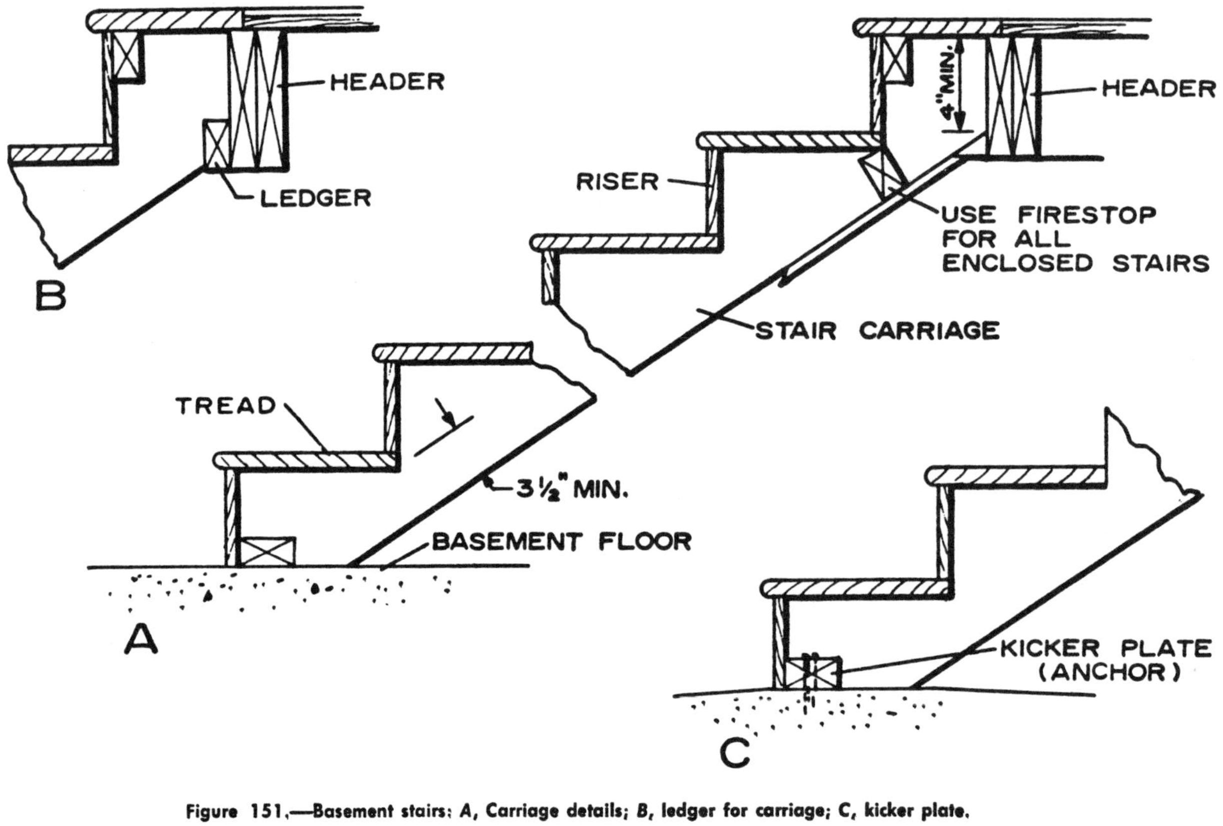

Figure 151.—Basement stairs: *A*, Carriage details; *B*, ledger for carriage; C, kicker plate.

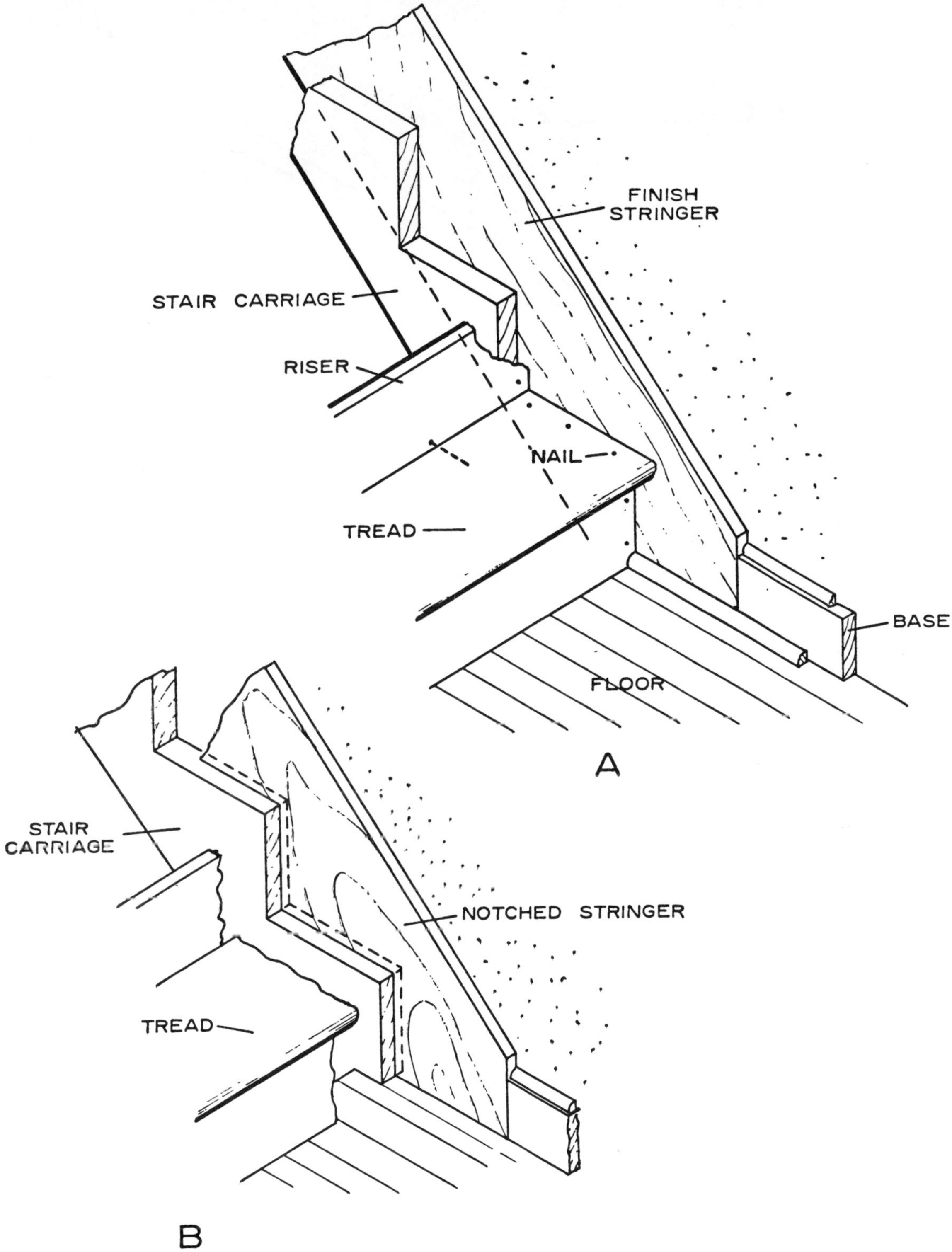

Figure 152.—Enclosed stairway details: *A*, With full stringer; *B*, with notched stringer.

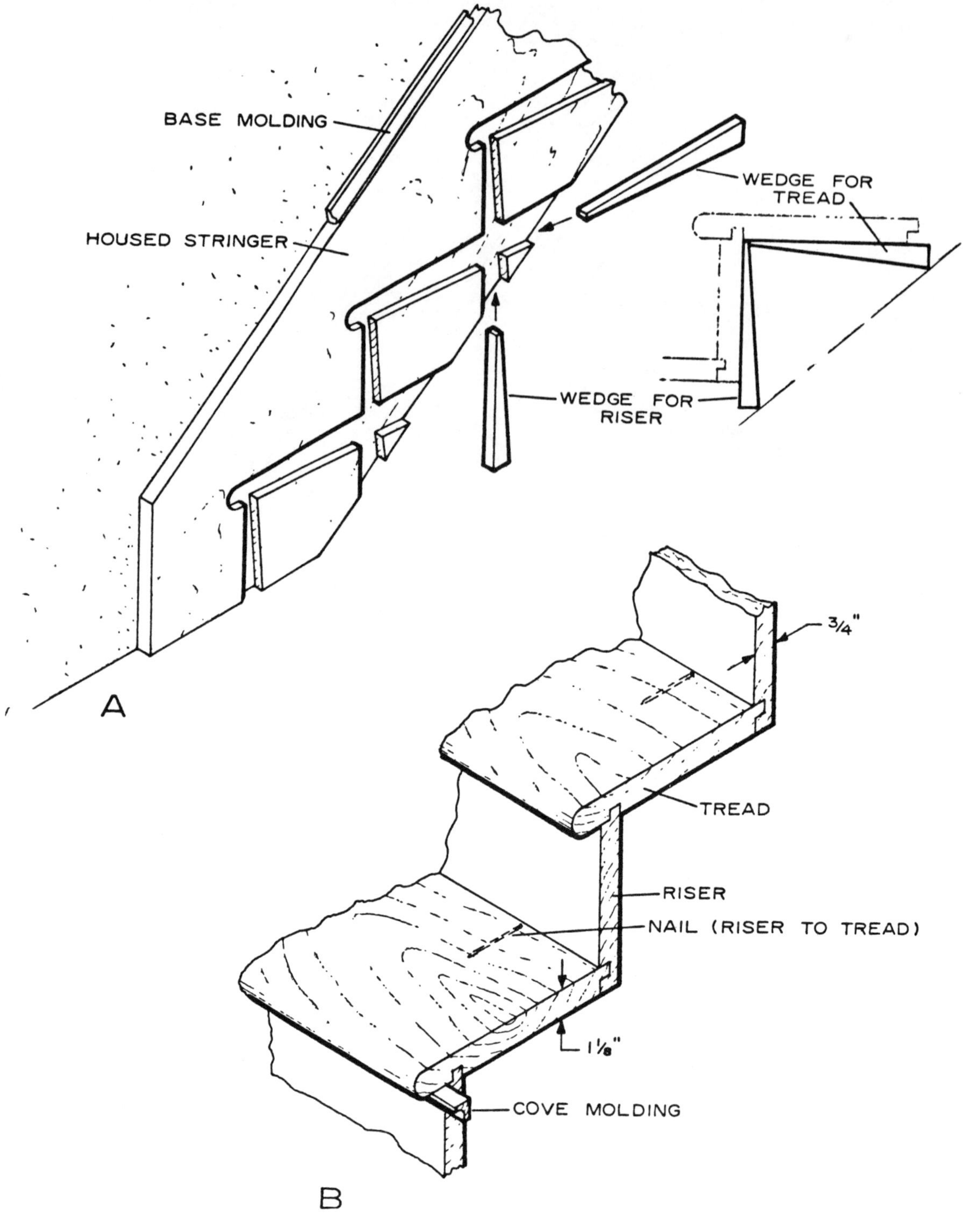

Figure 153.—Main stair detail with: *A*, Housed stringer; *B*, combination of treads and risers.

An open main stairway with its railing and balusters ending in a *newel* post can be very decorative and pleasing in the traditional house interior. It can also be translated to a contemporary stairway design and again result in a pleasing feature.

The main stairway differs from the other types previously described because of: (a) The housed stringers which replace the rough plank carriage; (b) the routed and grooved treads and risers; (c) the decorative railing and balusters in open stairways; and (d) the wood species, most of which can be given a natural finish.

The supporting member of the finished main stairway is the housed stringer (fig. 153,*A*). One is used on each side of the stairway and fastened to the plastered or finished walls. They are routed to fit both the tread and riser. The stair is assembled by means of hardwood wedges which are spread with glue and driven under the ends of the treads and in back of the risers. Assembly is usually done from under and the rear side of the stairway. In addition, nails are used to fasten the riser to the tread between the ends of the step (fig. 153,*B*). When treads and risers are wedged and glued into housed stringers, the maximum allowable width is usually 3 feet 6 inches. For wider stairs, a notched carriage is used between the housed stringers.

When stairs are open on one side, a railing and balusters are commonly used. Balusters may be fastened to the end of the treads which have a finished return (fig. 154). The balusters are also fastened to a railing which is terminated at a newel post. Balusters may be turned to form doweled ends, which fit into drilled holes in the treads and the railing. A stringer and appropriate moldings are used to complete the stairway trim.

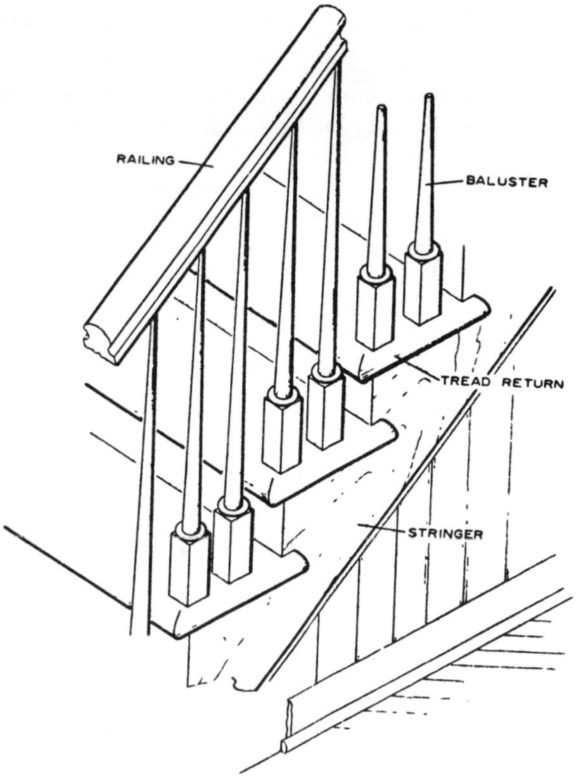

Figure 154.—Details of open main stairway.

Attic Folding Stairs

Where attics are used primarily for storage and where space for a fixed stairway is not available, hinged or folding stairs are often used and may be purchased ready to install. They operate through an opening in the ceiling of a hall and swing up into the attic space, out of the way when not in use. Where such stairs are to be installed, the attic floor joists should be designed for limited floor loading. One common size of folding stairs requires only a 26- by 54-inch rough opening. These openings should be framed out as described for normal stair openings.

Exterior Stairs

Proportioning of risers and treads in laying out porch steps or approaches to terraces should be as carefully considered as the design of interior stairways. Similar riser-to-tread ratios can be use; however, the riser used in principal exterior steps should normally be between 6 and 7 inches in height. The need for a good support or foundation for outside steps is often overlooked. Where wood steps are used, the bottom step should be concrete or supported by treated wood members. Where the steps are located over backfill or disturbed ground, the foundation should be carried down to undisturbed ground.

CHAPTER 24

FLASHING AND OTHER SHEET METAL WORK

In house construction, the *sheet-metal work* normally consists of flashing, gutters, and downspouts, and sometimes attic ventilators. Flashing (*11*) is often provided to prevent wicking action by joints between moisture-absorbent materials. It might also be used to provide protection from wind-driven rain or from action of melting snows. For instance, damage from ice dams is often the result of inadequate flashing. Thus, proper installation of these materials is important, as well as their selection and location.

Gutters are installed at the cornice line of a pitched-roof house to carry the rain or melted snow to the downspouts and away from the foundation area. They are especially needed for houses with narrow roof overhangs. Where positive rain disposal cannot be assured, downspouts should be connected with storm sewers or other drains. Poor drainage away from the wall is often the cause of wet basements and other moisture problems.

Materials

Materials most commonly used for sheet-metal work are galvanized metal, terneplate, aluminum, copper, and stainless steel. Near the seacoast, where the salt in the air may corrode galvanized sheet metal, copper or stainless steel is preferred for gutters, downspouts, and flashings. Molded wood gutters, cut from solid pieces of Douglas-fir or redwood, are also used in coastal areas because they are not affected by the corrosive atmosphere. Wood gutters can be attractive in appearance and are preferred by some builders.

Galvanized (zinc-coated) sheet metal is used in two weights of zinc coatings: 1.25 and 1.50 ounces per square foot (total weight of coating on both sides). When the lightly coated 1.25-ounce sheet is used for exposed flashing and for gutters and downspouts, 26-gage metal is required. With the heavier 1.50-ounce coating, a 28-gage metal is satisfactory for most metal work, except that gutters should be 26-gage.

Aluminum flashing should have a minimum thickness of 0.019 inch, the same as for roof valleys. Gutters should be made from 0.027-inch-thick metal and downspouts from 0.020-inch thickness. Copper for flashing and similar uses should have a minimum thickness of 0.020 inch (16 oz.). Aluminum is not normally used when it comes in contact with concrete or stucco unless it is protected with a coat of asphaltum or other protection against reaction with the alkali in the cement.

The types of metal fastenings, such as nails and screws, and the hangers and clips used with the various metals, are important to prevent corrosion or deterioration when unlike metals are used together. For aluminum, only aluminum or stainless steel fasteners should be used. For copper flashing, use copper nails and fittings. Galvanized sheet metal or terneplate should be fastened with galvanized or stainless-steel fasteners.

Flashing

Flashing should be used at the junction of a roof and a wood or masonry wall, at chimneys, over exposed doors and windows, at siding material changes, in roof valleys, and other areas where rain or melted snow may penetrate into the house.

Material Changes

One wall area which requires flashing is at the intersection of two types of siding materials. For example, a stucco-finish gable end and a wood-siding lower wall should be flashed (fig. 155,*A*). A wood molding such as a drip cap separates the two materials and is covered by the flashing which extends behind the stucco. The flashing should extend at least 4 inches above the intersection. When sheathing paper is used, it should lap the flashing (fig. 155,*A*).

When a wood-siding pattern change occurs on the same wall, the intersection should also be flashed. A vertical board-sided upper wall with horizontal siding below usually requires some type of flashing (fig. 155,*B*). A small space above the molding provides a drip for rain. This will prevent paint peeling which could occur if the boards were in tight contact with the molding. A drip cap is sometimes used as a terminating molding (fig. 84). When the upper wall, such as a gable end, projects slightly beyond the lower wall (fig. 85), flashing is usually *not* required.

Doors and Windows

The same type of flashing shown in figure 155,*A* should be used over door and window openings exposed to driving rain. However, window and doorheads protected by wide overhangs in a single-story house with a hip roof do not ordinarily require such flashing. When building paper is used on the sidewalls, it should lap the top edge of the flashing. To protect the walls behind the window sill in a brick veneer exterior, flashing should extend under the masonry sill up to the underside of the wood sill.

Flat Roof

Flashing is also required at the junctions of an exterior wall and a flat or low-pitched built-up roof

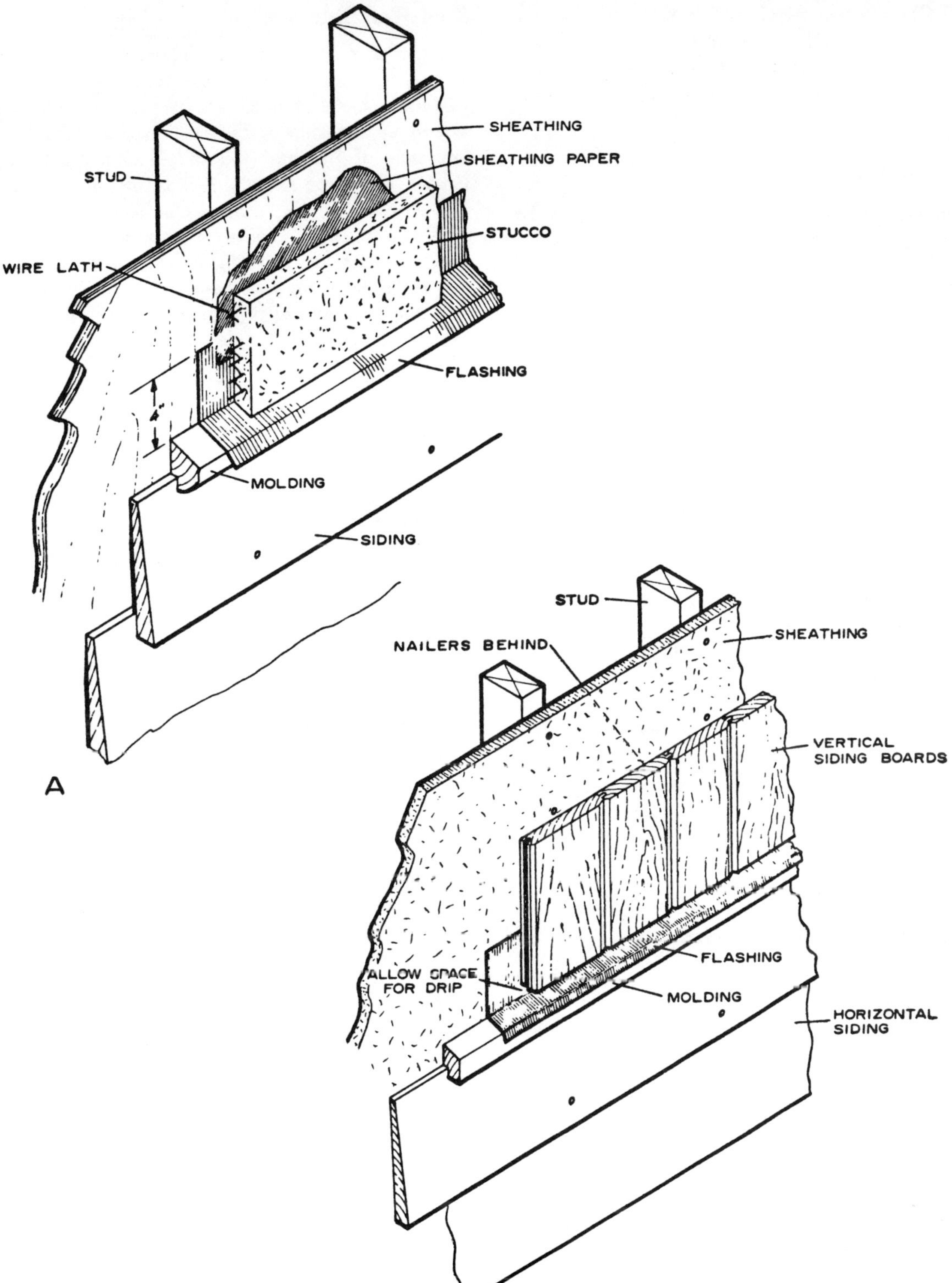

Figure 155.—Flashing at material changes: *A*, Stucco above, siding below; *B*, vertical siding above, horizontal below.

(fig. 71,*C*). When a metal roof is used, the metal is turned up on the wall and covered by the siding. A clearance of 2 inches should be allowed at the bottom of the siding for protection from melted snow and water.

Ridge and Roof

Ridge flashing should be used under a Boston ridge in wood shingle or shake roofs to prevent water entry (fig. 72,*B*). The flashing should extend about 3 inches on each side of the ridge and be nailed in place only at the outer edges. The ridge shingles or shakes, which are 6 to 8 inches wide, cover the flashing.

Stack vents and roof ventilators are provided with flashing collars which are lapped by the shingles on the upper side. The lower edge of the collar laps the shingles. Sides are nailed to the shingles and calked with a roofing mastic.

Valley

The valley formed by two intersecting rooflines is usually covered with metal flashing. Some building regulations allow the use of two thicknesses of mineral-surfaced roll roofing in place of the metal flashing. As an alternate, one 36-inch-wide strip of roll roofing with closed or woven asphalt shingles is also allowed. This type of valley is normally used only on roofs with a slope of 10 in 12 or steeper.

Widths of sheet-metal flashing for valleys should not be less than:

(a) 12 inches wide for roof slopes of 7 in 12 and over.

(b) 18 inches wide for 4 in 12 to 7 in 12 roof slopes.

(c) 24 inches wide for slopes less than 4 in 12.

The width of the valley between shingles should increase from the top to the bottom (fig. 156,*A*). The minimum open width at the top is 4 inches and should be increased at the rate of about ⅛ inch per foot. These widths can be chalklined on the flashing before shingles are applied.

When adjacent roof slopes vary, such as a low-slope porch roof intersecting a steeper main roof, a 1-inch crimped standing seam should be used (fig. 156,*B*). This will keep heavy rains on the steeper slopes from overrunning the valley and being forced under the shingles on the adjoining slope. Nails for the shingles should be kept back as far as possible to eliminate holes in the flashing. A ribbon of asphalt-roofing mastic is often used under the edge of the shingles. It is wise to use the wider valley flashings supplemented by a width of 15- or 30-pound asphalt felt where snow and ice dams may cause melting snow water to back under shingles.

Roof-Wall Intersections

When shingles on a roof intersect a vertical wall, shingle flashing is used at the junction. These tin or galvanized-metal shingles are bent at a 90° angle and extend up the side of the wall over the sheathing a minimum of 4 inches (fig. 157,*A*). When roofing felt is used under the shingle, it is turned up on the wall and covered by the flashing. One piece of flashing is used at each shingle course. The siding is then applied over the flashing, allowing about a 2-inch space between the bevel edge of the siding and the roof.

If the roof intersects a brick wall or chimney, the same type of metal shingle flashing is used at the end of each shingle course as described for the wood-sided wall. In addition, counterflashing or brick flashing is used to cover the shingle flashing (fig. 157,*B*). This counterflashing is often preformed in sections and is inserted in open mortar joints. Unless soldered together, each section should overlap the next a minimum of 3 inches with the joint calked. In laying up the chimney or the brick wall, the mortar is usually raked out for a depth of about 1 inch at flashing locations. Lead wedges driven into the joint above the flashing hold it in place. The joint is then calked to provide a watertight connection. In chimneys, this counterflashing is often preformed to cover one entire side.

Around small chimneys, chimney flashing often consists of simple counterflashing on each side. For single-flue chimneys, the shingle flashing on the high side should be carried up under the shingles. The vertical distance at top of the flashing and the upturned edge should be about 4 inches above the roof boards (fig. 158,*A*).

A wood *saddle* usually constructed on the high side of wide chimneys for better drainage, is made of a ridgeboard and post and sheathed with plywood or boards (fig. 158,*B*). It is then covered with metal, which extends up on the brick and under the shingles. Counterflashing at the chimney is then used (as previously described) by lead plugging and calking. A very wide chimney may contain a partial gable on the high side and be shingled in the same manner as the main roof.

Roof Edge

The cornice and the rake section of the roof are sometimes protected by a metal edging. This edging forms a desirable drip edge at the rake and prevents rain from entering behind the shingles (fig. 70,*B*).

At the eave line, a similar metal edging may be used to advantage (fig. 159,*A*). This edging, with the addition of a roll roof flashing (fig. 68,*B*,) will aid in resisting water entry from ice dams. Variations of it are shown in figure 159,*B* and *C*. They form a good drip edge and prevent or minimize the chance of rain

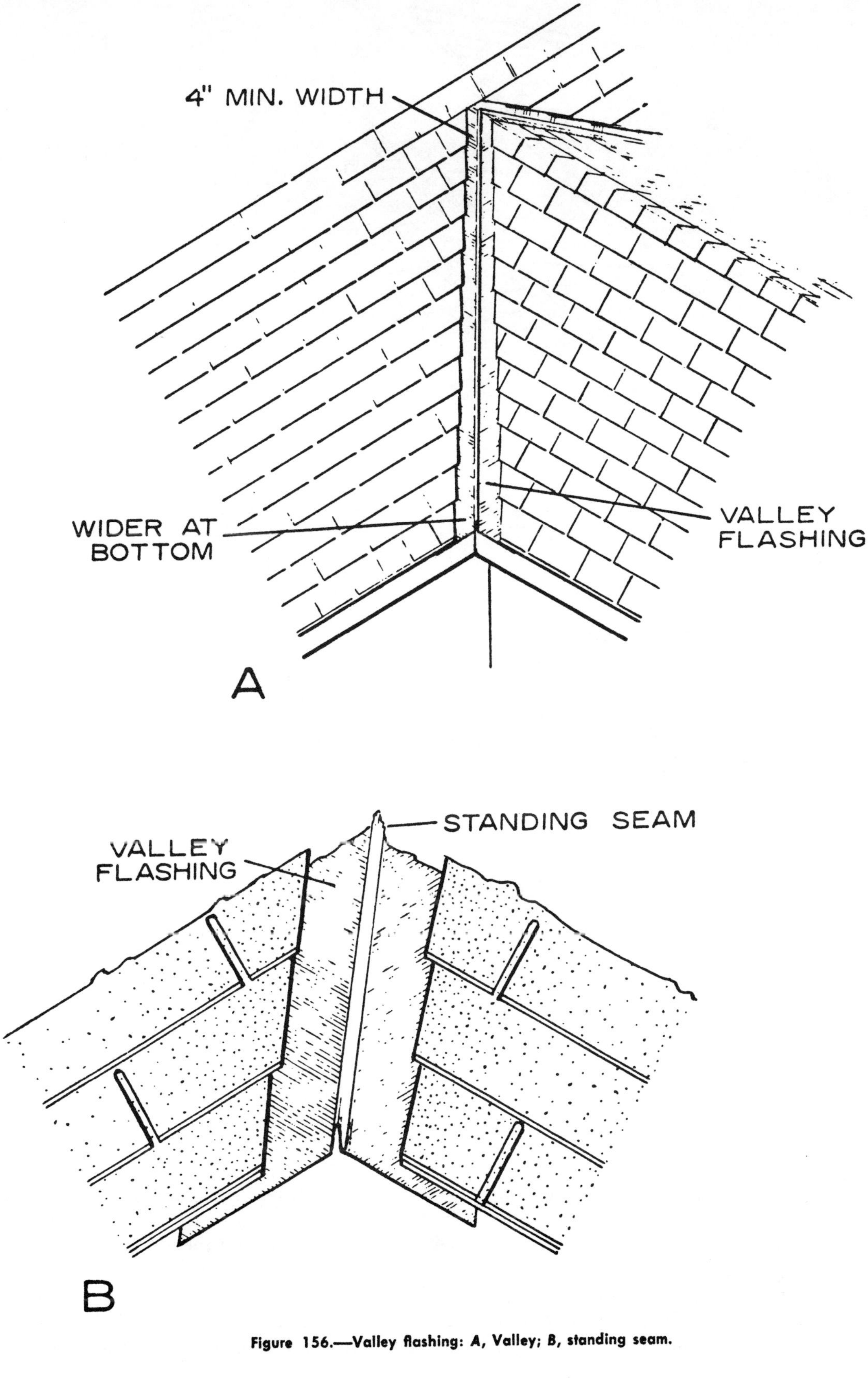

Figure 156.—Valley flashing: *A*, Valley; *B*, standing seam.

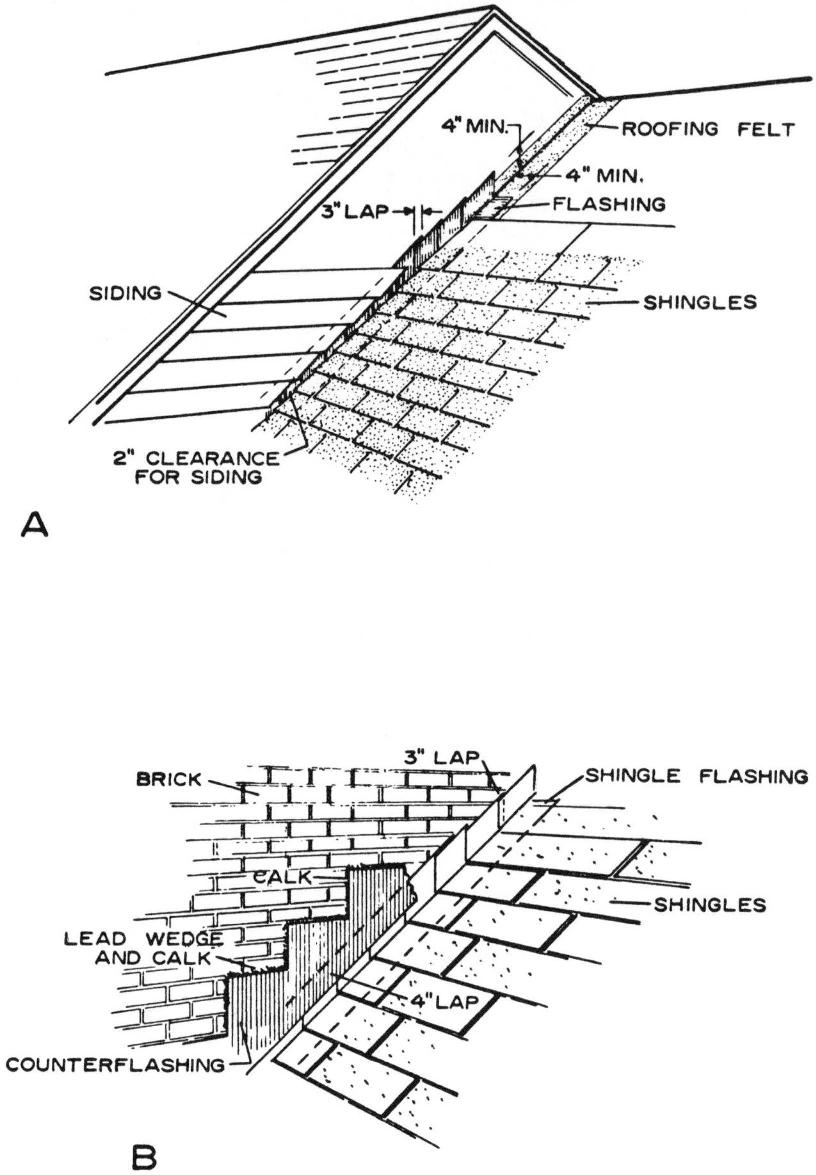

Figure 157.—Roof and wall intersection: *A*, Wood siding wall; *B*, brick wall.

being blown back under the shingles. This type of drip edge is desirable whether or not a gutter is used.

Gutters and Downspouts

Types

Several types of gutters are available to guide the rainwater to the downspouts and away from the foundation. Some houses have built-in gutters in the cornice. These are lined with sheet metal and connected to the downspouts. On flat roofs, water is often drained from one or more locations and carried through an inside wall to an underground drain. All downspouts connected to an underground drain should contain basket strainers at the junction of the gutter.

Perhaps the most commonly used gutter is the type hung from the edge of the roof or fastened to the edge of the cornice facia. Metal gutters may be the half-round (fig. 160,*A*) or the formed type (fig. 160,*B*) and may be galvanized metal, copper, or aluminum. Some have a factory-applied enamel finish.

Downspouts are round or rectangular (fig. 160, *C* and *D*), the round type being used for the half-round gutters. They are usually corrugated to provide extra stiffness and strength. Corrugated patterns are less likely to burst when plugged with ice.

Wood gutters have a pleasing appearance and are fastened to the facia board rather than being carried by hangers as are most metal gutters. The wood should

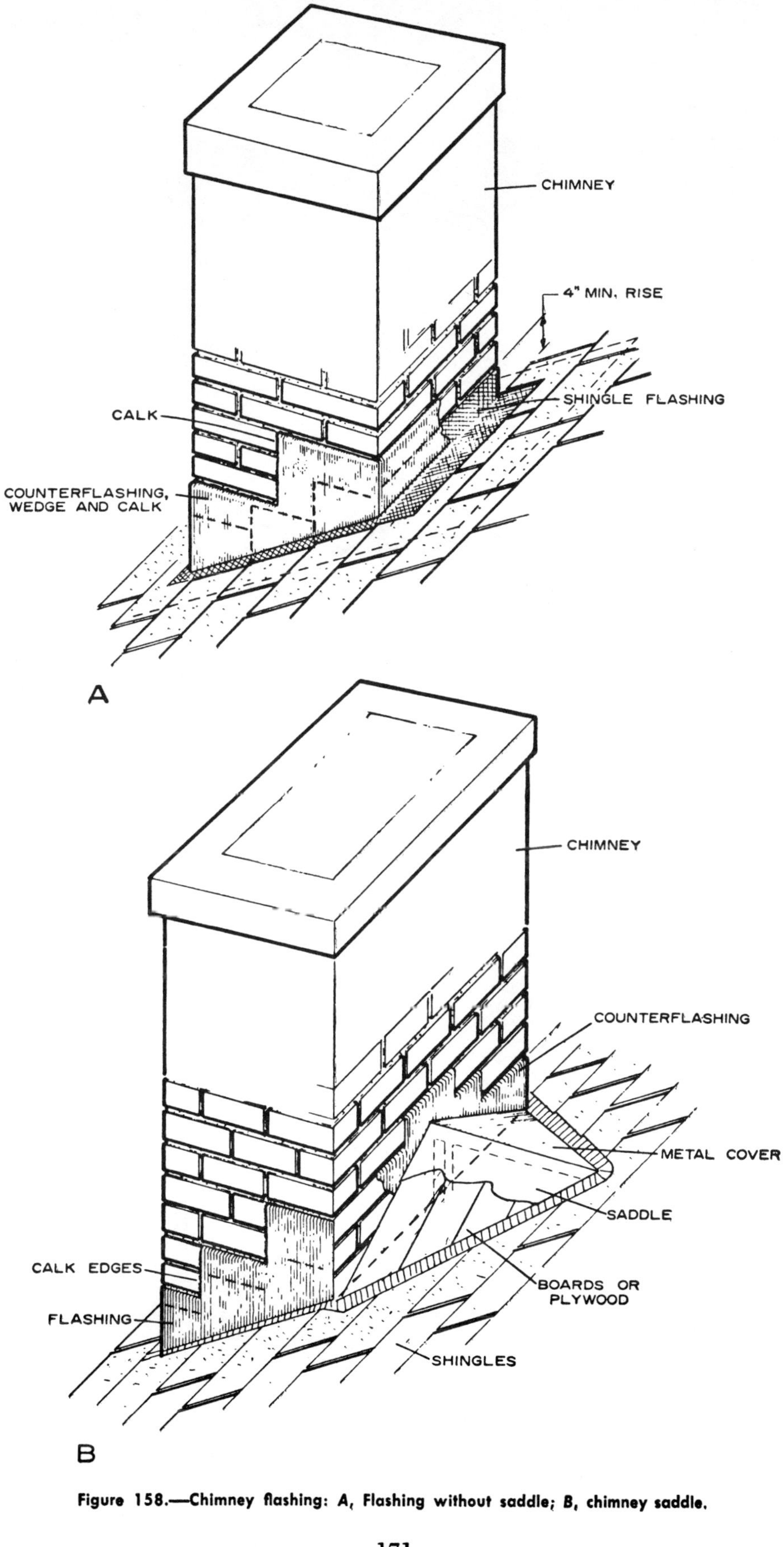

Figure 158.—Chimney flashing: *A*, Flashing without saddle; *B*, chimney saddle.

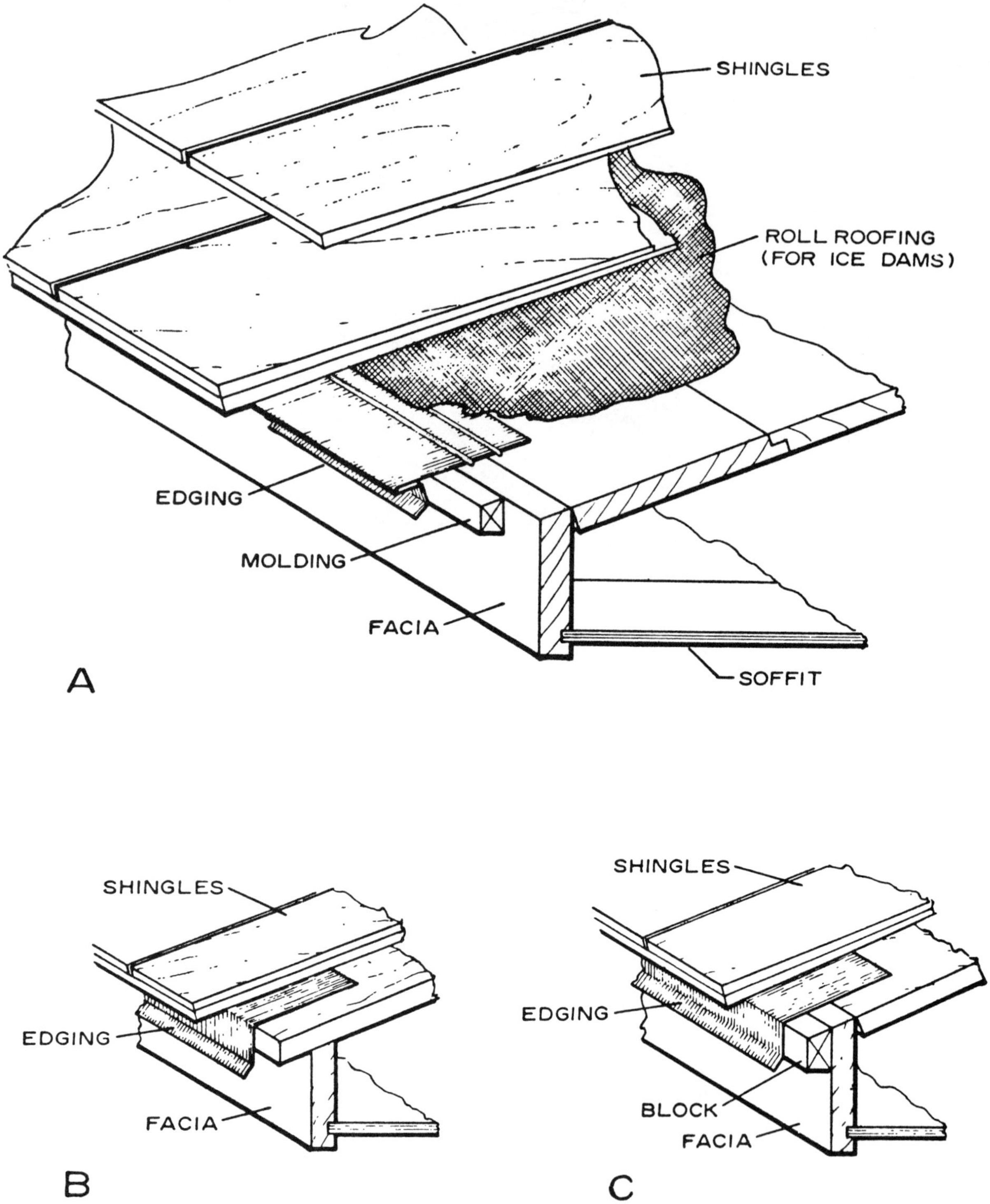

Figure 159.—Cornice flashing: ***A,*** **Formed flashing;** ***B,*** **flashing without wood blocking; C, flashing with wood blocking.**

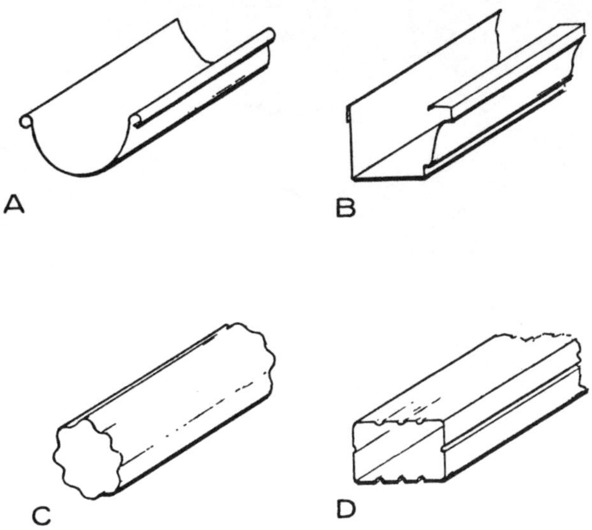

Figure 160.—Gutters and downspouts: *A*, Half-round gutter; *B*, formed gutter; *C*, round downspout; *D*, rectangular downspout.

be clear and free of knots and preferably treated, unless made of all heartwood from such species as redwood, western redcedar, and cypress. Continuous sections should be used wherever possible. When splices are necessary, they should be square-cut butt joints fastened with dowels or a spline. Joints should be set in white lead or similar material. When untreated wood gutters are used, it is good practice to brush several generous coats of water-repellent preservative on the interior.

Size

The size of gutters should be determined by the size and spacing of the downspouts used. One square inch of downspout is required for each 100 square feet of roof. When downspouts are spaced up to 40 feet apart, the gutter should have the same area as the downspout. For greater spacing, the width of the gutter should be increased.

Installation

On long runs of gutters, such as required around a hip-roof house, at least four downspouts are desirable. Gutters should be installed with a slight pitch toward the downspouts. Metal gutters are often suspended from the edge of the roof with hangers (fig. 161,*A*). Hangers should be spaced 48 inches apart when made of galvanized steel and 30 inches apart when made of copper or aluminum. Formed gutters may be mounted on furring strips, but the gutter should be reinforced with wrap-around hangers at 48-inch intervals. Gutter splices, downspout connections, and corner joints should be soldered or provided with watertight joints.

Wood gutters are mounted on the facia using furring blocks spaced 24 inches apart (fig. 161,*B*). Rustproof screws are commonly used to fasten the gutters to the blocks and facia backing. The edge shingle should be located so that the drip is near the center of the gutter.

Downspouts are fastened to the wall by straps or hooks (fig. 162,*A*). Several patterns of these fasteners allow a space between the wall and downspout. One common type consists of a galvanized metal strap with

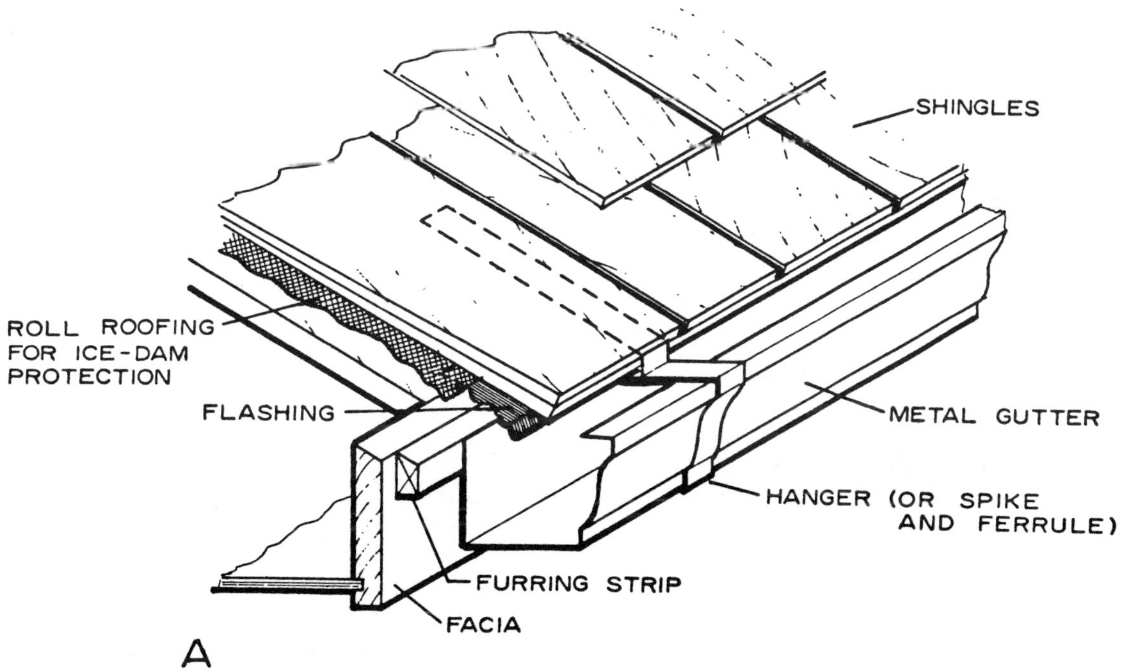

Figure 161a.—Gutter installation: *A*, Formed metal gutter.

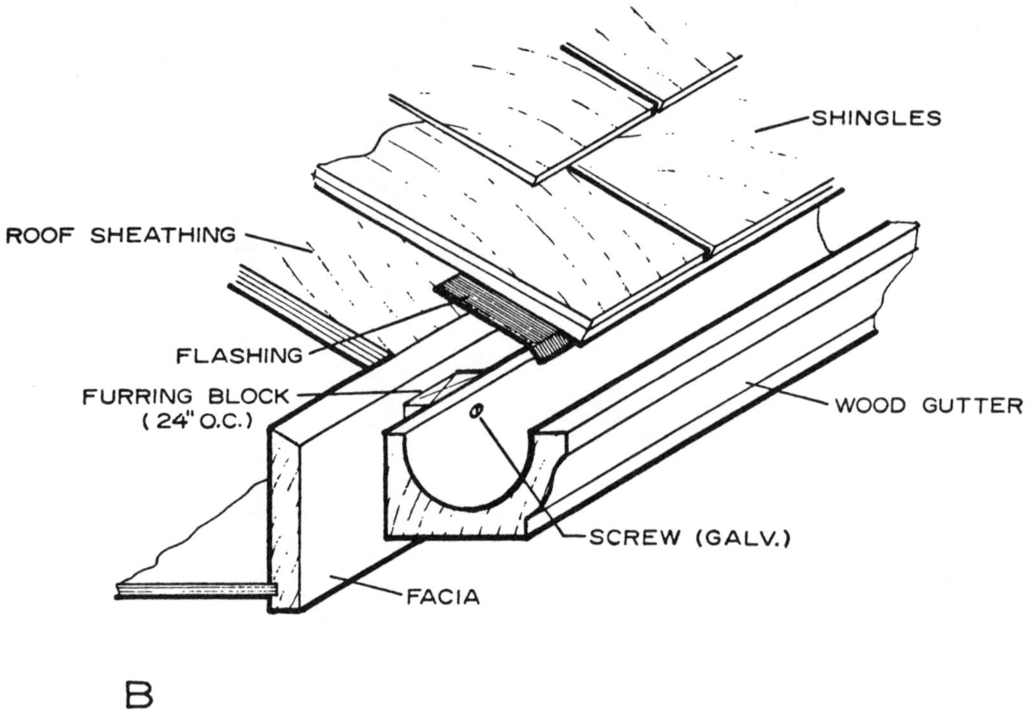

Figure 161b.—Gutter installation: *B*, Wood gutter.

a spike and spacer collar. After the spike is driven through the collar and into the siding and backing stud, the strap is fastened around the pipe. Downspouts should be fastened at the top and bottom. In addition, for long downspouts a strap or hook should be used for every 6 feet of length.

An elbow should be used at the bottom of the downspout, as well as a splash block, to carry the water away from the wall. However, a vitrified tile line is sometimes used to carry the water to a storm sewer. In such installations, the splash block is not required (fig. 162,*B*).

CHAPTER 25

PORCHES AND GARAGES

An attached porch or garage which is in keeping with the house design usually adds to overall pleasing appearance. Thus, any similar attachments to the house after it has been built should also be in keeping structurally and architecturally with the basic design. In such additions, the connections of the porch or garage to the main house should be by means of the framing members and roof sheathing. Rafters, ceiling joists, and studs should be securely attached by nailing to the house framing.

When additions are made to an existing house, the siding or other finish is removed so that framing members can be easily and correctly fastened to the house. In many instances, the siding can be cut with a skill saw to the outline of the addition and removed only where necessary. When concrete foundations, piers, or slabs are added, they should also be structurally correct. Footings should be of sufficient size, the bottoms located below the frostline, and the foundation wall anchored to the house foundation when possible.

Porches

There are many types and designs of porches, some with roof slopes continuous with the roof of the house itself. Other porch roofs may have just enough pitch to provide drainage. The fundamental construction principles, however, are somewhat alike no matter what type is built. Thus, a general description, together with several construction details, can apply to several types.

Figure 163 shows the construction details of a

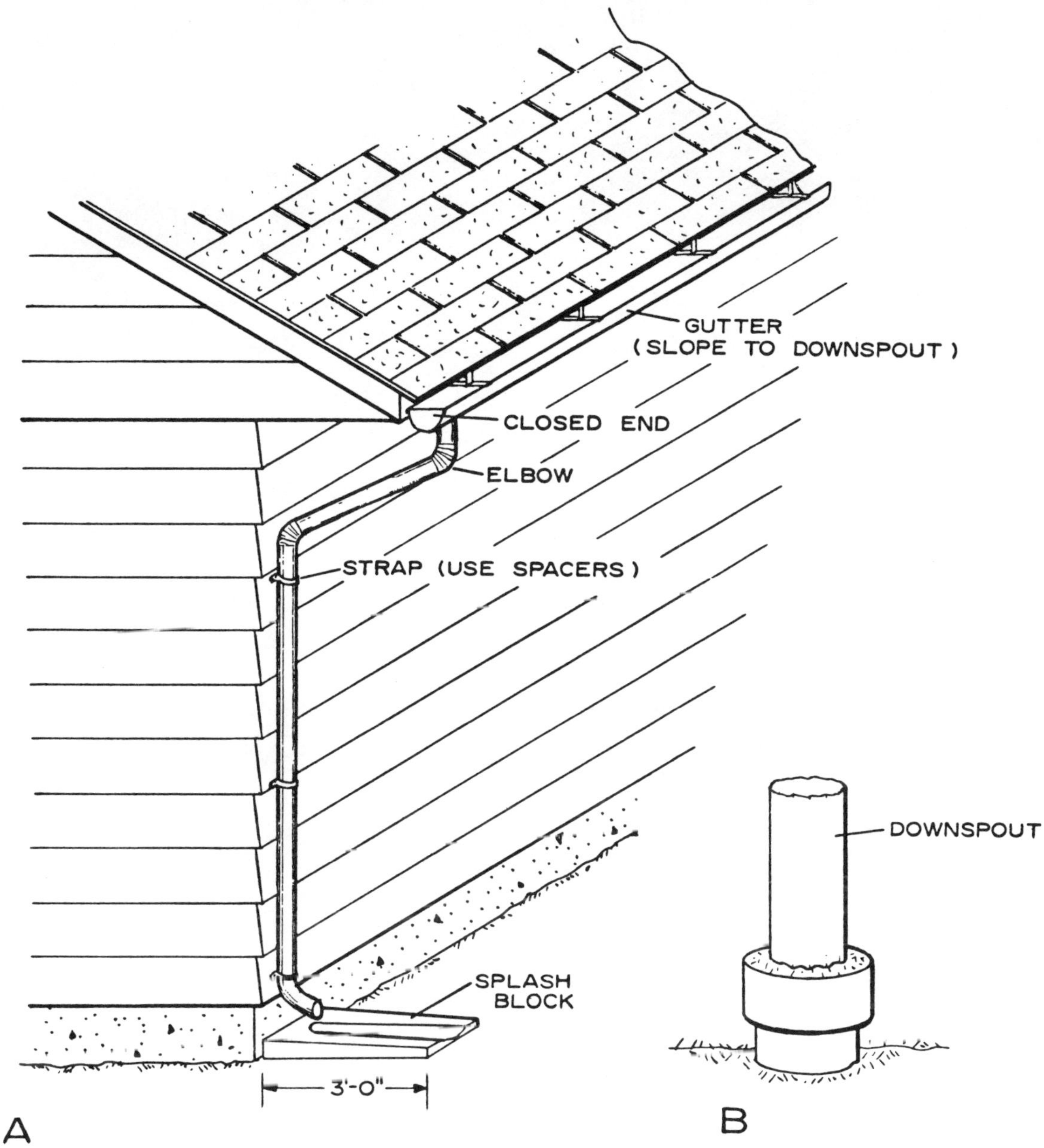

Figure 162.—Downspout installation: *A*, Downspout with splash block; *B*, drain to storm sewer.

typical flat-roofed porch with a concrete slab floor. An attached porch can be open or fully enclosed; or it can be constructed with a concrete slab floor, insulated or uninsulated. A porch can also be constructed using wood floor framing over a crawl space (fig. 164). Most details of such a unit should comply with those previously outlined for various parts of the house itself.

Porch Framing and Floors

Structural framing for the floors and walls should comply with the details given in Chapter 5, "Floor Framing," and Chapter 6, "Wall Framing." General details of the ceiling and roof framing are covered in Chapter 7, "Ceiling and Roof Framing."

Porch floors, whether wood or concrete, should have sufficient slope away from the house to provide good drainage. Weep holes or drains should be provided in any solid or fully sheathed perimeter wall. Open wood balusters with top and bottom railings should be constructed so that the bottom rail is free of the floor surface.

Floor framing for wood floor construction should be at least 18 inches above the soil. The use of a soil cover of polyethylene or similar material under a partially open or a closed porch is good practice.

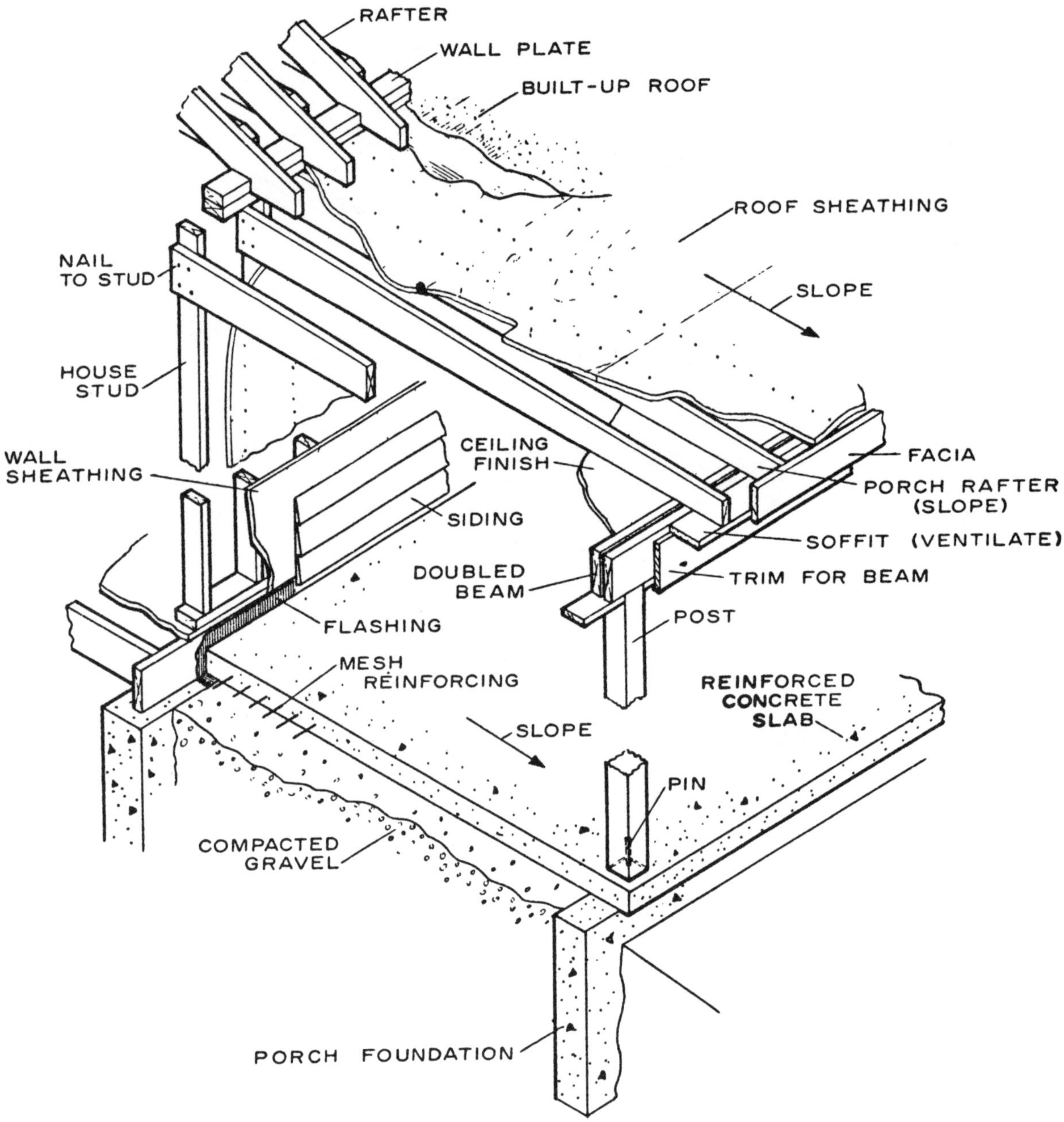

Figure 163.—Details of porch construction for concrete slab.

Slats or grillwork used around an open crawl space should be made with a removable section for entry in areas where termites may be present. (See Chapter 29, "Protection Against Decay and Termites.") A fully enclosed crawl-space foundation should be vented or have an opening to the basement.

Wood species used for finish porch floor should have good decay and wear resistance, be nonsplintering, and be free from warping. Species commonly used are cypress, Douglas-fir, western larch, southern pine, and redwood. Only treated material should be used where moisture conditions are severe.

Porch Columns

Supports for enclosed porches usually consist of fully framed stud walls. The studs are doubled at openings and at corners. Because both interior and exterior finish coverings are used, the walls are constructed much like the walls of the house. In open or partially open porches, however, solid or built-up posts or columns are used. A more finished or cased column is often made up of doubled 2 by 4's which are covered with 1- by 4-inch casing on two opposite sides and 1- by 6-inch finish casing on the other sides (fig. 165,*A*). Solid posts, normally 4 by 4 inches in

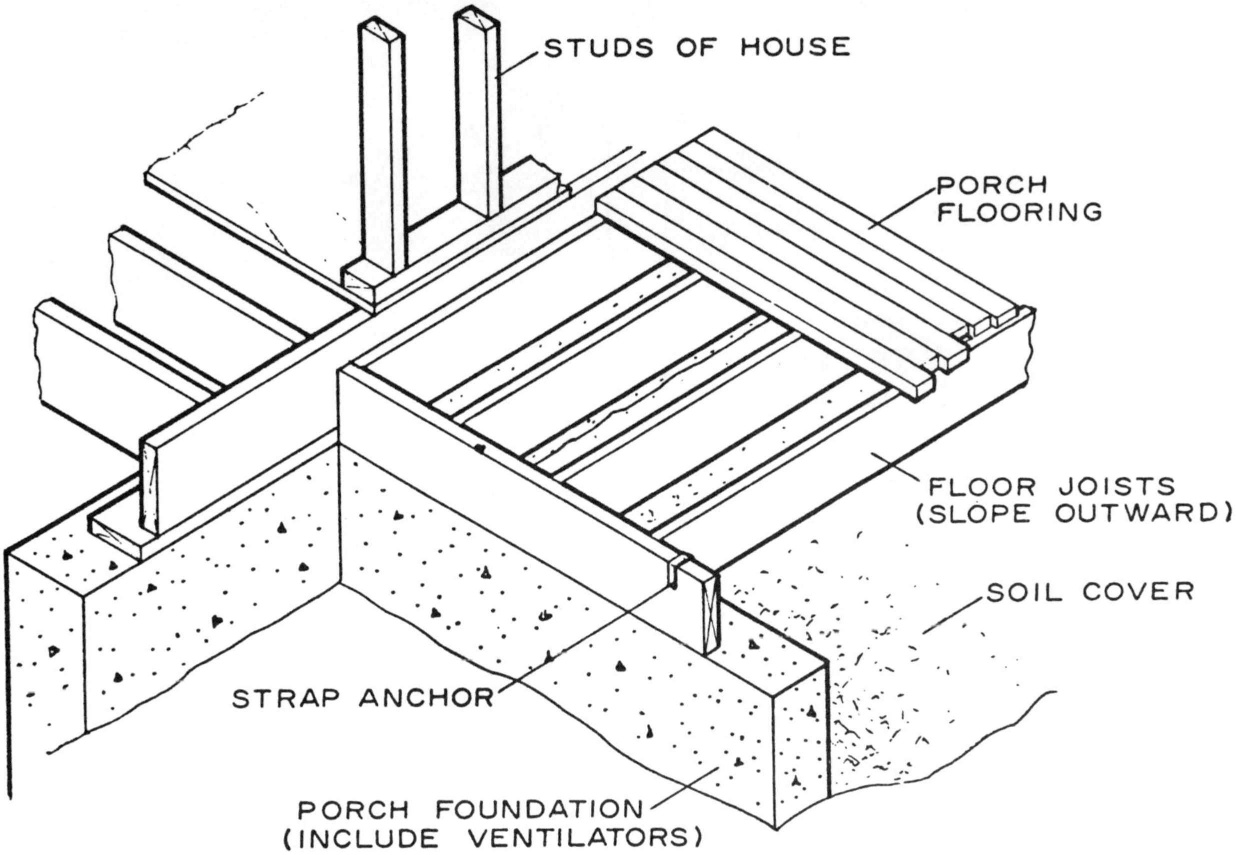

Figure 164.—Porch floor with wood framing.

size, are used mainly for open porches. An open railing may be used between posts.

A formal design of a large house entrance often includes the use of round built-up columns topped by *Doric* or *Ionic* capitals. These columns are factory-made and ready for installation at the building site.

The base of posts or columns in open porches should be designed so that no pockets are formed to retain moisture and encourage decay. In single posts, a steel pin may be used to locate the post and a large galvanized washer or similar spacer used to keep the bottom of the post above the concrete or wood floor (fig. 165,*B*). The bottom of the post should be treated to minimize moisture penetration. Often single posts of this type are made from a decay-resistant wood species. A cased post can be flashed under the base molding (fig. 165,*C*). Post anchors which provide connections to the floor and to the post are available commercially, as are post caps.

Balustrade

A porch *balustrade* usually consists of one or two railings with *balusters* between them. They are designed for an open porch to provide protection and to improve the appearance. There are innumerable combinations and arrangements of them. A closed balustrade may be used with screens or combination windows above (fig. 166,*A*). A balustrade with decorative railings may be used for an open porch (fig. 166,*B*). This type can also be used with full-height removable screens.

All balustrade members that are exposed to water and snow should be designed to shed water. The top of the railing should be tapered and connections with balusters protected as much as possible (fig. 167,*A*). Railings should not contact a concrete floor but should be blocked to provide a small space beneath. When wood must be in contact with the concrete, it should be treated to resist decay.

Connection of the railing with a post should be made in a way that prevents moisture from being trapped. One method provides a small space between the post and the end of the railing (fig. 167,*B*). When the railing is treated with paint or water-repellent preservative, this type connection should provide good service. Exposed members, such as posts, balusters, and railings, should be all-heart-

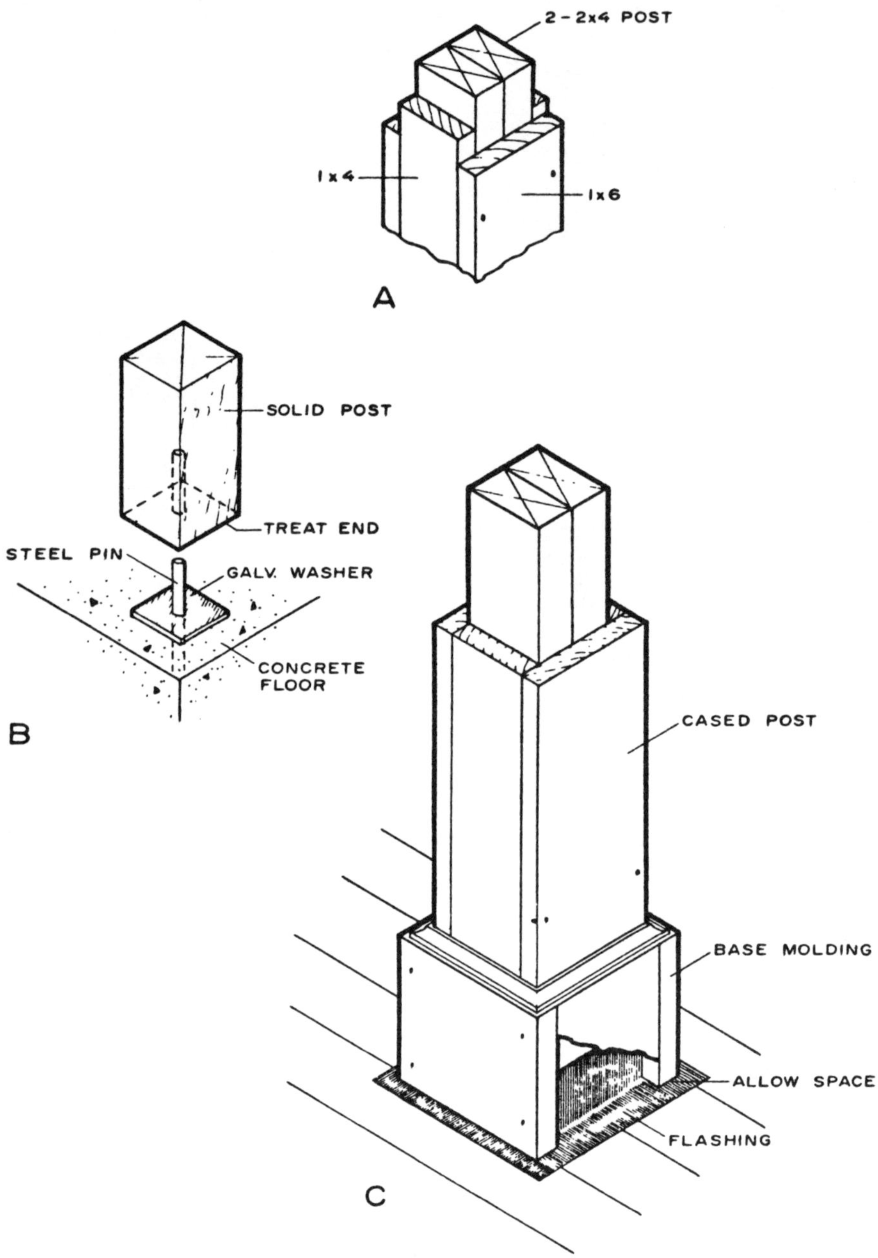

Figure 165.—Post details: *A*, Cased post; *B*, pin anchor and spacer; *C*, flashing at base.

wood stock of decay-resistant or treated wood to minimize decay.

Garages

Garages can be classified as attached, detached, basement, or carport. The selection of a garage type is often determined by limitations of the site and the size of the lot. Where space is not a limitation, the attached garage has much in its favor. It may give better architectural lines to the house, it is warmer during cold weather, and it provides covered protection to passengers, convenient space for storage, and a short, direct entrance to the house.

Building regulations often require that detached garages be located away from the house toward the rear of the lot. Where there is considerable slope to a lot, basement garages may be desirable, and generally such garages will cost less than those above grade.

Carports are car-storage spaces, generally attached to the house, that have roofs and often no sidewalls. To improve the appearance and utility of this type of structure, storage cabinets are often used on a side and at the end of the carport.

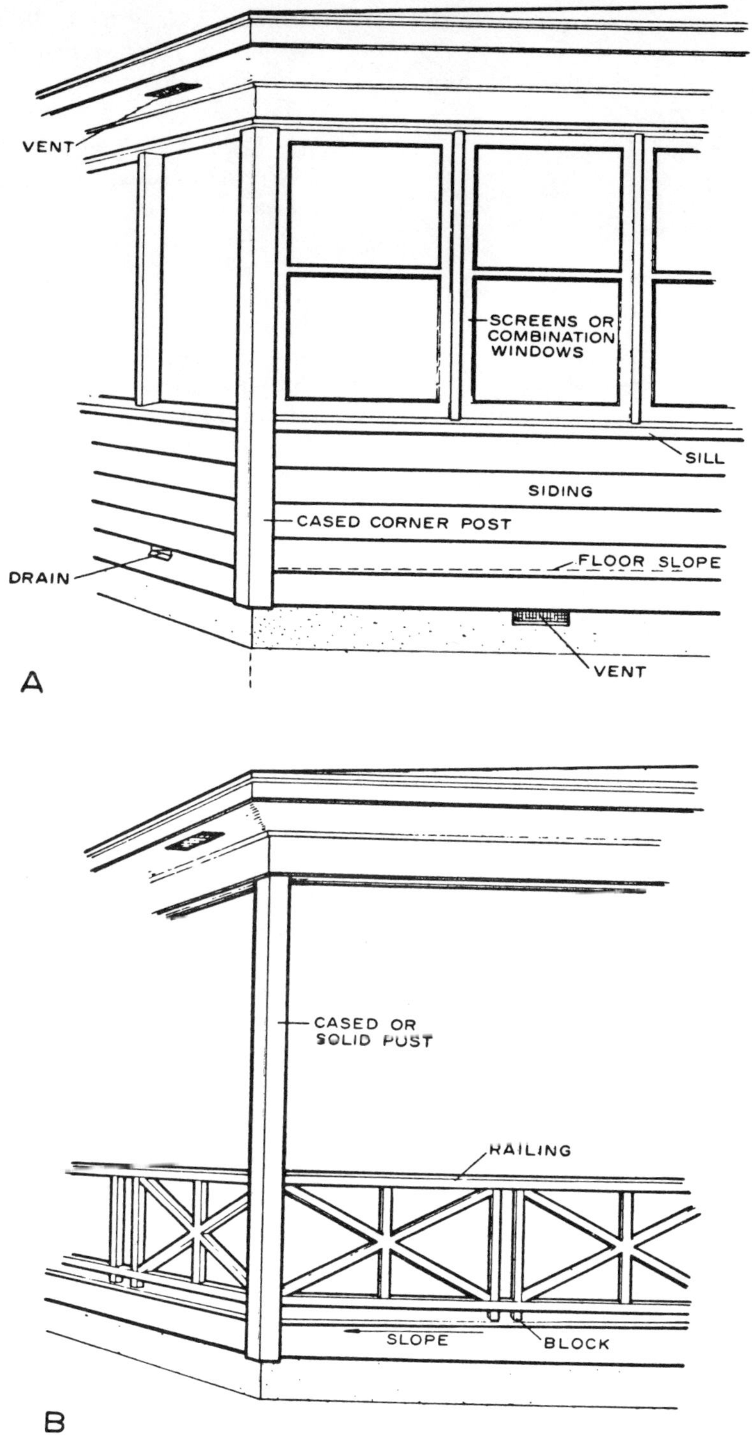

Figure 166.—Types of balustrades: *A*, Closed; *B*, open.

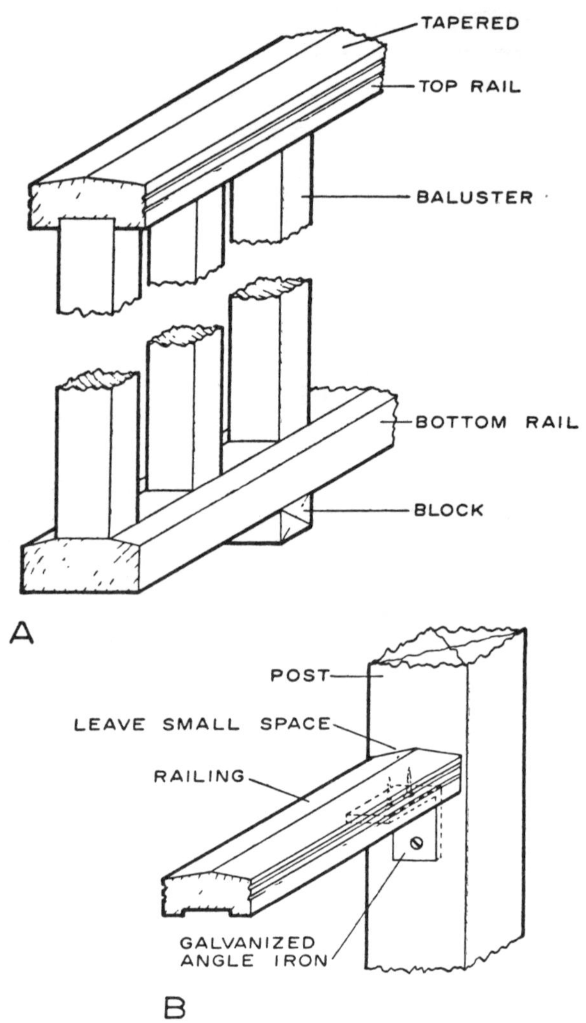

Figure 167.—Railing details: *A*, Balustrade assembly; *B*, rail-to-post connection.

Size

It is a mistake to design the garage too small for convenient use. Cars vary in size from the small import models to the large foreign and domestic sedans. Many popular models are now up to 215 inches long, and the larger and more expensive models are usually over 230 inches—almost 20 feet in length. Thus, while the garage need not necessarily be designed to take all sizes with adequate room around the car, it is wise to provide a minimum distance of 21 to 22 feet between the inside face of the front and rear walls. If additional storage or work space is required at the back, a greater depth is required.

The inside width of a single garage should never be less than 11 feet with 13 feet much more satisfactory.

The minimum outside size for a single garage, therefore, would be 14 by 22 feet. A double garage should be not less than 22 by 22 feet in outside dimensions to provide reasonable clearance and use. The addition of a shop or storage area would increase these minimum sizes.

For an attached garage, the foundation wall should extend below the frostline and about 8 inches above the finish-floor level. It should be not less than 6 inches thick, but is usually more because of the difficulty of trenching this width. The sill plate should be anchored to the foundation wall with anchor bolts spaced about 8 feet apart, at least two bolts in each sill piece. Extra anchors may be required at the sides of the main door. The framing of the sidewalls and roof and the application of the exterior covering material of an attached garage should be similar to that of the house.

The interior finish of the garage is often a matter of choice. The studs may be left exposed or covered with some type of sheet material or they may be plastered. Some building codes require that the wall between the house and the attached garage be made of fire-resistant material. Local building regulations and fire codes should be consulted before construction is begun.

If fill is required below the floor, it should preferably be sand or gravel well-compacted and tamped. If other types of soil fill are used, it should be wet down so that it will be well compacted and can then be well-tamped and time allowed before pouring. Unless these precautions are taken, the concrete floor will likely settle and crack.

The floor should be of concrete not less than 4 inches thick and laid with a pitch of about 2 inches from the back to the front of the garage. The use of wire reinforcing mesh is often advisable. The garage floor should be set about 1 inch above the drive or apron level. It is desirable at this point to have an expansion joint between the garage floor and the driveway or apron.

Garage Doors

The two overhead garage doors most commonly used are the sectional and the single-section swing types. The swing door (fig. 168,*A*) is hung with side and overhead brackets and an overhead track, and must be moved outward slightly at the bottom as it is opened. The sectional type (fig. 168,*B*), in four or five horizontal hinged sections, has a similar track extending along the sides and under the ceiling framing, with a roller for the side of each section. It is opened by lifting and is adaptable to automatic electric opening with remote control devices. The standard desirable size for a single door is 9 feet in width by $6\frac{1}{2}$ or 7 feet in height. Double doors are usually 16 by $6\frac{1}{2}$ or 7 feet in size.

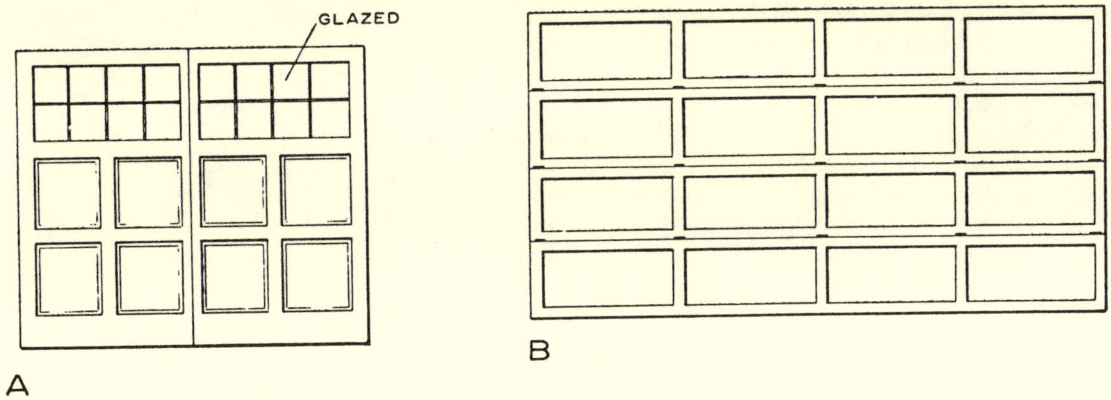

Figure 168.—Garage doors: *A*, One-section swing; *B*, sectional.

Doors vary in design, but those most often used are the panel type with solid stiles and rails and panel fillers. A glazed panel section is often included. Clearance above the top of the door required for overhead doors is usually about 12 inches. However, low-headroom brackets are available when such clearance is not possible.

The header beam over garage doors should be designed for the snow load which might be imposed on the roof above. In wide openings, this may be a steel I-beam or a built-up wood section. For spans of 8 or 9 feet, two doubled 2 by 10's of high-grade Douglas-fir or similar species are commonly used when only snow loads must be considered. If floor loads are also imposed on the header, a steel I-beam or wide-flange beam is usually selected.

CHAPTER 26

CHIMNEYS AND FIREPLACES

Chimneys are generally constructed of masonry units supported on a suitable foundation. A chimney must be structurally safe, capable of producing sufficient draft for the fireplace, and capable of carrying away harmful gases from the fuel-burning equipment and other utilities. Lightweight, prefabricated chimneys that do not require masonry protection or concrete foundations are now accepted for certain uses by fire underwriters. Make certain, however, they are approved and listed by Underwriters' Laboratories, Inc.

Fireplaces should not only be safe and durable but should be so constructed that they provide sufficient draft and are suitable for their intended use. From the standpoint of heat-production efficiency, which is estimated to be only 10 percent, they might be considered a luxury. However, they add a decorative note to a room and a cheerful atmosphere. Improved heating efficiency and the assurance of a correctly proportioned fireplace can usually be obtained by the installation of a factory-made circulating fireplace. This metal unit, enclosed by the masonry, allows air to be heated and circulated throughout the room in a system separate from the direct heat of the fire.

Chimneys

The chimney should be built on a concrete footing of sufficient area, depth, and strength for the imposed load. The footing should be below the frostline. For houses with a basement, the footings for the walls and fireplace are usually poured together and at the same elevation.

The size of the chimney depends on the number of flues, the presence of a fireplace, and the design of the house. The house design may include a room-wide brick or stone fireplace wall which extends through the roof. While only two or three flues may be required for heating units and fireplaces, several "false" flues may be added at the top for appearance. The flue sizes are made to conform to the width and length of a brick so that full-length bricks can be used to enclose the flue lining. Thus an 8- by 8-inch flue lining (about 8½ by 8½ in. in outside dimensions) with the minimum 4-inch thickness of surrounding masonry will use six standard bricks for each course (fig. 169,*A*). An 8- by 12-inch flue lining (8½ by 13 in. in outside dimensions) will be enclosed by seven bricks at each course (fig. 169,*B*), and a 12- by 12-inch flue (13 by 13 in. in outside dimension) by eight bricks (fig. 169,*C*), and so on. Each fireplace should have a separate flue and, for best performance, flues should be separated by a 4-inch-wide brick spacer (withe) between them (fig. 170,*A*).

The greater the difference in temperature between chimney gases and outside atmosphere, the better the draft. Thus, an interior chimney will have better draft because the masonry will retain heat longer. The height of the chimney as well as the size of the flue are important factors in providing sufficient draft.

The height of a chimney above the roofline usually depends upon its location in relation to the ridge. The top of the extending flue liners should not be less than 2 feet above the ridge or a wall that is within 10 feet (fig. 170,*B*). For flat or low-pitched roofs, the chimney should extend at least 3 feet above the highest point of the roof. To prevent moisture from entering between the brick and flue lining, a concrete cap is usually poured over the top course of brick (fig. 170,*C*). Precast or stone caps with a cement wash are also used.

Flashing for chimneys is illustrated in figure 158. Masonry chimneys should be separated from wood framing, subfloor, and other combustible materials. Framing members should have at least a 2-inch clearance and should be firestopped at each floor with asbestos or other types of noncombustible material (fig. 171). Subfloor, roof sheathing, and wall sheathing should have a ¾-inch clearance. A cleanout door is included in the bottom of the chimney where there are fireplaces or other solid fuel-burning equipment as well as at the bottom of other flues. The cleanout door for the furnace flue is usually located just below the smokepipe thimble with enough room for a soot pocket.

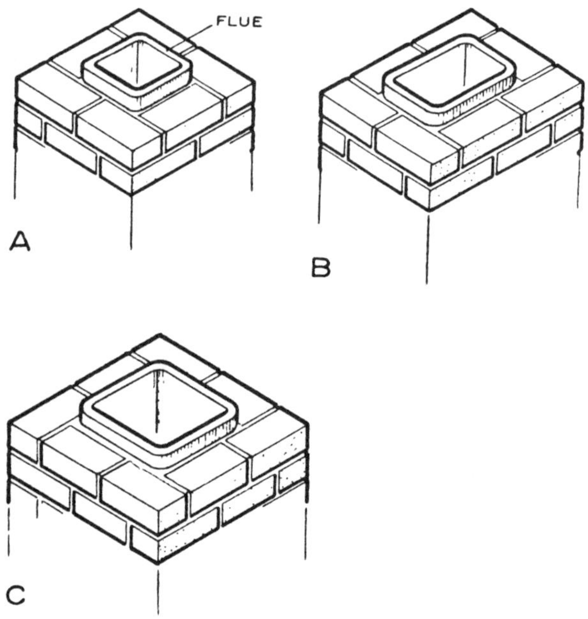

Figure 169.—Brick and flue combinations: *A*, 8- by 8-inch flue lining; *B*, 8- by 12-inch flue lining; *C*, 12- by 12-inch flue lining.

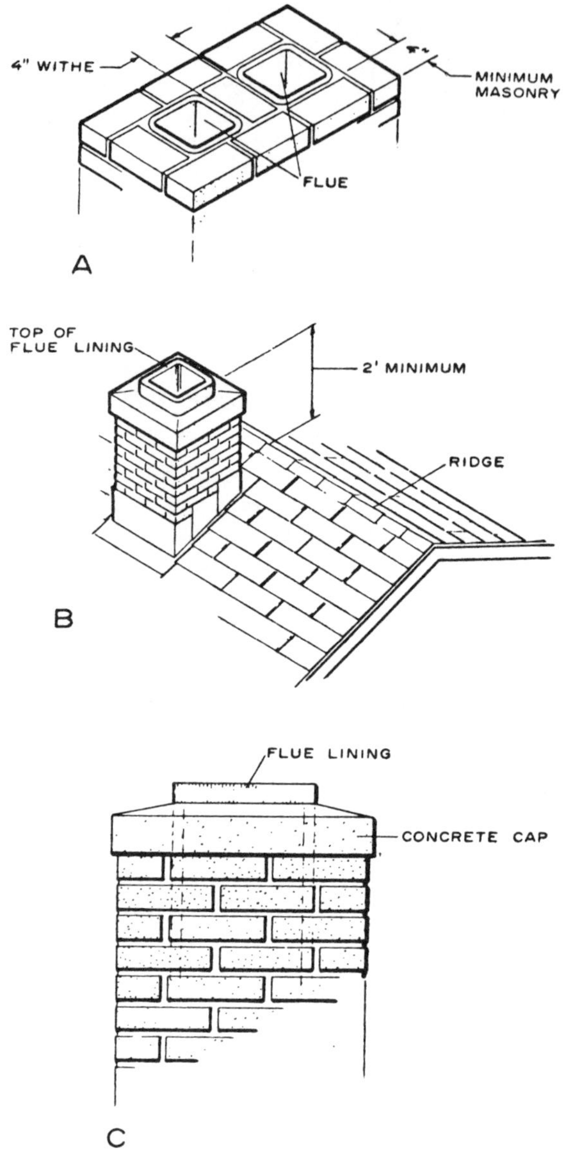

Figure 170.—Chimney details: *A*, Spacer between flues; *B*, height of chimneys; *C*, chimney cap.

Flue Linings

Rectangular fire-clay flue linings (previously described) or round vitrified tile are normally used in all chimneys. Vitrified (glazed) tile or a stainless-steel lining is usually required for gas-burning equipment. Local codes outline these specific requirements. A fireplace chimney with at least an 8-inch-thick masonry wall ordinarily does not require a flue lining. However, the cost of the extra brick or masonry and the labor involved are most likely greater than the

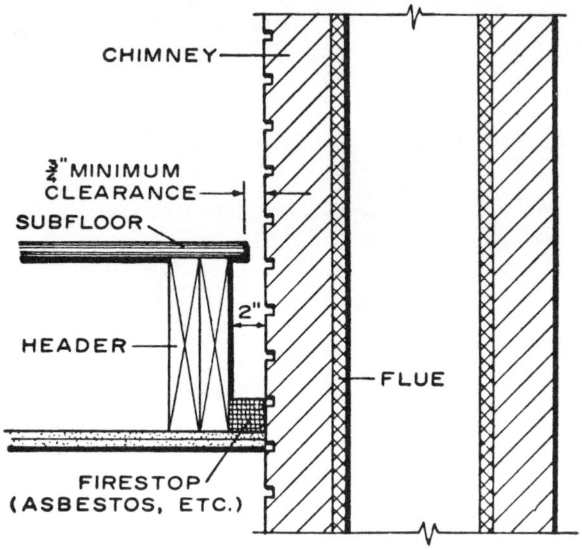

Figure 171.—Clearances for wood construction.

cost of flue lining. Furthermore, a well-installed flue lining will result in a safer chimney.

Flue liners should be installed enough ahead of the brick or masonry work, as it is carried up, so that careful bedding of the mortar will result in a tight and smooth joint. When diagonal offsets are necessary, the flue liners should be beveled at the direction change in order to have a tight joint. It is also good practice to stagger the joints in adjacent tile.

Flue lining is supported by masonry and begins at least 8 inches below the thimble for a connecting smoke or vent pipe from the furnace. In fireplaces, the flue liner should start at the top of the throat and extend to the top of the chimney.

Rectangular flue lining is made in 2-foot lengths and in sizes of 8 by 8, 8 by 12, 12 by 12, 12 by 16, and up to 20 by 20 inches. Wall thicknesses of the flue lining vary with the size of the flue. The smaller sizes have a 5/8-inch-thick wall, and the larger sizes vary from 3/4 to 1 3/8 inches in thickness. Vitrified tiles, 8 inches in diameter, are most commonly used for the flues of the heating unit, although larger sizes are also available. This tile has a bell joint.

Fireplaces

A fireplace adds to the attractiveness of the house interior, but one that does not "draw" properly is a detriment, not an asset. By following several rules on the relation of the fireplace opening size to flue area, depth of the opening, and other measurements, satisfactory performance can be assured. Metal circulating fireplaces, which form the main outline of the opening and are enclosed with brick, are designed for proper functioning when flues are the correct size.

One rule which is often recommended is that the depth of the fireplace should be about two-thirds the height of the opening. Thus, a 30-inch-high fireplace would be 20 inches deep from the face to the rear of the opening.

The flue area should be at least one-tenth of the open area of the fireplace (width times height) when the chimney is 15 feet or more in height. When less than 15 feet, the flue area in square inches should be one-eighth of the opening of the fireplace. This height is measured from the throat to the top of the chimney. Thus, a fireplace with a 30-inch width and 24-inch height (720 sq. in.) would require an 8- by 12-inch flue, which has an inside area of about 80 square inches. A 12- by 12-inch flue liner has an area of about 125 square inches, and this would be large enough for a 36- by 30-inch opening when the chimney height is 15 feet or over.

The back width of the fireplace is usually 6 to 8 inches narrower than the front. This helps to guide the smoke and fumes toward the rear. A vertical back-wall of about a 14-inch height then tapers toward the upper section or "throat" of the fireplace (fig. 172). The area of the throat should be about 1 1/4 to 1 1/3 times the area of the flue to promote better draft. An adjustable damper is used at this area for easy control of the opening.

The *smoke shelf* (top of the throat) is necessary to prevent back drafts. The height of the smoke shelf should be 8 inches above the top of the fireplace opening (fig. 172). The smoke shelf is concave to retain any slight amount of rain that may enter.

Steel angle iron is used to support the brick or masonry over the fireplace opening. The bottom of the inner hearth, the sides, and the back are built of a heat-resistant material such as firebrick. The outer hearth should extend at least 16 inches out from the face of the fireplace and be supported by a reinforced concrete slab (fig. 172). This outer hearth is a precaution against flying sparks and is made of noncombustible materials such as glazed tile. Other fireplace details of clearance, framing of the wall, and cleanout opening and ash dump are also shown. Hangers and brackets for fireplace screens are often built into the face of the fireplace.

Fireplaces with two or more openings (fig. 173) require much larger flues than the conventional fireplace. For example, a fireplace with two open adjacent faces (fig. 173,*A*) would require a 12- by 16-inch flue for a 34- by 20- by 30-inch (width, depth, and height, respectively) opening. Local building regulations usually cover the proper sizes for these types of fireplaces.

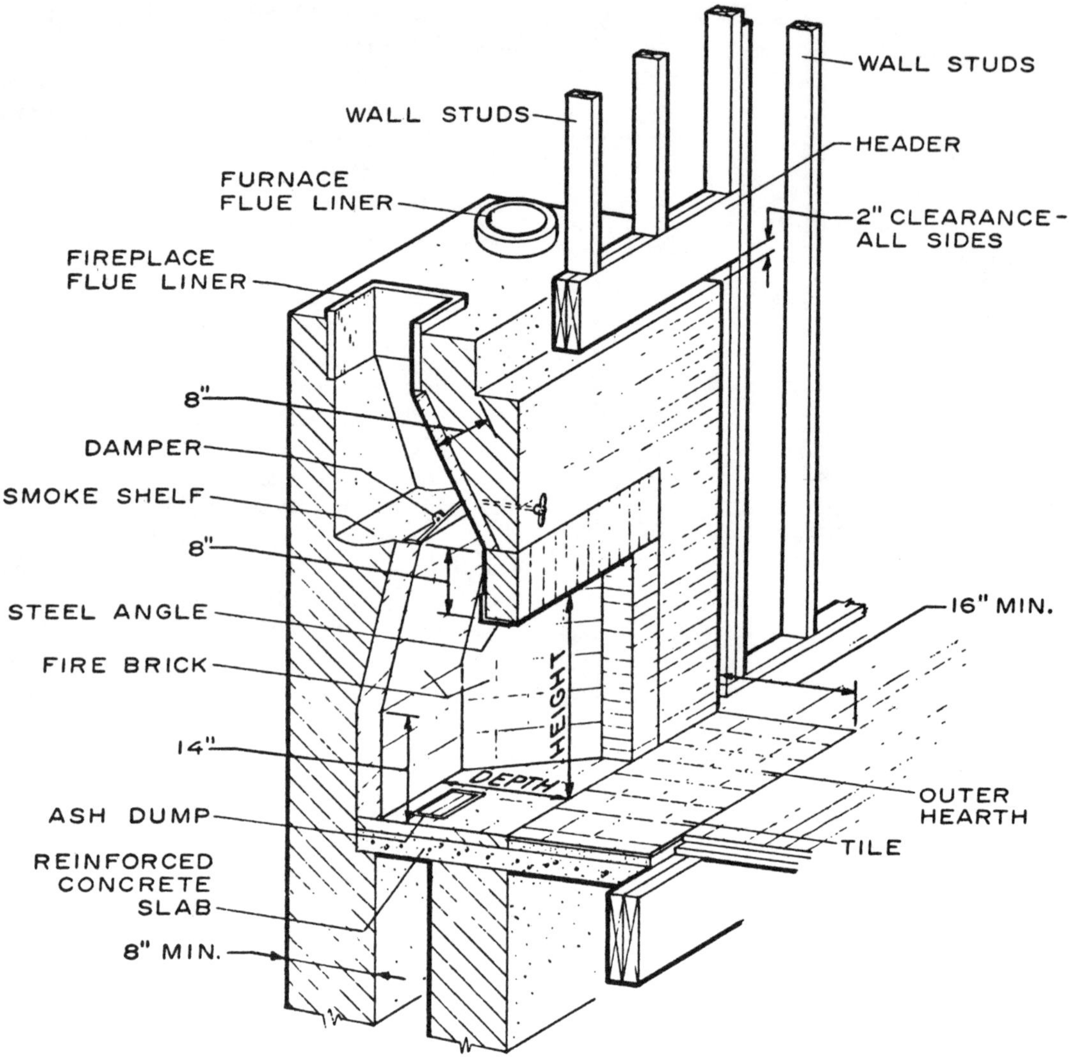

Figure 172.—Masonry fireplace.

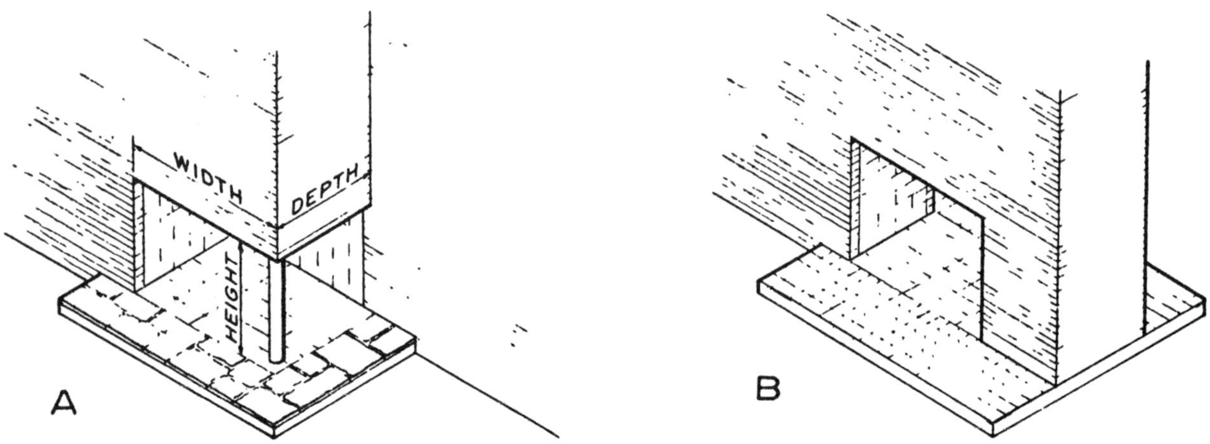

Figure 173.—Dual-opening fireplace: *A*, Adjacent opening; *B*, through fireplace.

CHAPTER 27

DRIVEWAYS, WALKS, AND BASEMENT FLOORS

A new home is not complete until driveways and walks have been installed so that landscaping can be started. Landscaping includes final grading, planting of shrubs and trees, and seeding or sodding of lawn areas. Because the automobile is an important element in American life, the garage is usually a prominent part of house design. This in turn establishes the location of driveways and walks.

Concrete and bituminous pavement are most commonly used in the construction of walks and drives, especially in areas where snow removal is important. In some areas of the country, a gravel driveway and a flagstone walk may be satisfactory and would reduce the cost of improvements.

Basements are normally finished with a concrete floor of some type, whether or not the area is to contain habitable rooms. These floors are poured after all improvements such as sewer and waterlines have been connected. Concrete slabs should *not* be poured on recently filled areas.

Driveways

The grade, width, and radius of curves in a driveway are important factors in establishing a safe entry to the garage. Driveways for attached garages, which are located near the street on relatively level property, need only be sufficiently wide to be adequate. Driveways that have a grade more than 7 percent (7-ft. rise in 100 ft.) should have some type of pavement to prevent wash. Driveways that are long and require an area for a turnaround should be designed carefully. Figure 174 shows a drive way and turnaround which allow the driver to back out of a single or double garage into the turn and proceed to the street or highway in a forward direction. This, in areas of heavy traffic, is much safer than having to back into the street or roadway. A double garage should be serviced by a wider entry and turnaround.

Driveways that are of necessity quite steep should have a near-level area in front of the garage for safety, from 12 to 16 feet long.

Two types of paved driveways may be used, (a) the more common *slab* or full-width type and (b) the *ribbon* type. When driveways are fairly long or steep, the full-width type is the most practical. The ribbon driveway is cheaper and perhaps less conspicuous. because of the grass strip between the concrete runners. However, it is not always practical for all locations.

The width of the single-slab type drive should be 9 feet for the modern car, although 8 feet is often considered minimum (fig. 175*A*,). When the driveway is also used as a walk, it should be at least 10 feet wide to allow for a parked car as well as a walkway. The width should be increased by at least 1 foot at curves. The radius of the drive at the curb should be at least 5 feet (fig. 175,*A*). Relatively short double driveways should be at least 18 feet wide, and 2 feet wider when they are also to be used as a walk from the street.

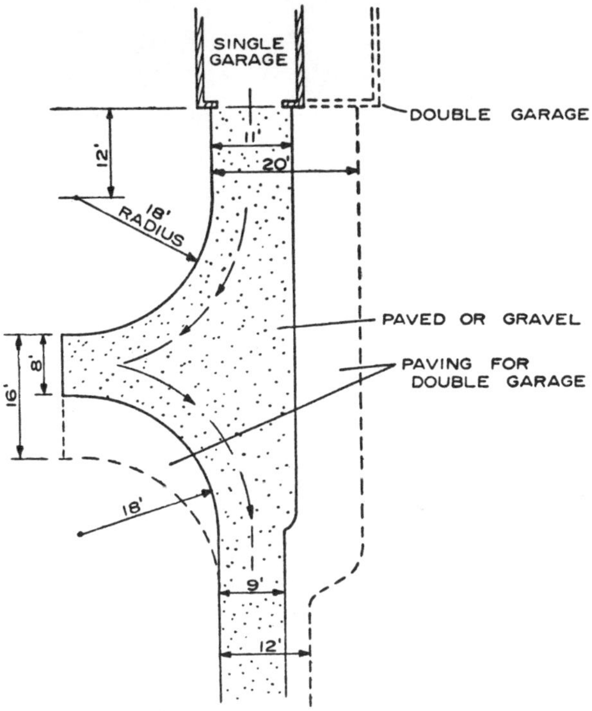

Figure 174.—Driveway turnaround.

The concrete strips in a ribbon driveway should be at least 2 feet in width and located so that they are 5 feet on center (fig. 175,*B*). When the ribbon is also used as a walk, the width of strips should be increased to at least 3 feet. This type of driveway is not practical if there is a curve or turn involved or the driveway is long.

Pouring a concrete driveway over an area that has been recently filled is poor practice unless the fill, preferably gravel, has settled and is well tamped. A gravel base is not ordinarily required on sandy undisturbed soil but should be used under all other conditions. Concrete should be about 5 inches thick. A 2 by 6 is often used for a side form. These members establish the elevation and alinement of the driveway and are used for striking off the concrete. Under most conditions, the use of steel reinforcing is good practice. Steel mesh, 6 by 6 inches in size,

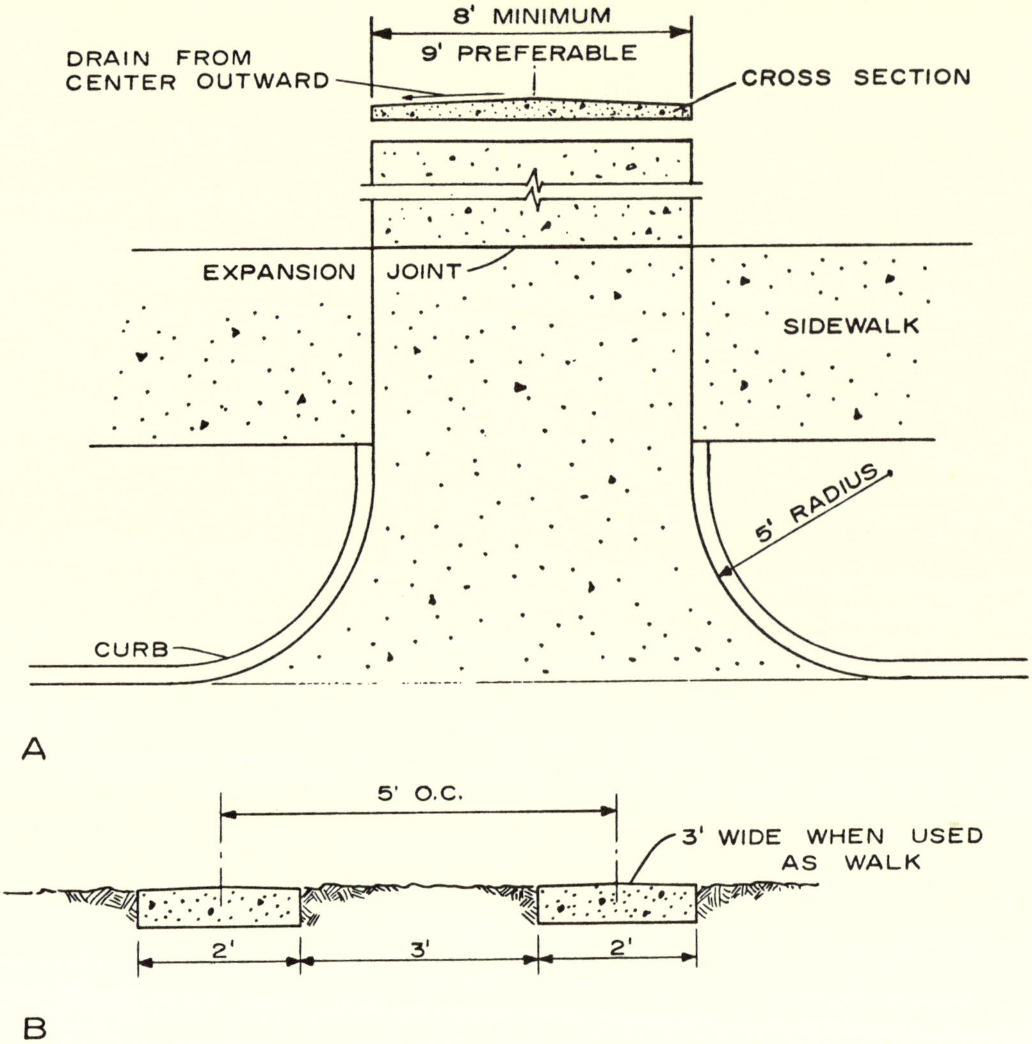

Figure 175.—Driveway details: ***A*****, Single-slab driveway;** ***B*****, ribbon-type driveway.**

will normally prevent or minimize cracking of the concrete. Expansion joints should be used (a) at the junction of the driveway with the public walk or curb, (b) at the garage slab, and (c) about every 40 feet on long driveways. A 5- or 5½-bag commercial mix is ordinarily used for driveways. However, a 5½- to 6-bag mix containing an air-entraining mixture should be used in areas having severe winter climates.

Contraction joints should be provided at 10- to 12-foot intervals. These crosswise grooves, cut into the partially set concrete, will allow the concrete to open along these lines during the cold weather rather than in irregular cracks in other areas.

Blacktop driveways, normally constructed by paving contractors, should also have a well-tamped gravel or crushed rock base. Top should be slightly crowned for drainage.

Sidewalks

Main sidewalks should extend from the front entry to the street or front walk or to a driveway leading to the street. A 5 percent grade is considered maximum for sidewalks; any greater slope usually requires steps. Walks should be at least 3 feet wide.

Concrete sidewalks should be constructed in the same general manner as outlined for concrete driveways. They should not be poured over filled areas unless they have settled and are very well tamped. This is especially true of the areas near the house after basement excavation backfill has been completed.

The minimum thickness of the concrete over normal undisturbed soil is usually 4 inches. As described for concrete driveways, contraction joints should be used and spaced on 4-foot centers.

When slopes to the house are greater than a 5 percent grade, stairs or steps should be used. This may be accomplished with a ramp sidewalk, a flight of stairs at a terrace, or a continuing sidewalk (fig. 176,*A*). These stairs have 11-inch treads and 7-inch risers when the stair is 30 inches or less in height. When the rise is more than 30 inches, the tread is 12 inches and the riser 6 inches. For a moderately uniform slope, a stepped ramp may be satisfactory (fig. 176,*B*). Generally, the rise should be about 6 to 6½ inches and the length between risers sufficient for two or three normal paces.

Walks can also be made of brick, flagstone, or other types of stone. Brick and stone are often placed directly over a well-tamped sand base. However, this system is not completely satisfactory where freezing of the soil is possible. For a more durable walk in cold climates, the brick or stone topping is embedded in a freshly laid reinforced concrete base (fig. 177).

As in all concrete sidewalks and curbed or uncurbed driveways, a slight crown should be included in the walk for drainage. Joints between brick or stone may be filled with a cement mortar mix or with sand.

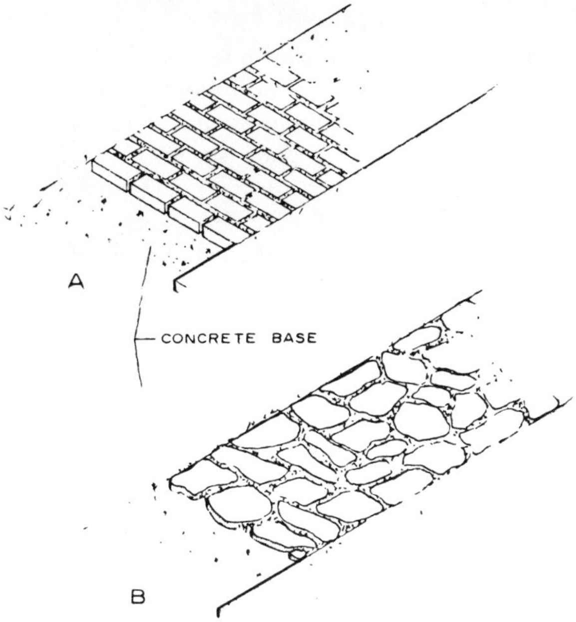

Figure 177.—Masonry paved walks: *A*, Brick; *B*, flagstone.

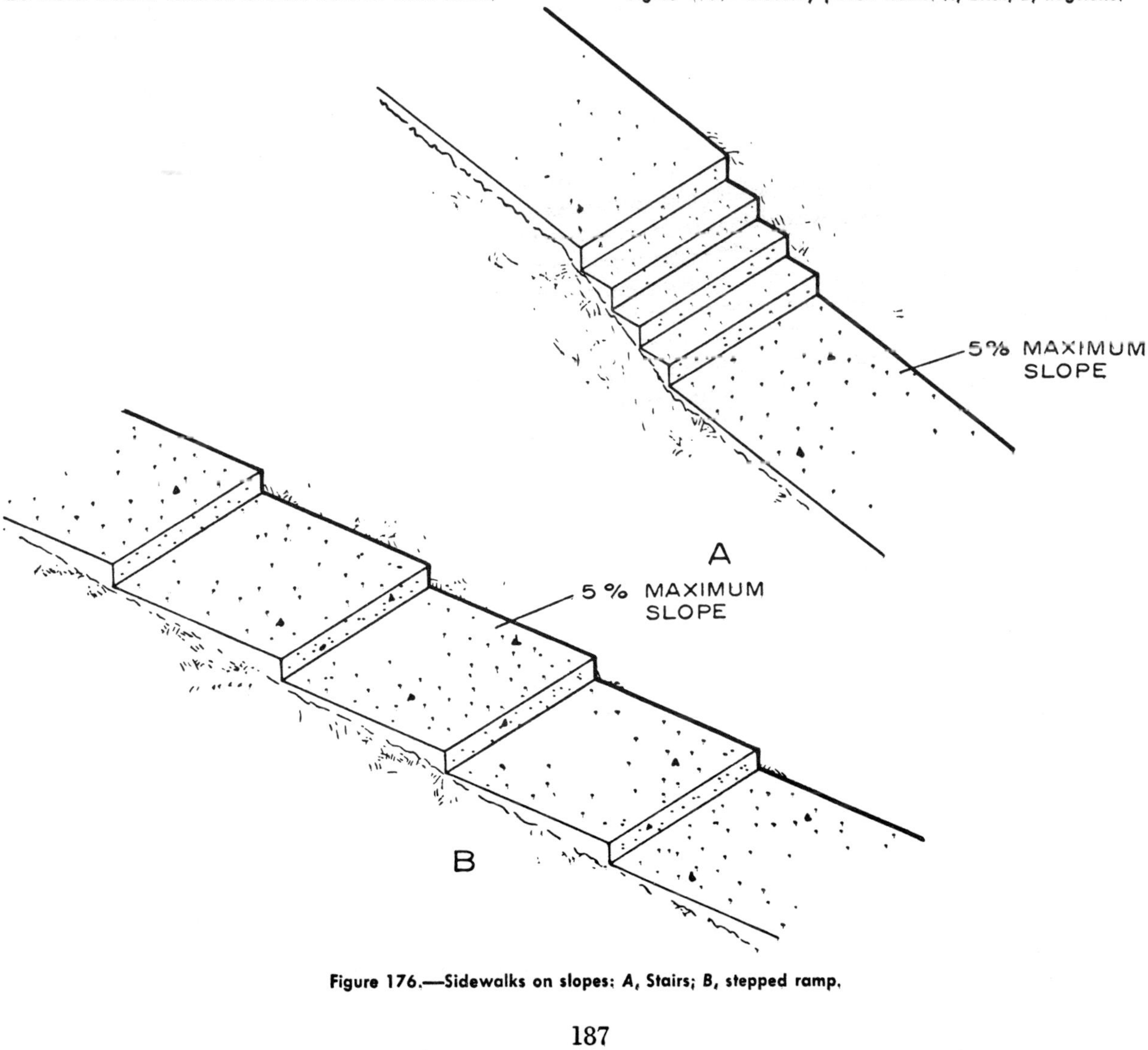

Figure 176.—Sidewalks on slopes: *A*, Stairs; *B*, stepped ramp.

Basement Floors

Basement floor slabs should be no less than $3\frac{1}{2}$ inches thick and sloped toward the floor drains. A 2 by 4 ($3\frac{1}{2}$ in. wide) is often used on edge for form work. At least one floor drain should be used, and for large floor areas, two are more satisfactory. One should be located near the laundry areas.

For a dry basement floor. the use of a polyethylene film or similar vapor barrier under the concrete slab is usually justified. However, basement areas or multi-level floors used only for utility or storage do not require a vapor barrier unless soil conditions are adverse. When finished rooms have concrete floors, the use of a vapor barrier is normally required. Details of basement floor or concrete-slab construction are outlined in Chapter 4, "Concrete Floor Slabs on Ground," and in Chapter 18. "Basement Rooms."

CHAPTER 28

PAINTING AND FINISHING

Wood and wood products in a variety of species, grain patterns, texture, and colors are available for use as exterior and interior surfaces. These wood surfaces can be finished quite effectively by several different methods. Painting, which totally obscures the wood grain, is used to achieve a particular color decor. Penetrating-type preservatives and pigmented stains permit some or all of the wood grain and texture to show and provide a special color effect as well as a natural or rustic appearance. The type of finish, painted or natural. often depends on the wood to be finished.

Effect of Wood Properties

Wood surfaces that shrink and swell the least are best for painting. For this reason, vertical- or edge-grained surfaces are far better than flat-grained surfaces of any species. Also, because the swelling of wood is directly proportional to density, low-density species are preferred over high-density species. However, even high-swelling and dense wood surfaces with flat grain have been stabilized with a resin-treated paper overlay, such as overlaid exterior plywood and lumber, to make them excellent for painting.

Medium-density fiberboard products fabricated with a uniform, low-density surface for exterior use are often painted. but little is known of their long-time performance. The most widely used species for exterior siding to be painted are vertical-grained western redcedar and redwood. These species are classified in Group I. those woods easiest to keep painted (table 10). Other species in Group I are excellent for painting but are not generally available in all parts of the country.

Species that are not normally cut as vertical-grained lumber. are high in density (swelling), or have defects such as knots or pitch are classified in Groups II through V, depending upon their general paint-holding characteristics. Many species in Groups II through IV are commonly painted, particularly the pines, Douglas-fir, and spruce; but these species generally require more care and attention than the species in Group I. Resinous species should be thoroughly kiln dried at temperatures that will effectively set the pitch.

The properties of wood that detract from its paintability do not necessarily affect the finishing of such boards naturally with penetrating preservatives and stains. These finishes penetrate into wood without forming a continuous film on the surface. Therefore, they will not blister, crack, or peel even if excessive moisture penetrates into wood. One way to further improve the performance of penetrating finishes is to leave the wood surface rough sawn. Allowing the high-density, flat-grained wood surfaces of lumber and plywood to weather several months also roughens the surface and improves it for staining. Rough-textured surfaces absorb more of the preservative and stain, insuring a more durable finish.

Natural Finishes for Exterior Wood

Weathered Wood

The simplest of natural finishes for wood is natural weathering. Without paint or treatment of any kind, wood surfaces change in color and texture in a few months or years, and then may stay almost unaltered for a long time if the wood does not decay. Generally, the dark-colored woods become lighter and the light-colored woods become darker. As weathering continues, all woods become gray, accompanied by degradation of the wood cells at the surface. Unfinished wood will wear away at the rate of about $\frac{1}{4}$ inch in 100 years.

The appearance of weathered wood is affected by dark-colored spores and mycelia of fungi or mildew on the surface, which give the wood a dark gray, blotchy, and unsightly appearance. Highly-colored wood extractives in such species as western redcedar and redwood also influence the color of weathered wood. The dark brown color may persist for a long time in areas not exposed to the sun and where the

TABLE 10.—*Characteristics of woods for painting and finishing (omission in the table indicate inadequate data for classification)*

Wood	Ease of keeping well-painted I—easiest V—most exacting [1]	Weathering: Resistance to cupping 1—best 4—worst	Weathering: Conspicuousness of checking 1—least 2—most	Appearance: Color of heartwood (sapwood is always light)	Appearance: Degree of figure on flat-grained surface
SOFTWOODS					
Cedar:					
Alaska	I	1	1	Yellow	Faint
California incense	I			Brown	Do.
Port-Orford	I		1	Cream	Do.
Western redcedar	I	1	1	Brown	Distinct
White	I	1		Light brown	Do.
Cypress	I	1	1	do	Strong
Redwood	I	1	1	Dark brown	Distinct
Pine:					
Eastern white	II	2	2	Cream	Faint
Sugar	II	2	2	do	Do.
Western white	II	2	2	do	Do.
Ponderosa	III	2	2	do	Distinct
Fir, commercial white	III	2	2	White	Faint
Hemlock	III	2	2	Pale brown	Do.
Spruce	III	2	2	White	Do.
Douglas-fir (lumber and plywood)	IV	2	2	Pale red	Strong
Larch	IV	2	2	Brown	Do.
Pine:					
Norway	IV	2	2	Light brown	Distinct
Southern (lumber and plywood)	IV	2	2	do	Strong
Tamarack	IV	2	2	Brown	Do.
HARDWOODS					
Alder	III			Pale brown	Faint
Aspen	III	2	1	do	Do.
Basswood	III	2	2	Cream	Do.
Cottonwood	III	4	2	White	Do.
Magnolia	III	2		Pale brown	Do.
Poplar	III	2	1	do	Do.
Beech	IV	4	2	do	Do.
Birch	IV	4	2	Light brown	Do.
Gum	IV	4	2	Brown	Do.
Maple	IV	4	2	Light brown	Do.
Sycamore	IV			Pale brown	Do.
Ash	V or III	4	2	Light brown	Distinct
Butternut	V or III			do	Faint
Cherry	V or III			Brown	Do.
Chestnut	V or III	3	2	Light brown	Distinct
Walnut	V or III	3	2	Dark brown	Do.
Elm	V or IV	4	2	Brown	Do.
Hickory	V or IV	4	2	Light brown	Do.
Oak, white	V or IV	4	2	Brown	Do.
Oak, red	V or IV	4	2	do	Do.

[1] Woods ranked in group V for *ease of keeping well-painted* are hardwoods with large pores that need filling with wood filler for durable painting. When so filled before painting, the second classification recorded in the table applies.

extractives are not removed by rain.

With naturally weathered wood, it is important to avoid the unsightly effect of rusting nails. Iron nails rust rapidly and produce a severe brown or black discoloration. Because of this, only aluminum or stainless steel nails should be used for natural finishes.

Water-Repellent Preservatives

The natural weathering of wood may be modified by treatment with water-repellent finishes that contain a preservative (usually pentachlorophenol), a small amount of resin, and a very small amount of a water repellent which frequently is wax or waxlike in nature. The treatment, which penetrates the wood surface, retards the growth of mildew, prevents water staining of the ends of boards, reduces warping, and protects species that have a low natural resistance to decay. A clear, golden tan color can be achieved on such popular sidings as smooth or rough-sawn western redcedar and redwood.

The preservative solution can be easily applied by dipping, brushing, or spraying. All lap and butt joints, edges, and ends of boards should be liberally treated. Rough surfaces will absorb more solution than smoothly planed surfaces and be more durable.

The initial application to smooth surfaces is usually short-lived. When the surfaces start to show a blotchy discoloration due to extractives or mildew, clean them with detergent solution and re-treat following thorough drying. During the first 2 to 3 years, the finish may have to be applied every year or so. After weathering to uniform color, the treatments are more durable and need refinishing only when the surface becomes unevenly colored.

Pigmented colors can also be added to the water-repellent preservative solutions to provide special color effects. Two to six fluid ounces of colors-in-oil or tinting colors can be added to each gallon of treating solution. Light-brown colors which match the natural color of the wood and extractives are preferred. The addition of pigment to the finish helps to stabilize the color and increases the durability of the finish. In applying pigmented systems, a complete course of siding should be finished at one time to prevent lapping.

Pigmented Penetrating Stains

The pigmented penetrating stains are semitransparent, permitting much of the grain pattern to show through, and penetrate into the wood without forming a continuous film on the surface. Therefore, they will not blister, crack, or peel even if excessive moisture enters the wood.

Penetrating stains are suitable for both smooth and rough-textured surfaces; however, their performance is markedly improved if applied to rough-sawn, weathered, or rough-textured wood. They are especially effective on lumber and plywood that does not hold paint well, such as flat-grained surfaces of dense species. One coat of penetrating stains applied to smooth surfaces may last only 2 to 4 years, but the second application, after the surface has roughened by weathering, will last 8 to 10 years. A finish life of close to 10 years can be achieved initially by applying two coats of stain to rough-sawn surfaces. Two-coat staining is usually best for the highly adsorptive rough-sawn or weathered surfaces to reduce lapping or uneven stain application. The second coat should always be applied the same day as the first and before the first dries.

An effective stain of this type is the Forest Products Laboratory natural finish (*13*). This finish has a linseed oil vehicle; a fungicide, pentachlorophenol, that protects the oil from mildew; and a water repellent, paraffin wax, that protects the wood from excessive penetration of water. Durable red and brown iron oxide pigments simulate the natural colors of redwood and cedar. A variety of colors can be achieved with this finish, but the more durable ones are considered to be the red and brown iron oxide stains.

Paints for Exterior Wood

Of all the finishes, paints provide the most protection for wood against weathering and offer the widest selection of colors. A nonporous paint film retards penetration of moisture and reduces discoloration by wood extractives, paint peeling, and checking and warping of the wood. Paint, however, is *not* a preservative; it will not prevent decay if conditions are favorable for fungal growth. Original and maintenance costs are usually higher for a paint finish than for a water-repellent preservative or penetrating stain finish.

The durability of paint coatings on exterior wood is affected both by variables in the wood surface and type of paint.

Application

Exterior wood surfaces can be very effectively painted by following a simple 3-step procedure:

Step 1. Water-repellent preservative treatment.—make sure wood siding and trim have been treated with water-repellent preservative to protect them against the entrance of rain and heavy dew at joints. If treated exterior woodwork was not installed, treat it by brushing or spraying in place. Care should be taken to brush well into lap and butt joints, especially retreating cut ends. Allow 2 warm, sunny days for adequate drying of the treatment before painting.

Step 2. Primer.—New wood should be given three coats of paint. The first, or prime, coat is the most important and should be applied soon after the woodwork is erected; topcoats should be applied within 2 days to 2 weeks. Use a nonporous linseed oil primer free of zinc pigments (Federal Specification TT-P-25). Apply enough primer to obscure the wood grain. Many painters tend to spread primer too thinly. For best results, follow the spreading rates recommended by the manufacturer, or approximately 400 to 450 square feet per gallon for a paint that is about 85 percent solids by weight. A properly applied coat of a nonporous house paint primer will greatly reduce moisture blistering, peeling, and staining of paint by wood extractives.

The wood primer is *not* suitable for galvanized iron.

Allow such surfaces to weather for several months and then prime with an appropriate primer, such as a linseed oil or resin vehicle pigmented with metallic zinc dust (about 80 pct.) and zinc oxide (about 20 pct.).

Step 3. Finish Coats.—Keep the following points in mind when applying topcoats over the primer on new wood and galvanized iron:

(1) Use two coats of a wood-quality latex, alkyd, or oil-base house paint over the nonporous primer. This is particularly important for areas that are fully exposed to the weather, such as the south side of a house.

(2) To avoid future separation between coats of paint, or intercoat peeling, apply the first topcoat within 2 weeks after the primer and the second within 2 weeks of the first.

(3) To avoid temperature blistering, *do not* apply oil-base paints on a cool surface that will be heated by the sun within a few hours. Follow the sun around the house. Temperature blistering is most common with thickly applied paints of dark colors. The blisters usually show up in the last coat of paint and occur within a few hours to 1 or 2 days after painting. They do not contain water.

(4) To avoid the wrinkling, fading, or loss of gloss of oil-base paint, and streaking of latex paints, do not paint in the evenings of cool spring and fall days when heavy dews are frequent before the surface of the paint has dried.

Repainting

(1) Repaint only when the old paint has worn thin and no longer protects the wood. Faded or dirty paint can often be freshened by washing. Where wood surfaces are exposed, spot prime with a zinc-free linseed oil primer before applying the finish coat. Too-frequent repainting produces an excessively thick film that is more sensitive to the weather and also likely to crack abnormally across the grain of the paint. The grain of the paint is in the direction of the last brush strokes. Complete paint removal is the only cure for cross-grain cracking.

(2) Use the same brand and type of paint originally applied for the topcoat. A change is advisable only if a paint has given trouble. When repainting with latex paint, apply a nonporous, oil-base primer overall before applying the latex paint.

(3) To avoid intercoat peeling, which indicates a weak bond between coats of paint, clean the old painted surface well and allow no more than 2 weeks between coats in two-coat repainting. Do not repaint sheltered areas, such as eaves and porch ceilings, every time the weathered body of the house is painted. In repainting sheltered areas, wash the old paint surface with trisodium phosphate or detergent solution to remove surface contaminants that will interfere with adhesion of the new coat of paint. Following washing, rinse sheltered areas with large amounts of water and let dry thoroughly before repainting. When intercoat peeling does occur, complete paint removal is the only satisfactory procedure.

Blistering and Peeling

When too much water gets into paint or wood, the paint may blister and peel. The moisture blisters normally appear first and peeling follows. But sometimes the paint peels without blistering. At other times the blisters go unnoticed. Moisture blisters usually contain water when they form, or soon afterward, and eventually dry out. Small blisters may disappear completely on drying; however, fairly large ones may leave a rough spot on the surface. If the blistering is severe, the paint may peel.

New, thin coatings are more likely to blister because of too much moisture under them than old, thick coatings. The older and thicker coatings are too rigid to stretch, as they must do to blister, and so are more prone to cracking and peeling.

House construction features that will *minimize* water damage of outside paint are: (a) Wide roof overhang, (b) wide flashing under shingles at roof edges, (c) effective vapor barriers, (d) adequate eave troughs and properly hung downspouts, and (e) adequate ventilation of the house. If these features are lacking in a new house, persistent blistering and peeling may occur.

Discoloration by Extractives

Water-soluble color extractives occur naturally in western redcedar and redwood. It is to these substances that the heartwood of these two species owes its attractive color, good stability, and natural decay resistance. Discoloration occurs when the extractives are dissolved and leached from the wood by water. When the solution of extractives reaches the painted surface, the water evaporates, leaving the extractives as a reddish-brown stain. The water that gets behind the paint and causes moisture blisters also causes migration of extractives. The discoloration produced by water wetting the siding from the back side frequently forms a rundown or streaked pattern.

The emulsion paints and the so-called "breather" or low-luster oil paints are more porous than conventional oil paints. If these are used on new wood without a good oil primer, or if any paint is applied too thinly on new wood (a skimpy two-coat paint job, for example), rain or even heavy dew can penetrate the coating and reach the wood. When the water dries from the wood, the extractives are brought to the surface of the paint. Discoloration of paint by this process forms a diffused pattern.

On rough surfaces, such as shingles, machine-grooved shakes, and rough-sawn lumber sidings, it is difficult to obtain an adequately thick coating on the high points. Therefore, extractive staining is more likely to occur on such surfaces by water penetrating through the coating. But the reddish-brown extractives will be less conspicuous if dark-colored paints are used.

Effect of Impregnated Preservatives on Painting

Wood treated with the water-soluble preservatives in common use can be painted satisfactorily after it is redried. The coating may not last quite as long as it would have on untreated wood, but there is no vast difference. Certainly, a slight loss in durability is not enough to offer any practical objection to using treated wood where preservation against decay is necessary, protection against weathering desired. and appearance of painted wood important. When such treated wood is used indoors in textile or pulpmills. or other places where the relative humidity may be above 90 percent for long periods. paint may discolor or preservative solution exude. Coal-tar creosote or other dark oily preservatives tend to stain through paint unless the treated wood has been exposed to the weather for many months before it is painted.

Wood treated with oilborne, chlorinated phenols can be painted only when the solvent oils have evaporated completely from the treated wood. If volatile solvents that evaporate rapidly are used for the treating solution, such as in water-repellent preservatives, painting can be done only after the treated wood has dried.

Finishes for Interior Woodwork

Interior finishing differs from exterior chiefly in that interior woodwork usually requires much less protection against moisture, more exacting standards of appearance, and a greater variety of effects. Good interior finishes used indoors should last much longer than paint coatings on exterior surfaces. Veneered panels and plywood, however, present special finishing problems because of the tendency of these wood constructions to surface check.

Opaque Finishes

Interior surfaces may be painted with the materials and by the procedures recommended for exterior surfaces. As a rule, however, smoother surfaces, better color, and a more lasting sheen are demanded for interior woodwork, especially the wood trim; therefore, enamels or semigloss enamels rather than paints are used.

Before enameling, the wood surface should be extremely smooth. Imperfections, such as planer marks, hammer marks, and raised grain, are accentuated by enamel finish. Raised grain is especially troublesome on flat-grained surfaces of the heavier softwoods because the hard bands of summerwood are sometimes crushed into the soft springwood in planing, and later are pushed up again when the wood changes in moisture content. It is helpful to sponge softwoods with water, allow them to dry thoroughly, and then sandpaper them lightly with sharp sandpaper before enameling. In new buildings, woodwork should be allowed adequate time to come to its equilibrium moisture content before finishing.

For hardwoods with large pores, such as oak and ash, the pores must be filled with wood filler before the priming coat is applied. The priming coat for all woods may be the same as for exterior woodwork, or special priming paints may be used. Knots in the white pines. ponderosa pine, or southern yellow pine should be shellacked or sealed with a special knot sealer after the priming coat is dry. A coat of knot sealer is also sometimes necessary over wood of white pines and ponderosa pine to prevent pitch exudation and discoloration of light-colored enamels by colored matter apparently present in the resin of the heartwood of these species.

One or two coats of enamel undercoat are next applied; this should completely hide the wood and also present a surface that can easily be sandpapered smooth. For best results, the surface should be sandpapered before applying the finishing enamel; however, this operation is sometimes omitted. After the finishing enamel has been applied, it may be left with its natural gloss, or rubbed to a dull finish. When wood trim and paneling are finished with a flat paint, the surface preparation is not nearly as exacting.

Transparent Finishes

Transparent finishes are used on most hardwood and some softwood trim and paneling, according to personal preference. Most finishing consists of some combination of the fundamental operations of staining, filling, sealing. surface coating, or waxing. Before finishing, planer marks and other blemishes of the wood surface that would be accentuated by the finish must be removed.

Both softwoods and hardwoods are often finished without staining, especially if the wood is one with a pleasing and characteristic color. When used, however. stain often provides much more than color alone because it is absorbed unequally by different parts of the wood; therefore, it accentuates the natural variations in grain. With hardwoods, such emphasis of the grain is usually desirable; the best stains for the purpose are dyes dissolved either in water or in oil. The water stains give the most pleasing results, but raise the grain of the wood and require an extra sanding operation after the stain is dry.

The most commonly used stains are the "non-grain-raising" ones which dry quickly, and often approach the water stains in clearness and uniformity of color. Stains on softwoods color the springwood more strongly than the summerwood, reversing the natural gradation in color in a manner that is often garish. Pigment-oil stains, which are essentially thin paints, are less subject to this objection, and are therefore more suitable for softwoods. Alternatively, the softwood may be coated with clear sealer before applying the pigment-oil stain to give more nearly uniform coloring.

In hardwoods with large pores, the pores must be filled before varnish or lacquer is applied if a smooth coating is desired. The filler may be transparent and without effect on the color of the finish, or it may be colored to contrast with the surrounding wood.

Sealer (thinned out varnish or lacquer) is used to prevent absorption of subsequent surface coatings and prevent the bleeding of some stains and fillers into surface coatings, especially lacquer coatings. Lacquer sealers have the advantage of being very fast drying.

Transparent surface coatings over the sealer may be of gloss varnish, semigloss varnish, nitrocellulose lacquer, or wax. Wax provides a characteristic sheen without forming a thick coating and without greatly enhancing the natural luster of the wood. Coatings of a more resinous nature, especially lacquer and varnish, accentuate the natural luster of some hardwoods and seem to permit the observer to look down in the wood. Shellac applied by the laborious process of *French polishing* probably achieves this impression of depth most fully, but the coating is expensive and easily marred by water. Rubbing varnishes made with resins of high refractive index for light are nearly as effective as shellac. Lacquers have the advantages of drying rapidly and forming a hard surface, but require more applications than varnish to build up a lustrous coating.

Varnish and lacquer usually dry with a highly glossy surface. To reduce the gloss, the surfaces may be rubbed with pumice stone and water or polishing oil. Waterproof sandpaper and water may be used instead of pumice stone. The final sheen varies with the fineness of the powdered pumice stone, coarse powders making a dull surface and fine powders a bright sheen. For very smooth surfaces with high polish, the final rubbing is done with rotten-stone and oil. Varnish and lacquer made to dry to semigloss are also available.

Flat oil finishes are currently very popular. This type of finish penetrates the wood and forms no noticeable film on the surface. Two coats of oil are usually applied, which may be followed with a paste wax. Such finishes are easily applied and maintained but are more subject to soiling than a film-forming type of finish.

Filling Porous Hardwoods Before Painting

For finishing purposes, the hardwoods may be classified as follows:

Hardwoods with large pores	*Hardwoods with small pores*
Ash	Alder, red
Butternut	Aspen
Chestnut	Basswood
Elm	Beech
Hackberry	Cherry
Hickory	Cottonwood
Khaya (African mahogany)	Gum
Mahogany	Magnolia
Oak	Maple
Sugarberry	Poplar
Walnut	Sycamore

Birch has pores large enough to take wood filler effectively when desired, but small enough as a rule to be finished satisfactorily without filling.

Hardwoods with small pores may be finished with paints, enamels, and varnishes in exactly the same manner as softwoods. Hardwoods with large pores require wood filler before they can be covered smoothly with a film-forming finish. Without filler, the pores not only appear as depressions in the coating, but also become centers of surface imperfections and early failure.

Finishes for Floors

Interior Floors

Wood possesses a variety of properties that make it a highly desirable flooring material for home, industrial, and public structures. A variety of wood flooring products permit a wide selection of attractive and serviceable wood floors. Selection is available not only from a variety of different wood species and grain characteristics, but also from a considerable number of distinctive flooring types and patterns.

The natural color and grain of wood floors make them inherently attractive and beautiful. It is the function of floor finishes to enhance the natural beauty of wood, protect it from excessive wear and abrasion, and make the floors easier to clean. A complete finishing process may consist of four steps: Sanding the surface, applying a filler for certain woods, applying a stain to achieve a desired color effect, and applying a finish. Detailed procedures and specified materials depend largely on the species of wood used and individual preference in type of finish.

Careful sanding to provide a smooth surface is essential for a good finish because any irregularities or roughness in the base surface will be magnified by the finish. The production of a satisfactory surface requires sanding in several steps with progressively finer sandpaper, usually with a machine, unless the area is small. The final sanding is usually done with a 2/0 grade paper. When sanding is complete, all dust must be removed by vacuum cleaner or tack rag. Steel wool should not be used on floors unprotected by finish because minute steel particles left in the wood may later cause staining or discoloration.

A filler is required for wood with large pores, such as oak and walnut, if a smooth, glossy, varnish finish is desired. A filler may be paste or liquid, natural or colored. It is applied by brushing first across the grain and then by brushing with the grain. Surplus filler must be removed immediately after the glossy wet appearance disappears. Wipe first across the grain to pack the filler into the pores; then complete the wiping with a few light strokes with the grain. Filler should be allowed to dry thoroughly before the finish coats are applied.

Stains are sometimes used to obtain a more nearly uniform color when individual boards vary too much in their natural color. Stains may also be used to accent the grain pattern. If the natural color of the wood is acceptable, staining is omitted. The stain should be an oil-base or a non-grain-raising stain. Stains penetrate wood only slightly; therefore, the finish should be carefully maintained to prevent wearing through the stained layer. It is difficult to renew the stain at worn spots in a way that will match the color of the surrounding area.

Finishes commonly used for wood floors are classified either as sealers or varnishes. Sealers, which are usually thinned out varnishes, are widely used in residential flooring. They penetrate the wood just enough to avoid formation of a surface coating of appreciable thickness. Wax is usually applied over the sealer; however, if greater gloss is desired, the sealed floor makes an excellent base for varnish. The thin surface coat of sealer and wax needs more frequent attention than varnished surfaces. However, rewaxing or resealing and waxing of high traffic areas is a relatively simple maintenance procedure.

Varnish may be based on *phenolic*, *alkyd*, *epoxy*, or *polyurethane* resins. They form a distinct coating over the wood and give a lustrous finish. The kind of service expected usually determines the type of varnish. Varnishes especially designed for homes, schools, gymnasiums, and other public buildings are available. Information on types of floor finishes can be obtained from the flooring associations or the individual flooring manufacturers.

Durability of floor finishes can be improved by keeping them waxed. Paste waxes generally give the best appearance and durability. Two coats are recommended and, if a liquid wax is used, additional coats may be necessary to get an adequate film for good performance.

Porches and Decks

Exposed flooring on porches and decks is commonly painted. The recommended procedure of treating with water-repellent preservative and primer is the same as for wood siding. After the primer, an undercoat and matching coat of porch and deck enamel should be applied.

Many fully exposed rustic-type decks are effectively finished with only water-repellent preservative or a penetrating-type pigmented stain (*13*). Because these finishes penetrate and form no film on the surface, they de not crack and peel. They may need more frequent refinishing than painted surfaces, but this is easily done because there is no need for laborious surface preparation as when painted surfaces start to peel.

Moisture-excluding Effectiveness of Coatings

The protection afforded by coatings in excluding moisture vapor from wood depends on a number of variables. Among them are film thickness, absence of defects and voids in the film, type of pigment, type of vehicle, volume ratio of pigment to vehicle, vapor pressure gradient across the film, and length of exposure period.

The relative effectiveness of several typical treating and finishing systems for wood in retarding adsorption of water vapor at 97 percent relative humidity is compared in table 11. Perfect protection, or no adsorption of water, would be represented by 100 percent effectiveness; complete lack of protection (as with unfinished wood) by 0 percent.

Paints which are porous, such as the latex paints and low-luster or breather-type oil-base paints formulated at a pigment volume concentration above 40 percent, afford little protection against moisture vapor. These porous paints also permit rapid entry of water and so provide little protection against dew and rain unless applied over a nonporous primer.

CHAPTER 29

PROTECTION AGAINST DECAY AND TERMITES

Wood used under conditions where it will always be dry, or even where it is wetted briefly and rapidly redried, will not decay. However, all wood and wood products in construction use are susceptible to decay if kept wet for long periods under temperature conditions favorable to the growth of decay organisms. Most

TABLE 11.—*Some typical values of moisture-excluding effectiveness of coatings after 2 weeks' exposure of wood initially conditioned from 80° F. and 65 percent relative humidity to 80° F. and 97 percent relative humidity*

Coatings: Type	Number of coats	Effectiveness	Coatings: Type	Number of coats	Effectiveness
		Pct.			*Pct.*
INTERIOR COATINGS			EXTERIOR COATINGS		
Uncoated wood		0	Water-repellent preservative	1	0
Latex paint	2	0	FPL natural finish (stain)	1	0
Floor seal	2	0	Exterior latex paint	2	3
Floor seal plus wax	2	10	House paint primer:	1	20
Linseed oil	1	1	Plus latex paint	2	22
Do.	2	5	Plus titanium-zinc linseed oil paint (low luster oil base) (30 pct. PVC) [1]	1	65
Do.	3	21			
Furniture wax	3	8	Titanium-alkyd oil:		
Phenolic varnish	1	5	30 pct. PVC [1]	1	45
Do.	2	49	40 pct. PVC [1]	1	3
Do.	3	73	50 pct. PVC [1]	1	0
Semigloss enamel	2	52	Aluminum powder in long oil phenolic varnish	1	39
Cellulose lacquer	3	73	Do.	2	88
Lacquer enamel	3	76	Do.	3	95
Shellac	3	87			

[1] PVC (pigment volume concentration) is the volume of pigment, in percent, in the nonvolatile portion of the paint.

of the wood as used in a house is not subjected to such conditions. There are places where water can work into the structure, but such places can be protected. Protection is accomplished by methods of design and construction, by use of suitable materials, and in some cases by using treated material.

Wood is also subject to attack by termites and some other insects. Termites can be grouped into two main classes—*subterranean* and *dry-wood.* Subterranean termites are important in the northernmost States where serious damage is confined to scattered, localized areas of infestation (fig. 178). Buildings may be fully protected against subterranean termites by incorporating comparatively inexpensive protection measures during construction. The Formosan subterranean termite has recently (1966) been discovered in several locations in the South. It is a serious pest because its colonies contain large numbers of the worker caste and cause damage rapidly. Though presently in localized areas, it could spread to other areas. Controls are similar to those for other subterranean species. Dry-wood termites are found principally in Florida, southern California, and the Gulf Coast States. They are more difficult to control, but the damage is less serious than that caused by subterranean termites.

Wood has proved itself through the years to be desirable and satisfactory as a building material. Damage from decay and termites has been small in proportion to the total value of wood in residential structures, but it has been a troublesome problem to many homeowners. With changes in building-design features and use of new building materials, it becomes pertinent to restate the basic safeguards to protect buildings against both decay and termites.

Decay

Wood decay is caused by certain fungi that can utilize wood for food. These fungi, like the higher plants, require air, warmth, food, and moisture for growth. Early stages of decay caused by these fungi may be accompanied by a discoloration of the wood. Paint also may become discolored where the underlying wood is rotting. Advanced decay is easily recognized because the wood has then undergone definite changes in properties and appearance. In advanced stages of building decay, the affected wood generally is brown and crumbly, and sometimes may be comparatively white and spongy. These changes may not be apparent on the surface, but the loss of sound wood inside often is reflected by sunken areas on the surface or by a "hollow" sound when the wood is tapped with a hammer. Where the surrounding atmosphere is very damp, the decay fungus may grow out on the surface—appearing as white or brownish growths in patches or strands or in special cases as vine-like structures.

Fungi grow most rapidly at temperatures of about 70° to 85° F. Elevated temperatures such as those used in kiln-drying of lumber kill fungi, but low temperatures, even far below zero, merely cause them to remain dormant.

Moisture requirements of fungi are within definite

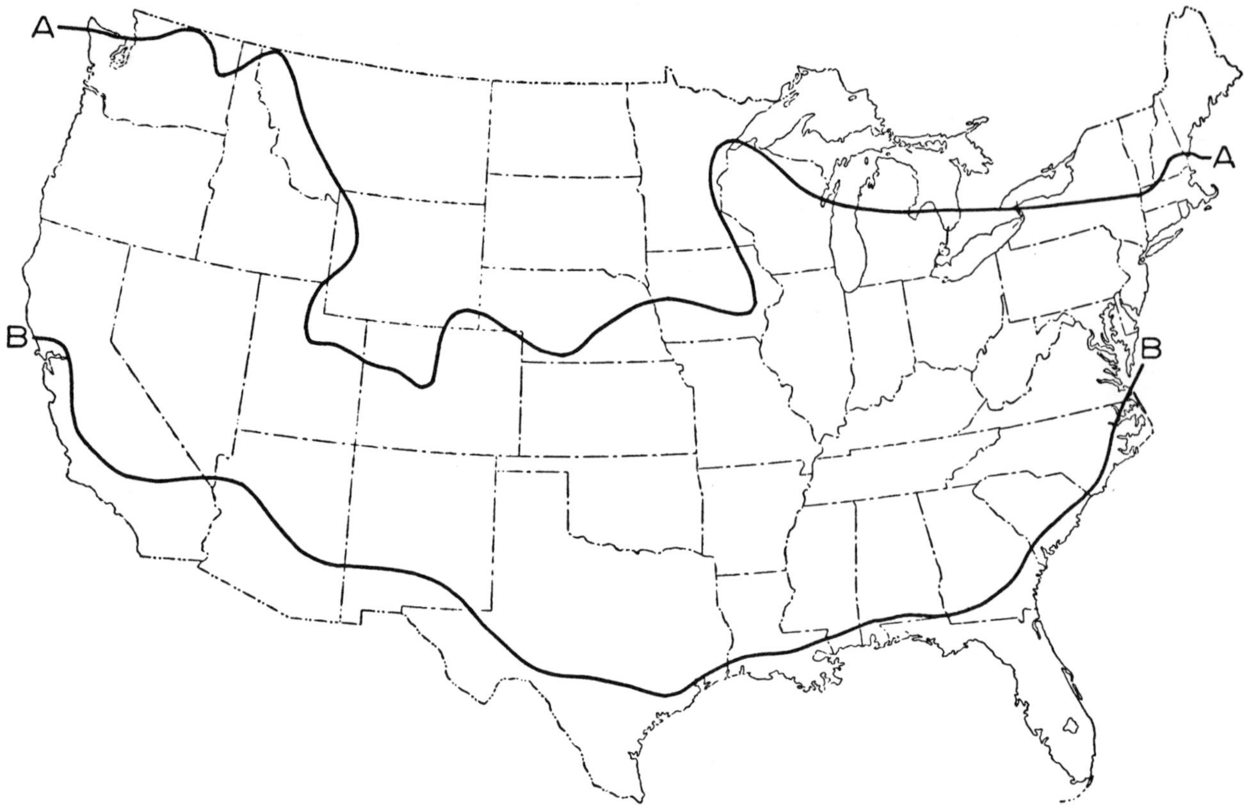

Figure 178—The northern limit of damage in the United States by subterranean termites, Line *A*; by dry-wood or nonsubterranean termites, line *B*.

limitations. Wood-destroying fungi will not become established in dry wood. A moisture content of 20 percent (which can be determined with an electrical moisture meter) is safe. Moisture contents greater than this are practically never reached in wood sheltered against rain and protected, if necessary, against wetting by condensation or fog. Decay can be permanently arrested by simply taking measures to dry out the infected wood and to keep it dry. Brown crumbly decay, in the dry condition, is sometimes called "dry rot," but this is a misnomer. Such wood must necessarily be damp if rotting is to occur.

The presence of mold or stain fungi should serve as a warning that conditions are or have been suitable for decay fungi. Heavily molded or stained lumber, therefore, should be examined for evidence of decay. Furthermore, such discolored wood is not entirely satisfactory for exterior millwork because it has greater water absorptiveness than bright wood.

The natural decay resistance of all common native species of wood lies in the heartwood. When untreated, the sapwood of all species has low resistance to decay and usually has a short life under decay-producing conditions. Of the species of wood commonly used in house construction, the heartwood of baldcypress, redwood, and the cedars is classified as being highest in decay resistance. All-heartwood, quality lumber is becoming more and more difficult to obtain, however, as increasing amounts of timber are cut from the smaller trees of second-growth stands. In general, when substantial decay resistance is needed in load-bearing members that are difficult and expensive to replace, appropriate preservative-treated wood is recommended.

Subterranean Termites

Subterranean termites are the most destructive of the insects that infest wood in houses. The chance of infestation is great enough to justify preventive measures in the design and construction of buildings in areas where termites are common.

Subterranean termites are common throughout the southern two-thirds of the United States except in mountainous and extremely dry areas.

One of the requirements for subterranean-termite life is the moisture available in the soil. These termites become most numerous in moist, warm soil containing an abundant supply of food in the form of wood or other cellulosic material. In their search for additional food (wood), they build earthlike shelter tubes over foundation walls or in cracks in the walls, or on pipes or supports leading from the soil to the house. These tubes are from 1/4 to 1/2 inch or more in width and flattened, and serve to protect the termites in their travels between food and shelter.

Since subterranean termites eat the interior of the wood, they may cause much damage before they are discovered. They honeycomb the wood with definite tunnels that are separated by thin layers of sound wood. Decay fungi, on the other hand, soften the wood and eventually cause it to shrink, crack, and crumble without producing anything like these continuous tunnels. When both decay fungi and subterranean termites are present in the same wood, even the layers between the termite tunnels will be softened.

Dry-wood Termites

Dry-wood termites fly directly to and bore into the wood instead of building tunnels from the ground as do the subterranean termites. Dry-wood termites are common in the tropics, and damage has been recorded in the United States in a narrow strip along the Atlantic Coast from Cape Henry, Va., to the Florida Keys, and westward along the coast of the Gulf of Mexico to the Pacific Coast as far as northern California (fig. 178). Serious damage has been noted in southern California and in localities around Tampa, Miami, and Key West, Fla. Infestations may be found in structural timber and other woodwork in buildings, and also in furniture, particularly where the surface is not adequately protected by paint or other finishes.

Dry-wood termites cut across the grain of the wood and excavate broad pockets, or chambers, connected by tunnels about the diameter of the termite's body. They destroy both springwood and the usually harder summerwood, whereas subterranean termites principally attack springwood. Dry-wood termites remain hidden in the wood and are seldom seen, except when they make dispersal flights.

Safeguards Against Decay

Except for special cases of wetting by condensation or fog, a dry piece of wood, when placed off the ground under a tight roof with wide overhang, will stay dry and never decay. This principle of "umbrella protection," when applied to houses of proper design and construction, is a good precaution. The use of dry lumber in designs that will keep the wood dry is the simplest way to avoid decay in buildings.

Most of the details regarding wood decay have been included in earlier chapters, but they are given here as a reminder of their relationship to protection from decay and termites.

Untreated wood should *not* come in contact with the soil. It is desirable that the foundation walls have a clearance of at least 8 inches above the exterior finish grade, and that the floor construction have a clearance of 18 inches or more from the bottom of the joists to the ground in basementless spaces. The foundation should be accessible at all points for inspection. Porches that prevent access should be isolated from the soil by concrete or from the building proper by metal barriers or aprons (fig. 179).

Steps and stair carriages, posts, wallplates, and sills should be insulated from the ground with concrete or masonry. Sill plates and other wood in contact with concrete near the ground should be protected by a moistureproof membrane, such as heavy roll roofing or 6-mil polyethylene. Girder and joists openings in masonry walls should be big enough to assure an air space around the ends of these members.

Design Details

Surfaces like steps, porches, door and window frames, roofs, and other protections should be sloped to promote runoff of water (Ch. 25, "Porches and Garages"). Noncorroding flashing should be used around chimneys, windows, doors, or other places where water might seep in (Ch. 24, "Flashing and Sheet Metal"). Roofs with considerable overhang give added protection to the siding and other parts of the house. Gutters and downspouts should be placed and maintained to divert water away from the buildings. Porch columns and screen rails should be shimmed above the floor to allow quick drying, or posts should slightly overhang raised concrete bases (Ch. 25, "Porches and Garages").

Exterior steps, rails, and porch floors exposed to rain need protection from decay, particularly in warm, damp parts of the country. Pressure treatment of the wood in accordance with the recommendation of Federal Specification TT–W–571 provides a high degree of protection against decay and termite attack (*12*). Where the likelihood of decay is relatively small or where pressure-treated wood is not readily obtainable, on-the-job application of water-repellent preservatives by dipping or soaking has been found to be worthwhile. The wood should by dry, cut to final dimensions, and then dipped or soaked in the preservative solution (*19*). Soaking is the best of these nonpressure methods, and the ends of the boards should be soaked for a minimum of 3 minutes. It is important to protect the end grain of wood at joints, for this area absorbs water easily and is the most common infection point. The edges of porch flooring should be coated with thick white lead or other durable coating as it is laid.

Leaking pipes should be remedied immediately to prevent damage to the house, as well as to guard against possible decay.

Green or Partially Seasoned Lumber

Construction lumber that is green or partially seasoned may be infected with one or more of the stain-

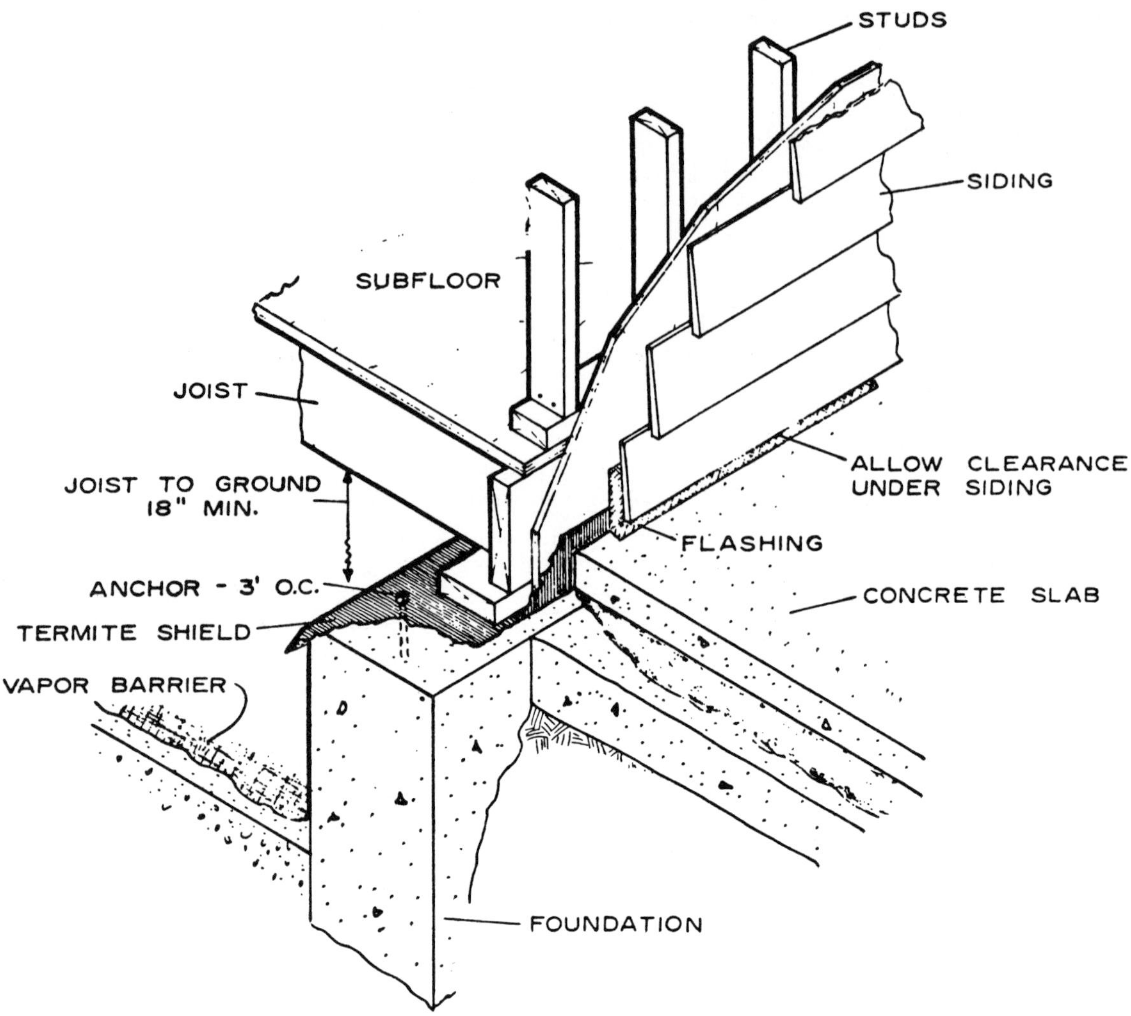

Figure 179.—Metal shield used to protect wood at porch slab.

ing, molding, or decay fungi and should be avoided. Such wood may contribute to serious decay in both the substructure and exterior parts of buildings. If wet lumber must be used, or if wetting occurs during construction, the wood should not be fully enclosed or painted until thoroughly dried.

Water Vapor from the Soil

Crawl spaces of houses built on poorly-drained sites may be subjected to high humidity. During the winter when the sills and outer joists are cold, moisture condenses on them and, in time, the wood absorbs so much moisture that it is susceptible to attack by fungi. Unless this moisture dries out before temperatures favorable for fungus growth are reached, considerable decay may result. However, this decay may progress so slowly that no weakening of the wood becomes apparent for a few years. Placing a layer of 45 pound or heavier roll roofing or a 6-mil sheet of polyethylene over the soil to keep the vapor from getting into the crawl space would prevent such decay. This might be recommended for all sites where, during the cold months, the soil is wet enough to be compressed in the hand.

If the floor is uninsulated, there is an advantage in closing the foundation vents during the coldest months from the standpoint of fuel savings. However, unless the crawl space is used as a heat plenum chamber, insulation is usually located between floor joists. The vents could then remain open. Crawl-space vents can be very small when soil covers are used; only 10 percent of the area required without covers (Ch. 16, "Ventilation").

Water Vapor from Household Activities

Water vapor is also given off during cooking, washing, and other household activities. This vapor can pass through walls and ceilings during very cold weather and condense on sheathing, studs, and rafters, causing condensation problems. A vapor barrier of an approved type is needed on the warm side of walls. (See section on "Vapor Barriers," Ch. 15.) It is also important that the attic space be ventilated (*17*), as previously discussed in Chapter 16, "Ventilation."

Water Supplied by the Fungus Itself

In the warmer coastal areas principally, some substructure decay is caused by a fungus that provides its own needed moisture by conducting it through a vinelike structure from moist ground to the wood. The total damage caused by this water-conducting fungus is not large, but in individual instances it tends to be unusually severe. Preventive and remedial measures depend on getting the soil dry and avoiding untreated wood "bridges" such as posts between ground and sills or beams.

Safeguards Against Termites

The best time to provide protection against termites is during the planning and construction of the building. The first requirement is to remove all woody debris like stumps and discarded form boards from the soil at the building site before and after construction. Steps should also be taken to keep the soil under the house as dry as possible.

Next, the foundation should be made impervious to subterranean termites to prevent them from crawling up through hidden cracks to the wood in the building above. Properly reinforced concrete makes the best foundation, but unit-construction walls or piers capped with at least 4 inches of reinforced concrete are also satisfactory. No wood member of the structural part of the house should be in contact with the soil.

The best protection against subterranean termites is by treating the soil near the foundation or under an entire slab foundation. The effective soil treatments are water emulsions of aldrin (0.5 pct.), chlordane (1.0 pct.), dieldrin (0.5 pct.), and heptachlor (0.5 pct.). The rate of application is 4 gallons per 10 linear feet at the edge and along expansion joints of slabs or along a foundation. For brick or hollow-block foundations, the rate is 4 gallons per 10 linear feet for each foot of depth to the footing. One to 1½ gallons of emulsion per 10 square feet of surface area is recommended for overall treatment before pouring concrete slab foundations. Any wood used in secondary appendages, such as wall extensions, decorative fences, and gates, should be pressure-treated with a good preservative.

In regions where dry-wood termites occur, the following measures should be taken to prevent damage:

1. All lumber, particularly secondhand material, should be carefully inspected before use. If infected, discard the piece.
2. All doors, windows (especially attic windows), and other ventilation openings should be screened with metal wire with not less than 20 meshes to the inch.
3. Preservative treatment in accordance with Federal Specification TT–W–571 ("Wood Preservatives: Treating Practices," available through GSA Regional Offices) can be used to prevent attack in construction timber and lumber.
4. Several coats of house paint will provide considerable protection to exterior woodwork in buildings. All cracks, crevices, and joints between exterior wood members should be filled with a mastic calking or plastic wood before painting.
5. The heartwood of foundation-grade redwood, particularly when painted, is more resistant to attack than most other native commercial species.

Pesticides used improperly can be injurious to man, animals, and plants. Follow the directions and heed all precautions on the labels.

Store pesticides in original containers—out of reach of children and pets—and away from foodstuff.

Apply pesticides selectively and carefully. Do not apply a pesticide when there is danger of drift to other areas. Avoid prolonged inhalation of a pesticide spray or dust. When applying a pesticide it is advisable that you be fully clothed.

After handling a pesticide, do not eat, drink or smoke until you have washed. In case a pesticide is swallowed or gets in the eyes, follow the first aid treatment given on the label, and get prompt medical attention. If the pesticide is spilled on your skin or clothing, remove clothing immediately and wash skin thoroughly.

Dispose of empty pesticide containers by wrapping them in several layers of newspaper and placing them in your trash can.

It is difficult to remove all traces of a herbicide (weed killer) from equipment. Therefore, to prevent injury to desirable plants do not use the same equipment for insecticides and fungicides that you use for a herbicide.

NOTE: Registrations of pesticides are under constant review by the U. S. Department of Agriculture. Use only pesticides that bear the USDA registration number and carry directions for home and garden use.

CHAPTER 30

PROTECTION AGAINST FIRE

Fire hazards exist to some extent in nearly all houses. Even though the dwelling is of the best fire-resistant construction, hazards can result from occupancy and the presence of combustible furnishings and other contents.

The following tabulation showing the main causes of fires in one- and two-family dwellings is based on an analysis (16) made of 500 fires by the National Fire Protection Association.

Cause of fire	*Percent of Total*
Heating equipment	23.8
Smoking materials	17.7
Electrical	13.8
Children and matches	9.7
Mishandling of flammable liquids	9.2
Cooking equipment	4.9
Natural gas leaks	4.4
Clothing ignition	4.2
Combustibles near heater	3.6
Other miscellaneous	8.7
	100.0

Fire-protection engineers generally recognize that a majority of fires begin in the contents, rather than in the building structure itself. Proper housekeeping and care with smoking, matches, and heating devices can reduce the possibility of fires. Other precautions to reduce the hazards of fires in dwellings—such as fire stops, spacing around heating units and fireplaces, and protection over furnaces—are recommended elsewhere in this handbook.

Fire Stops

Fire stops are intended to prevent drafts that foster movement of hot combustible gases from one area of the building to another during a fire. Exterior walls of wood-frame construction should be fire-stopped at each floor level (figs. 26, 35, and 38), at the top-story ceiling level, and at the foot of the rafters.

Fire stops should be of noncombustible materials or wood not less than 2 inches in nominal thickness. The fire stops should be placed horizontally and be well fitted to completely fill the width and depth of the spacing. This applies primarily to balloon-type frame construction. Platform walls are constructed with top and bottom plates for each story (fig. 31). Similar fire stops should be used at the floor and ceiling of interior stud partitions, and headers should be used at the top and bottom of stair carriages (fig. 151).

Noncombustible fillings should also be placed in any spacings around vertical ducts and pipes passing through floors and ceilings, and self-closing doors should be used on shafts, such as clothes chutes.

When cold-air return ducts are installed between studs or floor joists, the portions used for this purpose should be cut off from all unused portions by tight-fitting stops of sheet metal or wood not less than 2 inches in nominal thickness. These ducts should be constructed of sheet metal or other materials no more flammable than 1-inch (nominal) boards.

Fire stops should also be placed vertically and horizontally behind any wainscoting or paneling applied over furring, to limit the formed areas to less than 10 feet in either dimension.

With suspended ceilings, vertical panels of noncombustible materials from lumber of 2-inch nominal thickness or the equivalent, should be used to subdivide the enclosed space into areas of less than 1,000 square feet. Attic spaces should be similarly divided into areas of less than 3,000 square feet.

Chimney and Fireplace Construction

The fire hazards within home construction can be reduced by insuring that chimney and fireplace constructions are placed in proper foundations and properly framed and enclosed. (See Ch. 26, "Chimneys and Fireplaces.") In addition, care should be taken that combustibles are not placed too close to the areas of high temperature. Combustible framing should be no closer than 2 inches to chimney construction; however, when required, this distance can be reduced to ½ inch, provided the wood is faced with a ¼-inch-thick asbestos sheeting.

For fireplace construction, wood should not be placed closer than 4 inches from the backwall nor within 8 inches of either side or top of the fireplace opening. When used, wood mantels should be located at least 12 inches from the fireplace opening.

Heating Systems

Almost 25 percent of fires are attributed to faulty construction or to improper use of heating equipment, and the greater proportion of these fires originate in the basement. Combustible products should not generally be located nearer than 24 inches from a hot air, hot water, or steam heating furnace; however, this distance can be reduced in the case of properly insulated furnaces or when the combustible materials are protected by gypsum board, plaster, or other materials with low flame spread. Most fire-protection agencies limit to 170° F. the temperature to which combustible wood products should be exposed for long periods of time, although experimentally, ignition does not occur until much higher temperatures have been reached.

In confining a fire to the basement of a home, added protection can be obtained with gypsum board, asbestos board, or plaster construction on the basement ceiling, either as the exterior surface or as backings for decorative materials. These ceiling surfaces are frequently omitted to reduce costs, but particular attention should be given to protection of the wood members directly above and near the furnace.

Flame Spread and Interior Finish

In some areas of a building, flame-spread ratings are assigned to limit spread of fire on wall and ceiling surfaces. Usually, these requirements do *not* apply to private dwellings because of their highly combustible content, particularly in furnishings and drapes usually found in this type of structure.

To determine the effect of the flammability of wall linings on the fire hazards within small rooms, burnout tests have been made at the Forest Products Laboratory in Madison, Wis. For this purpose, an 8- by 12- by 8-foot high room was furnished with an average amount of furniture and combustible contents. This room was lined with various wall panel products, plywood, fiber insulation board, plaster on fiberboard lath, and gypsum wallboard. When a fire was started in the center of this room, the time to reach the *critical* temperature (when temperature rise became very rapid) or the flashover temperature (when everything combustible burst into flames) was not significantly influenced by either combustible or noncombustible wall linings. In the time necessary to reach these critical temperatures (usually less than 10 minutes) the room would already be unsafe for human occupancy.

Similar recent tests in a long corridor, partially ventilated, showed that the "flashover" condition would develop for 60 to 70 feet along a corridor ceiling within 5 to 7 minutes from the burning of a small amount of combustible contents. This "flashover" condition developed in approximately the same time, whether combustible or noncombustible wall linings were used, and before any appreciable flame spread along wall surfaces.

Wood paneling, treated with fire-retardant chemicals or fire-resistant coatings as listed by the Underwriters' Laboratory, Inc. or other recognized testing laboratories, can also be used in areas where flame-spread resistance is especially critical. Such treatments, however, are not considered necessary in dwellings, nor can the extra cost of treatment be justified.

Fire-resistant Walls

Whenever it is desirable to construct fire-resistant walls and partitions in attached garages and heating rooms, information on fire resistance ratings using wood and other materials is readily available through local code authorities. Wood construction assemblies can provide ½ hour to 2 hour fire resistance under recognized testing methods, depending on the covering material.

CHAPTER 31

METHODS OF REDUCING BUILDING COSTS

The average homebuilder is interested in reducing the overall cost of his house but not at the expense of its livability or resale value. This is often somewhat difficult to do for a single custom-built house.

Operators of large housing developments often build hundreds of houses each year. Because of their need for huge volumes of materials, they buy direct from the manufacturer. They also develop the building sites from large sections of land. Much of the work, such as installation of the roofing, application of gypsum board interiors, and painting, is done by subcontractors. Their own crews are specialists, each crew becoming proficient in its own phase of the work. Central shops are established where all material is cut to length and often preassembled before being trucked to the site. These methods reduce the cost of the individual house in a large building project, but few of them can be applied to an individual house built by the owner.

If a home builder pays attention to various construction details and to the choice of materials, however, this information will usually aid in reducing costs. The following suggestions are intended as possible ways for the owner to lower the cost of his house.

Design

The first area where costs of the house may be reduced somewhat is during the design stages. However, such details should not affect the architectural lines or appearance of the house, but rather the room arrangement and other factors. The following design elements might be considered before final plans are chosen:

1. The size of the house, width and length, should be such that standard-length joists and rafters and standard spacings can be used without wasting material. The architect or contractor will have this informa-

tion available. Also reflected in the house size is the use of standard-width sheets of sheathing materials on the exterior as well as in the interior. Any waste or ripping required adds to both the labor and material costs.

The rooms should be arranged so that the plumbing, water, and heating lines are short and risers can serve more than one room. An "expandable" house may mean the use of a steeper pitched roof to provide space for future rooms in the attic area. It might also be desirable to include second-floor dormers in the original design. Additional rooms can thus be provided at a much lower cost than by adding to the side or rear of the house at a future date. Roughing in plumbing and heating lines to the second floor will also reduce future costs when the second floor is completed, yet not add appreciably to the original construction costs.

While a rectangular plan is the most economical from many standpoints, it should not always govern final design. A rectangular plan of the house proper, with a full basement, can be made more desirable by a garage or porch wing of a different size or alinement. Such attachments require only shallow footings, without the excavation necessary for basement areas.

2. The type of foundation to be used, such as slab, crawl space, or basement, is an important consideration. Base this selection on climatic conditions and needs of the family for storage, hobby, or recreation space. While space in the basement is not so desirable as in areas above grade, its cubic-foot cost is a great deal lower. The design of a slab-type house usually includes some additional space for heating, laundry, and storage. This extra area may often cost as much as a full basement. Many multilevel houses include habitable rooms over concrete slabs as well as a full-basement. Consult local architects or contractors for their opinions on the most desirable type of home in your area from the cost standpoint.

3. Many contemporary house designs include a flat or low-pitched roof which allows one type of member to serve as both ceiling joists and rafters. This generally reduces the cost compared to that of a pitched roof, both in materials and labor. However, all styles of houses are not adaptable to such a roof. Many contractors incur savings by using preassembled roof trusses for pitched roofs. Dealers who handle large quantities of lumber are usually equipped to furnish trusses of this type.

Pitched roofs are of *gable* or *hip* design, with the *gambrel* roof a variation of each. While the hip roof is somewhat more difficult to frame than the gable roof, it usually requires less trim and siding. Furthermore, painting is much simpler in the hip roof because of less wall area by elimination of the gable and because of accessibility. In the gambrel roof, which is adapted to two-story houses, roof shingles serve also as siding over the steep-pitched portions. Furthermore, a roof of this type provides a greater amount of headroom (perhaps the original purpose of this design) than the common gable type.

Choice of Materials

The type and grade of materials used in a house can vary greatly and savings can be effected in their choice. It is poor practice to use a low grade or an inferior material which could later result in excessive maintenance costs. On the other hand, it is not economical to use a material of too high a grade when not needed for strength or appearance (5).

Several points might be considered as a means of reducing costs. (Your contractor or lumber dealer who is familiar with these costs will aid you in your final selection.)

1. Consider the use of concrete blocks for foundation walls as opposed to the use of poured concrete. It is less costly to provide a good water-resistant surface on a poured wall than on a block wall. On the other hand, a common hollow concrete block has better insulating properties than a poured concrete wall of equal thickness. Costs often vary by areas.

2. If *precast* blocks are available, consider them for chimneys. These blocks are made to take flue linings of varied sizes and are laid up more rapidly than brick. Concrete block units are also used in laying up the base for a first-floor fireplace, rather than bricks. Prefabricated, lightweight chimneys that require no masonry may also save money.

3. Dimension material varies somewhat in cost by species and grades. Use the better grades for joists and rafters and the lower cost grades for studs. Do not use better grades of lumber than are actually needed. Conversely, grades that involve excessive cutting and selection would dissipate the saving by increased labor costs. Proper moisture content is an important factor.

4. Conventional items such as cabinets, moldings, windows, and other millwork, which are carried as stock or can be easily ordered, also reduce costs. Any special, nonstandard materials which require extra machine setups will be much more expensive. This need not restrict the homebuilder in his design, however, as there are numerous choices of millwork components from many manufacturers.

5. The use of a single material for wall and floor covering will provide a substantial saving. A combination subfloor underlayment of 5/8- or 3/4-inch tongued and grooved plywood will serve both as subfloor and as a base for resilient tile or similar material, as well as for carpeting. Panel siding consisting of 4-foot-wide full-height sheets of plywood or similar material serves both as sheathing and a finish siding. For example,

exterior particleboard with a painted finish and corner bracing on the stud wall may also qualify as a panel siding. Plywood may be obtained with a paper overlay, as well as rough sawn, striated, reverse board and batten, brushed, and other finishes.

6. In planning a truly low-cost house where each dollar is important, a crawl space design with the use of a treated wood post foundation is worth investigating. This construction utilizes treated wood foundation posts bearing on concrete footings. The post support floor beams upon which the floor joists rest. A variation of this design includes spacing of the beams on 48-inch centers and the use of $1\frac{1}{8}$-inch-thick tongue and groove plywood eliminating the need for joists as such.

7. Costs of exterior siding or other finish materials often vary a great deal. Many factory-primed sidings are available which require only finish coats after they are applied. A rough-sawn, low-grade cedar or similar species in board and batten pattern with a stained finish will often reduce the overall cost of exterior coverings. Many species and textures of plywood are available for the exterior. When these sheet materials are of the proper thickness and application, they might also serve as sheathing. Paintability of species is also important (*5*). Edge-grained boards or paper-overlaid plywood provide good bases for paint.

In applying all exterior siding and trim, galvanized or other rust-resistant nails reduce the need for frequent treatment or refinishing. Stainless steel or aluminum nails on siding having a natural finish are a must. Corrosion-resistant nails will add slightly to the cost but will save many dollars in reduced maintenance costs.

8. Interior coverage also deserves consideration. While gypsum board dry-wall construction may be lower in cost per square foot, it requires decorating before it can be considered complete; plaster walls do not require immediate decorating. These costs vary by areas, depending largely on the availability of the various trades. However, prefinished or plastic-faced gypsum board (available in a number of patterns) with a simple "V" joint or with a joint flap of the same covering, and the use of adhesive for application, will result in an economical wall and ceiling finish.

9. There are many cost-related considerations in the choice of flooring, trim, and other interior finish. Areas which will be fully carpeted do not require a finish floor. However, there is a trend to provide a finish floor under the carpeting. The replacement cost of the carpeting may be substantially greater than the cost of the original finish floor.

Species of trim, jambs, and other interior moldings vary from a relatively low-cost softwood to the higher cost hardwoods such as oak or birch. Softwoods are ordinarily painted, while the hardwoods have a natural finish or are lightly stained. The softwoods, though lower in cost, are less resistant to blows and impacts.

Another consideration is the selection of panel and flush doors. Flush doors can be obtained in a number of species and grades. Unselected gum, for example, might have a paint finish while the more costly woods are best finished with a varnish or sealer. Hollow-core flush doors are lower in cost and are satisfactory for interior use, but exterior flush doors should be solid core to better resist warping. The standard exterior panel door can be selected for many styles of architecture.

Construction

Methods of reducing construction costs are primarily based on reducing on-site labor time. The progressive contractor often accomplishes this in several ways, but the size of the operation generally governs the method of construction. A contractor might use two carpenter crews—one for framing and one for interior finishing. Close cooperation with the subcontractors—such as plumbers, plasterers, and electricians—avoids wasting time. Delivery of items when needed so that storage is not a problem also reduces on-site costs. Larger operators may preassemble components at a central shop to permit rapid on-site erection. While the small contractor building individual houses cannot always use the same cost-saving methods, he follows certain practices:

1. Power equipment, such as a radial-arm saw, skill saw, or an automatic nailer, aids in reducing the time required for framing and is used by most progressive contractors. Such equipment not only reduces assembly time for floor, wall, and roof framing and sheathing, but is helpful in applying siding and exterior and interior trim. For example, with a radial arm saw on the job, studs can be cut to length, headers and framing members prepared, and entire wall sections assembled on the floor and raised in place. Square cuts, equal lengths, and accurate layouts result in better nailing and more rigid joints.

2. Where a gypsum-board dry-wall finish is used, many contractors employ the horizontal method of application. This brings the taped joint below eye level, and large room-size sheets may be used. Vertical joints may be made at window or door openings. This reduces the number of joints to be treated and results in a better-looking wall.

3. Staining and painting of the exterior and interior surfaces and trim are important. For example, one cost study of interior painting indicated that prestaining of jambs, stops, casing, and other trim before application would result in a substantial saving. These are normally stained or sealed after they have been fitted and nailed.

4. During construction, the advantages of a simple

plan and the selection of an uncomplicated roof will be obvious. There will be less waste by cutting joists and rafters, and erection will be more rapid than on a house where intricate construction is involved.

CHAPTER 32

PROTECTION AND CARE OF MATERIALS AT THE BUILDING SITE

Many building contractors arrange for the materials needed for a house to be delivered just before construction begins. Perhaps the first load, after the foundation has been completed, would include all materials required for the wood-floor system. A second load, several days later, would provide the materials for framing and sheathing the walls, and a third load for roof and ceiling framing and roof sheathing. In this manner, storage of framing and sheathing materials on the site would not be as critical as when all materials were delivered at once. On the other hand, materials for factory-built or preassembled houses may be delivered in one large truckload, because a crew erects the house in a matter of hours. This practically eliminates the need for protection of materials on the site.

Protection Requirements

Unfortunately, the builder of a single house may not be able to have delivery coincide with construction needs. Thus, some type of protection may be required at the building site. This is especially true for such millwork items as window and door frames, doors, and moldings. Finished cabinets, floor underlayment, flooring, and other more critical items should be delivered only after the house is enclosed and can provide complete protection from the weather.

During fall, winter, and spring months, the interior of the house should be heated so that finished wood materials will not be affected. Exposure to damp and cold conditions will change the dimensions of such materials as flooring and cause problems if they are installed at too high a moisture content. Thus, care of the materials after they arrive at the site and the conditions to which they are exposed are important to most materials used in house construction.

Protection of Framing Materials

In normal construction procedures, after excavation is complete, some dimension lumber and sheathing materials are delivered on the job. After delivery, it is the builder's responsibility to protect these materials against wetting and other damage. Structural and framing materials in place on a house before it is enclosed may become wet during a storm, but exposed surfaces can dry out quickly in subsequent dry weather without causing damage.

Lumber should *not* be stored in tight piles without some type of protection. Rather, if lumber is not to be used for several days or a week, it should be unloaded on skids with a 6-inch clearance above the soil. The pile should then be covered with waterproof paper, canvas, or polyethylene so that it sheds water. However, the cover should allow air to circulate and not enclose the pile to the groundline. In a tight enclosure, moisture from the ground may affect the moisture content of lumber. The use of a polyethylene cover over the ground before lumber is piled will reduce moisture rise. The same type of protection should be given to sheathing grade plywood.

After the framing and the wall and roof sheathing have been completed, the exterior roof trim, such as the cornice and rake finish, is installed. During this period, the shingles may have been delivered. Asphalt shingles should be stored so that bundles lie flat without bending; curved or buckled shingles often result in a poor looking roof. Wood shingles can be stored with only moderate protection from rain.

Window and Door Frames

Window and exterior door frames should not be delivered until they can be installed. In normal construction procedures, these frames are installed after the roof is completed and roofing installed. Generally, window units are ready for installation with sash and weatherstrip in place, and all wood protected by a dip treatment with a water-repellent preservative. Such units are premium items and, even though so treated, should be protected against moisture or mechanical damage. If it not possible to install frames when they arrive, place them on a dry base in an upright position and cover them.

Siding and Lath

Siding materials can be protected by storing temporarily in the house or garage. Place them so they will not be stepped on and split. Wood bevel siding is usually bundled with the pieces face to face to protect the surfaces fom mechanical damage and soiling. Some manufacturers treat their siding with a water-repellent material and pack in bundles with an outer protective wrap. All siding materials that cannot be installed immediately should be protected against exposure to conditions that could appreciably change their moisture content.

Insulation and rock lath should be stored inside the house. These materials are generally not installed until the electrical, heating, and plumbing trades have completed the roughing-in phases of their work.

Plastering in Cold Weather

During winter months, and in colder spring and fall weather in northern areas of the country, the heating unit of a house should be in operation before plastering is started. In fact, if the wood-framing members are much above 15 percent moisture content, it is good practice to let them dry out somewhat before rock lath is applied. This normally presents no problem, as the plumber and electrician do the rough-in work shortly after the house is closed in. Heat will allow the plaster to dry more readily, but because much moisture is driven off during this period, windows should be opened slightly.

Interior Finish

Millwork, floor underlayment, flooring, and interior trim manufactured by reputable companies are normally shipped at a moisture content satisfactory for immediate use. However, if storage conditions at the lumber company or in an unheated house during the inclement seasons are not satisfactory, wood parts will pick up moisture. Results may not be apparent immediately. If material is installed at too high a moisture content, the following heating season openings will show up between flooring strips and poorly matched joints in the trim because members have dried out and shrunk.

In flooring, for instance, the recommended moisture content at installation varies from 10 percent in the damp southern States to 6 and 7 percent for other localities. In examining wood floors with objectionable cracks between the boards, it has been found that in most cases the material had picked up moisture *after* manufacture and *before* it was installed. As such material redries during the heating season, it shrinks and the boards separate. Some of the moisture pickup may occur before the flooring is delivered to the building, but often such pickup occurs after delivery and before laying.

In an unheated building under construction, the relative humidity will average much higher than that in an occupied house. Thus, the flooring and finish tend to absorb moisture. To prevent moisture pickup at the building and to dry out any excess moisture picked up between time of manufacture and delivery, the humidity must be reduced below that considered normal in an unheated house. This may be accomplished by maintaining a temperature above the outdoor temperature even during the warmer seasons.

Before any floor underlayment, flooring or interior finish is delivered, the outside doors and windows should be hung and the heating plant installed to supply heat. For warm-weather control, when the workmen leave at night, the thermostat should be set to maintain a temperature of 15°F. above the average outdoor temperature. In the morning when the workmen return, the thermostat can be set back so that the burner will not operate. During the winter, fall, and spring, the temperature should be kept at about 60° F.

Several days before flooring is to be laid, bundles should be opened and the boards spread about so that their surfaces can dry out evenly. This will permit the drying of moisture picked up before delivery. Wood wall paneling and floor underlayment should also be exposed to the heated conditions of the house so the material will approach the moisture content it reaches in service. Actually, exposure of all interior finish to this period of moisture adjustment is good practice. Supplying some heat to the house in damp weather, even during the summer months, will be justified by improved appearance and owner satisfaction.

CHAPTER 33

MAINTENANCE AND REPAIR

A well constructed house will require comparatively little maintenance if adequate attention was paid to details and to choice of materials, as presented in previous chapters. A house may have an outstanding appearance, but if construction details have not been correct, the additional maintenance that might be required would certainly be discouraging to the owner. This may mean only a little attention to some apparently unimportant details. For example, an extra $10 spent on corrosion-resistant nails for siding and trim may save $100 or more annually because of the need for less frequent painting. The use of edge-grained rather than flat-grained siding will provide a longer paint life, and the additional cost of the edge-grained boards then seems justified.

The following sections will outline some factors

relating to maintenance of the house and how to reduce or eliminate conditions that may be harmful as well as costly. These suggestions can apply to both new and old houses.

Basement

The basement of a poured block wall may be damp for some time after a new house has been completed. However, after the heating season begins, most of this dampness from walls and floors will gradually disappear if construction has been correct. If dampness or wet walls and floors persist, the owner should check various areas to eliminate any possibilities for water entry.

Possible sources of trouble:

1. Drainage at the downspouts. The final grade around the house should be away from the building and a splash block or other means provided to drain water away from the foundation wall.

2. Soil settling at the foundation wall and forming pockets in which water may collect. These areas should be filled and tamped so that surface water can drain away.

3. Leaking in a poured concrete wall at the form tie rods. These leaks usually seal themselves, but larger holes should be filled with a cement mortar or other sealer. Clean and slightly dampen the area first for good adhesion.

4. Concrete-block or other masonry walls exposed above grade often show dampness on the interior after a prolonged rainy spell. A number of waterproofing materials on the market will provide good resistance to moisture penetration when applied to the inner face of the basement wall. If the outside of below-grade basement walls is treated correctly during construction, waterproofing the interior walls is normally not required. (See Ch. 3, "Foundation Walls and Piers.")

5. There should be at least a 6-inch clearance between the bottom of the siding and the grass. This means that at least 8 inches should be allowed above the finish grade before sod is laid or foundation plantings made. This will minimize the chance of moisture absorption by siding, sill plates, or other adjacent wood parts. Shrubs and foundation plantings should also be kept away from the wall to improve air circulation and drying. In lawn sprinkling, it is poor practice to allow water to spray against the walls of the house.

6. Check areas between the foundation wall and the sill plate. Any openings should be filled with a cement mixture or a calking compound. This filling will decrease heat loss and also prevent entry of insects into the basement, as well as reduce air infiltration.

7. Dampness in the basement in the early summer months is often augmented by opening the windows for ventilation during the day to allow warm, moisture-laden outside air to enter. The lower temperature of the basement will cool the incoming air and frequently cause condensation to collect and drip from cold-water pipes and also collect on colder parts of the masonry walls and floors. To air out the basement, open the windows during the night.

Perhaps the most convenient method of reducing humidity in basement areas is with *dehumidifiers*. A mechanical dehumidifier is moderate in price and does a satisfactory job of removing moisture from the air during periods of high humidity. Basements containing living quarters and without air conditioners may require more than one dehumidifier unit. When they are in operation, all basement windows should be closed.

Crawl-space Area

Crawl-space areas should be checked as follows:

1. Inspect the crawl-space area annually, for termite activity. Termite tubes on the walls or piers are an indication of this. In termite areas, soil in the crawl space or under the concrete slab is normally treated with some type of chemical to prevent termite damage. A house should contain a termite shield under the wood sill with a 2-inch extension on the interior. It must be well installed to be effective. Examine the shield for proper projection, and also any cracks in the foundation walls, as such cracks form good channels for termites to enter (Ch. 29, "Protection Against Decay and Termites").

2. While in the crawl space, check exposed wood joists and beams for indications of excessive moisture. In older houses where soil covers had not been used in the past, signs of staining or decay may be present. Use a penknife to test questionable areas.

3. Soil covers should be used to protect wood members from ground moisture (Ch. 16, "Ventilation"). These may consist of plastic films, roll roofing, or other suitable materials. A small amount of ventilation is desirable to provide some air movement. If the crawl space is not presently covered, install a barrier for greater protection.

Roof and Attic

The roof and the attic area of both new and older houses might be inspected with attention to the following:

1. Humps which occur on an asphalt-shingle roof are often caused by movement of roofing nails which have been driven into knots, splits, or unsound wood. Remove such nails, seal the holes, and replace the nails with others driven into sound wood. Blind-nail such replacements so that the upper shingle tab covers the nailhead.

A line of buckled shingles across the roof of a relatively new house is often caused by shrinkage of wide roof boards. It is important to use sheathing boards not over 8 inches wide and at a moisture content not exceeding 12 to 15 percent. When moisture content is greater, boards should be allowed to dry out for several days before shingles are applied. Time and hot weather tend to reduce buckling. Plywood sheathing would eliminate this problem altogether.

2. A dirt streak down the gable end of a house with a close rake section can often be attributed to rain entering and running under the edge of the shingles. This results from insufficient shingle overhang or the lack of a metal roof edge, such as shown in figure 70,*B*. The addition of a flashing strip to form a drip edge will usually minimize this problem.

3. In winters with heavy snows, ice dams may form at the eaves, often resulting in water entering the cornice and walls of the house. The immediate remedy is to remove the snow on the roof for a short distance above the gutters and, if necessary, in the valleys. Additional insulation between heated rooms and roof space, and increased ventilation in the overhanging eaves to lower the general attic temperature, will help to decrease the melting of snow on the roof and thus minimize ice formation. Deep snow in valleys also sometimes forms ice dams that cause water to back up under shingles and valley flashing (fig. 68,*A* and *B*).

4. Roof leaks are often caused by improper flashing at the valley, ridge, or around the chimney. Observe these areas during a rainy spell to discover the source. Water may travel many feet from the point of entry before it drips off the roof members.

5. The attic ventilators are valuable year round; in summer, to lower the attic temperature and improve comfort conditions in the rooms below; in winter, to remove water vapor that may work through the ceiling and condense in the attic space and to minimize ice dam problems. The ventilators should be open both in winter and summer.

To check for sufficient ventilation during cold weather, examine the attic after a prolonged cold period. If nails protruding from the roof into the attic space are heavily coated with frost, ventilation is usually insufficient. Frost may also collect on the roof sheathing, first appearing near the eaves on the north side of the roof. Increase the size of the ventilators or place additional ones in the soffit area of the cornice. This will improve air movement and circulation. (See Ch. 16, "Ventilation," and figs. 99, 100, and 101 for proper size and location.)

Exterior Walls

One of the maintenance problems which sometimes occurs with a wood-sided house involves the exterior paint finish. Several reasons are known for peeling and poor adherence of paint. One of the major ones perhaps can be traced to moisture in its various forms. Paint quality and methods of application are other reasons. Another factor involves the species of wood and the direction of grain. Some species retain paint better than others, and edge grain provides a better surface for paint than flat grain. Chapter 28, "Painting and Finishing," covers correct methods of application, types of paint, and other recommendations for a good finish. Other phases of the exterior maintenance that the owner may encounter with his house are as follows:

1. In applying the siding, if bright steel nails have been used rather than galvanized, aluminum, stainless steel or other noncorrosive nails, rust spots may occur at the nailhead. These spots are quite common where nails are driven flush with the heads exposed. The spotting may be remedied somewhat, in the case of flush nailing, by setting the nailhead below the surface and puttying. The puttying should be preceded by a priming coat.

2. Brick and other types of masonry are not always waterproof, and continued rains may result in moisture penetration. Masonry veneer walls over a sheathed wood frame are normally backed with a waterproof sheathing paper to prevent moisture entry. When walls do not have such protection and the moisture problem persists, use a waterproof coating over the exposed masonry surfaces. Transparent waterproof materials can be obtained for this purpose.

3. Calking is usually required where a change in materials occurs on a vertical line, such as that of wood siding abutting against brick chimneys or walls. The wood should normally have a prime coating of paint for proper adhesion of the calking compound. Calking guns with cartridges are the best means of waterproofing these joints. Many permanent-type calking materials with a neoprene, elastomer or other type base are available.

4. Rainwater may work behind wood siding through butt and end joints and sometimes up under the butt edge by capillarity when joints are not tight. Setting the butt and end joints in white lead is an old-time custom that is very effective in preventing water from entering. Painting under the butt edges at the lap adds mechanical resistance to water ingress. However, moisture changes in the siding cause some swelling and shrinking that may break the paint film. Treating the siding with a water repellent before it is applied is effective in reducing capillary action. For houses already built, the water repellent could be applied under the butt edges of bevel siding or along the joints of drop siding and at all vertical joints. Such water repellents are often combined with a preservative and can be purchased at your local paint dealers as a water-repellent preservative. In-place application is often done with a plunger-type oil can. Excess repellent on the face of painted surfaces should be wiped off.

Interior

Plaster

The maintenance of plastered interior surfaces normally is no problem in a properly constructed house. However, the following points are worthy of attention:

1. Because of the curing (aging) period ordinarily required for plastered walls, it is not advisable to apply oil-base paints until at least 60 days after plastering is completed. Water-mix or resin-base paints may be applied without the necessity of an aging period.

2. In a newly constructed house, small plaster cracks may develop during or after the first heating season. Such cracks are ordinarily caused by the drying of framing members that had too high a moisture content when the plaster was applied. The cracks usually occur at interior corners and also above windows and doors because of shrinkage of the headers. For this reason, it is usually advisable to wait for a part of the heating season before painting plaster so that such cracks can be filled first.

3. Large plaster cracks in houses, new or old, often indicate a structural weakness in the framing or column footings. This may be due to excessive deflection or to settling of beam supports. Common areas for such defects might be along the center beam or around the basement stairway. In such cases, the use of an additional post and pedestal may be required. (See Ch. 5, "Floor Framing," for recommended methods of framing.)

Moisture on Windows

Moisture on inside surfaces of windows may often occur during the colder periods of the heating season. The following precautions and corrections should be observed during this time:

1. During cold weather, condensation and, in cold climates, frost will collect on the inner face of single-glazed windows. Water from the condensation or melting frost runs down the glass and soaks into the wood sash to cause stain, decay, and paint failure. The water may rust steel sash. To prevent such condensation, the window should be provided with a storm sash. Double glazing will also minimize this condensation. If it still presists on double-glazed windows, it usually indicates that the humidity is too high. If a humidifier is used, it should be turned off for a while or the setting lowered. Other moisture sources should also be reduced enough to remedy the problem. Increasing the inside temperature will also reduce surface condensation.

2. Occasionally, in very cold weather, frost may form on the inner surfaces of the storm windows. This may be caused by (a) loose-fitting window sash that allows moisture vapor from the house to enter the space between the window and storm sash, (b) high relative humidity in the living quarters, or (c) a combination of both. Generally, the condensation on storm sash does not create a maintenance problem, but it may be a nuisance. Weather-stripping the inner sash offers resistance to moisture flow and may prevent this condensation. Lower relative humidities in the house are also helpful.

Problems with Exterior Doors

Condensation may occur on the glass or even on the interior surface of exterior doors during periods of severe cold. Furthermore, warping may result. The addition of a tight-fitting storm or combination door will usually remedy both problems. A solid-core flush door or a panel door with solid stiles and rails is preferred over a hollow-core door to prevent or minimize this warping problem.

Openings in Flooring

Laying finish-strip flooring at too high a moisture content or laying individual boards with varying moisture contents may be a source of trouble to the homeowner. As the flooring dries out and reaches moisture equilibrium, spaces will form between the boards. These openings are often very difficult to correct. If the floor has a few large cracks, one expedient is to fit matching strips of wood between the flooring strips and glue them in place. In severe cases, it may be necessary to replace sections of the floor or to refloor the entire house.

Another method would be to cover the existing flooring with a thin flooring 5/16 or 3/8 inch thick. This would require removal of the base shoe, fitting the thin flooring around door jambs, and perhaps sawing off the door bottoms. (For proper methods of laying floors to prevent open joints in new houses, see Ch. 20, "Floor Coverings.")

Unheated Rooms

To lower fuel consumption and for personal reasons, some homeowners close off unused rooms and leave them unheated during the winter months. These factors of low temperatures and lack of heat, unfortunately, are conducive to trouble from condensation. Certain corrective or protective measures can be taken to prevent damage and subsequent maintenance expense, as follows:

1. Do not operate humidifiers or otherwise intentionally increase humidity in heated parts of the house.

2. Open the windows of unheated rooms during bright sunny days for several hours for ventilation. Ventilation will help draw moisture out of the rooms.

3. Install storm sash on all windows, including those in unheated rooms. This will materially reduce heat loss from both heated and unheated rooms and will minimize the condensation on the inner glass surfaces.

LITERATURE CITED

(1) American Institute of Timber Construction
1965. Standard for heavy timber roof decking. AITC 112. Washington, D.C.

(2) American Plywood Association
1964. Plywood truss designs, APA 64–650 11 pp., illus.

(3) ———
1967. Plywood in apartments. APA 67-310.

(4) American Society of Heating, Refrigerating, and Air-Conditioning Engineers
1967. ASHRAE handbook of fundamentals. 544 pp., illus.

(5) Anderson, L. O.
1967. Selection and use of wood products for home and farm building. U.S. Dep. Agr., Agr. Inform. Bull. 311, 41 pp., illus.

(6) ———
1969. Low-cost wood homes for rural America—construction manual. Agr. Handb. 364, 112 pp.

(7) ———, and Smith, W.
1965. Houses can resist hurricanes. U.S. Forest Serv. Res. Pap. FPL 33. Forest Prod. Lab., Madison, Wis., 44 pp. illus.

(8) Berendt, R.D., and Winzer, G.E.
1954. Sound insulation of wall, floor, and door constructions. Nat. Bur. of Stand. Monogr. No. 77, 49 pp.

(9) Federal Housing Administration
1961. Mat-formed particleboard for exterior use. FHA Use of Materials Bull. No. UM-32. June 19.

(10) ———
1963. A guide to impact noise control in multifamily dwellings. FHA No. 750, 86 pp.

(11) ———
1964. Minimum property standards for one and two family living units. FHA No. 300, 315 pp.

(12) Forest Products Laboratory, Forest Service, U.S. Dept. of Agriculture.
1955. Wood handbook. U.S. Dep. Agr., Agr. Handb. 72, 528 pp., illus.

(13) ———
1964. FPL natural finish. U.S. Forest Serv. Res. Note FPL–046, Madison, Wis., 5 pp.

(14) Insulation Board Institute
1963. Noise control with insulation board. Chicago, Ill., 15 pp., illus.

(15) Lewis, Wayne C.
1968. Thermal insulation from wood for buildings: Effects of moisture and its control. U.S.D.A. Forest Serv. Res. Pap. FPL 86. Forest Prod. Lab., Madison, Wis.

(16) National Fire Protection Association
1962. Occupancy fire record: One-and two-family dwellings. Fire Rec. Bull. FR56–2A, Boston, Mass., 20 pp., illus.

(17) Teesdale, L.V.
1955. Remedial measures for building condensation difficulties. U.S. Forest Prod. Lab. Rep. 1710. 15 pp., illus.

(18) U.S. Department of Commerce
1966. Softwood plywood—construction and industrial. U.S. Prod. Stand. PS 1–66, 28 pp., illus.

(19) Verrall, A.F.
1961. Brush, dip, and soak treatment with water-repellent preservatives. Forest Prod. J. 11(1): 23–26.

GLOSSARY OF HOUSING TERMS

Air-dried lumber. Lumber that has been piled in yards or sheds for any length of time. For the United States as a whole, the minimum moisture content of thoroughly air-dried lumber is 12 to 15 percent and the average is somewhat higher. In the South, air-dried lumber may be no lower than 19 percent.

Airway. A space between roof insulation and roof boards for movement of air.

Alligatoring. Coarse checking pattern characterized by a slipping of the new paint coating over the old coating to the extent that the old coating can be seen through the fissures.

Anchor bolts. Bolts to secure a wooden sill plate to concrete or masonry floor or wall.

Apron. The flat member of the inside trim of a window placed against the wall immediately beneath the stool.

Areaway. An open subsurface space adjacent to a building used to admit light or air or as a means of access to a basement.

Asphalt. Most native asphalt is a residue from evaporated petroleum. It is insoluble in water but soluble in gasoline and melts when heated. Used widely in building for waterproofing roof coverings of many types, exterior wall coverings, flooring tile, and the like.

Astragal. A molding, attached to one of a pair of swinging doors, against which the other door strikes.

Attic ventilators. In houses, screened openings provided to ventilate an attic space. They are located in the soffit area as inlet ventilators and in the gable end or along the ridge as outlet ventilators. They can also consist of power-driven fans used as an exhaust system. (See also **Louver.**)

Backband. A simple molding sometimes used around the outer edge of plain rectangular casing as a decorative feature.

Backfill. The replacement of excavated earth into a trench around and against a basement foundation.

Balusters. Usually small vertical members in a railing used between a top rail and the stair treads or a bottom rail.

Balustrade. A railing made up of balusters, top rail, and sometimes bottom rail, used on the edge of stairs, balconies, and porches.

Barge board. A decorative board covering the projecting rafter (fly rafter) of the gable end. At the cornice, this member is a facia board.

Base or baseboard. A board placed against the wall around a room next to the floor to finish properly between floor and plaster.

Base molding. Molding used to trim the upper edge of interior baseboard.

Base shoe. Molding used next to the floor on interior baseboard. Sometimes called a carpet strip.

Batten. Narrow strips of wood used to cover joints or as decorative vertical members over plywood or wide boards.

Batter board. One of a pair of horizontal boards nailed to posts set at the corners of an excavation, used to indicate the desired level, also as a fastening for stretched strings to indicate outlines of foundation walls.

Bay window. Any window space projecting outward from the walls of a building, either square or polygonal in plan.

Beam. A structural member transversely supporting a load.

Bearing partition. A partition that supports any vertical load in addition to its own weight.

Bearing wall. A wall that supports any vertical load in addition to its own weight.

Bed molding. A molding in an angle, as between the overhanging cornice, or eaves, of a building and the sidewalls.

Blind-nailing. Nailing in such a way that the nailheads are not visible on the face of the work—usually at the tongue of matched boards.

Blind stop. A rectangular molding, usually ¾ by 1-⅜ inches or more in width, used in the assembly of a window frame. Serves as a stop for storm and screen or combination windows and to resist air infiltration.

Blue stain. A bluish or grayish discoloration of the sapwood caused by the growth of certain moldlike fungi on the surface and in the interior of a piece, made possible by the same conditions that favor the growth of other fungi.

Bodied linseed oil. Linseed oil that has been thickened in viscosity by suitable processing with heat or chemicals. Bodied oils are obtainable in a great range in viscosity from a little greater than that of raw oil to just short of a jellied condition.

Boiled linseed oil. Linseed oil in which enough lead, manganese, or cobalt salts have been incorporated to make the oil harden more rapidly when spread in thin coatings.

Bolster. A short horizontal timber or steel beam on top of a column to support and decrease the span of beams or girders.

Boston ridge. A method of applying asphalt or wood shingles at the ridge or at the hips of a roof as a finish.

Brace. An inclined piece of framing lumber applied to wall or floor to stiffen the structure. Often used on walls as temporary bracing until framing has been completed.

Brick veneer. A facing of brick laid against and fastened to sheathing of a frame wall or tile wall construction.

Bridging. Small wood or metal members that are inserted in a diagonal position between the floor joists at midspan to act both as tension and compression members for the purpose of bracing the joists and spreading the action of loads.

Buck. Often used in reference to rough frame opening members. Door bucks used in reference to metal door frame.

Built-up roof. A roofing composed of three to five layers of asphalt felt laminated with coal tar, pitch, or asphalt. The top is finished with crushed slag or gravel. Generally used on flat or low-pitched roofs.

Butt joint. The junction where the ends of two timbers or other members meet in a square-cut joint.

Cant strip. A triangular-shaped piece of lumber used at the junction of a flat deck and a wall to prevent cracking of the roofing which is applied over it.

Cap. The upper member of a column, pilaster, door cornice, molding, and the like.

Casement frames and sash. Frames of wood or metal enclosing part or all of the sash, which may be opened by means of hinges affixed to the vertical edges.

Casing. Molding of various widths and thicknesses used to trim door and window openings at the jambs.

Cement, Keene's. A white finish plaster that produces an extremely durable wall. Because of its density, it excels for use in bathrooms and kitchens and is also used extensively for the finish coat in auditoriums, public buildings, and other places where walls may be subjected to unusually hard wear or abuse.

Checking. Fissures that appear with age in many exterior paint coatings, at first superficial, but which in time may penetrate entirely through the coating.

Checkrails. Meeting rails sufficiently thicker than a window to fill the opening between the top and bottom sash made by the parting stop in the frame of double-hung windows. They are usually beveled.

Collar beam. Nominal 1- or 2-inch-thick members connecting opposite roof rafters. They serve to stiffen the roof structure.

Column. In architecture: A perpendicular supporting member, circular or rectangular in section, usually consisting of a base, shaft, and capital. In engineering: A vertical structural compression member which supports loads acting in the direction of its longitudinal axis.

Combination doors or windows. Combination doors or windows used over regular openings. They provide winter insulation and summer protection and often have self-storing or removable glass and screen inserts. This eliminates the need for handling a different unit each season.

Concrete plain. Concrete either without reinforcement, or reinforced only for shrinkage or temperature changes.

Condensation. In a building: Beads or drops of water (and frequently frost in extremely cold weather) that accumulate on the inside of the exterior covering of a building when warm, moisture-laden air from the interior reaches a point where the temperature no longer permits the air to sustain the moisture it holds. Use of louvers or attic ventilators will reduce moisture condensation in attics. A vapor barrier under the gypsum lath or dry wall on exposed walls will reduce condensation in them.

Conduit, electrical. A pipe, usually metal, in which wire is installed.

Construction dry wall. A type of construction in which the interior wall finish is applied in a dry condition, generally in the form of sheet materials or wood paneling, as contrasted to plaster.

Construction, frame. A type of construction in which the structural parts are wood or depend upon a wood frame for support. In codes, if masonry veneer is applied to the exterior walls, the classification of this type of construction is usually unchanged.

Coped joint. See **Scribing.**

Corbel out. To build out one or more courses of brick or stone from the face of a wall, to form a support for timbers.

Corner bead. A strip of formed sheet metal, sometimes combined with a strip of metal lath, placed on corners before plastering to reinforce them. Also, a strip of wood finish three-quarters-round or angular placed over a plastered corner for protection.

Corner boards. Used as trim for the external corners of a house or other frame structure against which the ends of the siding are finished.

Corner braces. Diagonal braces at the corners of frame structure to stiffen and strengthen the wall.

Let-in brace. Nominal 1-inch-thick boards applied into notched studs diagonally.

Cut-in brace. Nominal 2-inch-thick members, usually 2 by 4's, cut in between each stud diagonally.

Cornerite. Metal-mesh lath cut into strips and bent to a right angle. Used in interior corners of walls and ceilings on lath to prevent cracks in plastering.

Cornice. Overhang of a pitched roof at the eave line, usually consisting of a facia board, a soffit for a closed cornice, and appropriate moldings.

Cornice return. That portion of the cornice that returns on the gable end of a house.

Counterflashing. A flashing usually used on chimneys at the roofline to cover shingle flashing and to prevent moisture entry.

Cove molding. A molding with a concave face used as trim or to finish interior corners.

Crawl space. A shallow space below the living quarters of a basementless house, normally enclosed by the foundation wall.

Cricket. A small drainage-diverting roof structure of single or double slope placed at the junction of larger surfaces that meet at an angle, such as above a chimney.

Cross-bridging. Diagonal bracing between adjacent floor joists, placed near the center of the joist span to prevent joists from twisting.

Crown molding. A molding used on cornice or wherever an interior angle is to be covered.

d. See **Penny.**

Dado. A rectangular groove across the width of a board or plank. In interior decoration, a special type of wall treatment.

Decay. Disintegration of wood or other substance through the action of fungi.

Deck paint. An enamel with a high degree of resistance to mechanical wear, designed for use on such surfaces as porch floors.

Density. The mass of substance in a unit volume. When expressed in the metric system, it is numerically equal to the specific gravity of the same substance.

Dewpoint. Temperature at which a vapor begins to deposit as a liquid. Applies especially to water in the atmosphere.

Dimension. See **Lumber dimension.**

Direct nailing. To nail perpendicular to the initial surface or to the junction of the pieces joined. Also termed **face nailing.**

Doorjamb, interior. The surrounding case into which and out of which a door closes and opens. It consists of two upright pieces, called side jambs, and a horizontal head jamb.

Dormer. An opening in a sloping roof, the framing of which projects out to form a vertical wall suitable for windows or other openings.

Downspout. A pipe, usually of metal, for carrying rainwater from roof gutters.

Dressed and matched (tongued and grooved). Boards or planks machined in such a maner that there is a groove on one edge and a corresponding tongue on the other.

Drier paint. Usually oil-soluble soaps of such metals as lead, manganese, or cobalt, which, in small proportions, hasten the oxidation and hardening (drying) of the drying oils in paints.

Drip. (a) A member of a cornice or other horizontal exterior-finish course that has a projection beyond the other parts for throwing off water. (b) A groove in the underside of a sill or drip cap to cause water to drop off on the outer edge instead of drawing back and running down the face of the building.

Drip cap. A molding placed on the exterior top side of a door or window frame to cause water to drip beyond the outside of the frame.

Dry-wall. Interior covering material, such as gypsum board or plywood, which is applied in large sheets or panels.

Ducts. In a house, usually round or rectangular metal pipes for distributing warm air from the heating plant to rooms, or air from a conditioning device or as cold air returns. Ducts are also made of asbestos and composition materials.

Eaves. The margin or lower part of a roof projecting over the wall.

Expansion joint. A bituminous fiber strip used to separate blocks or units of concrete to prevent cracking due to expansion as a result of temperature changes. Also used on concrete slabs.

Facia or fascia. A flat board, band, or face, used sometimes by itself but usually in combination with moldings, often located at the outer face of the cornice.

Filler (wood). A heavily pigmented preparation used for filling and leveling off the pores in open-pored woods.

Fire-resistive. In the absence of a specific ruling by the authority having jurisdiction, applies to materials for construction not combustible in the temperatures of ordinary fires and that will withstand such fires without serious impairment of their usefulness for at least 1 hour.

Fire-retardant chemical. A chemical or preparation of chemicals used to reduce flammability or to retard spread of flame.

Fire stop. A solid, tight closure of a concealed space, placed to prevent the spread of fire and smoke through such a space. In a frame wall, this will usually consist of 2 by 4 cross blocking between studs.

Fishplate. A wood or plywood piece used to fasten the ends of two members together at a butt joint with nails or bolts. Sometimes used at the junction of opposite rafters near the ridge line.

Flagstone (flagging or flags). Flat stones, from 1 to 4 inches thick, used for rustic walks, steps, floors, and the like.

Flashing. Sheet metal or other material used in roof and wall construction to protect a building from water seepage.

Flat paint. An interior paint that contains a high proportion of pigment and dries to a flat or lusterless finish.

Flue. The space or passage in a chimney through which smoke, gas, or fumes ascend. Each passage is called a flue, which together with any others and the surrounding masonry make up the chimney.

Flue lining. Fire clay or terra-cotta pipe, round or square, usually made in all ordinary flue sizes and in 2-foot lengths, used for the inner lining of chimneys with the brick or masonry work around the outside. Flue lining in chimneys runs from about a foot below the flue connection to the top of the chimney.

Fly rafters. End rafters of the gable overhang supported by roof sheathing and lookouts.

Footing. A masonry section, usually concrete, in a rectangular form wider than the bottom of the foundation wall or pier it supports.

Foundation. The supporting portion of a structure below the first-floor construction, or below grade, including the footings.

Framing, balloon. A system of framing a building in which all vertical structural elements of the bearing walls and partitions consist of single pieces extending from the top of the foundation sill plate to the roofplate and to which all floor joists are fastened.

Framing, platform. A system of framing a building in which floor joists of each story rest on the top plates of the story below or on the foundation sill for the first story, and the bearing walls and partitions rest on the subfloor of each story.

Frieze. In house construction, a horizontal member connecting the top of the siding with the soffit of the cornice.

Frostline. The depth of frost penetration in soil. This depth varies in different parts of the country. Footings should be placed below this depth to prevent movement.

Fungi, wood. Microscopic plants that live in damp wood and cause mold, stain, and decay.

Fungicide. A chemical that is poisonous to fungi.

Furring. Strips of wood or metal applied to a wall or other surface to even it and normally to serve as a fastening base for finish material.

Gable. In house construction, the portion of the roof above the eave line of a double-sloped roof.

Gable end. An end wall having a gable.

Gloss enamel. A finishing material made of varnish and sufficient pigments to provide opacity and color, but little or no pigment of low opacity. Such an enamel forms a hard coating with maximum smoothness of surface and a high degree of gloss.

Gloss (paint or enamel). A paint or enamel that contains a relatively low proportion of pigment and dries to a sheen or luster.

Girder. A large or principal beam of wood or steel used to support concentrated loads at isolated points along its length.

Grain. The direction, size, arrangement, appearance, or quality of the fibers in wood.

Grain, edge (vertical). Edge-grain lumber has been sawed parallel to the pith of the log and approximately at right angles to the growth rings; i.e., the rings form an angle of 45° or more with the surface of the piece.

Grain, flat. Flat-grain lumber has been sawed parallel to the pith of the log and approximately tangent to the growth rings, i.e., the rings form an angle of less than 45° with the surface of the piece.

Grain, quartersawn. Another term for edge grain.

Grounds. Guides used around openings and at the floorline to strike off plaster. They can consist of narrow strips of wood or of wide subjambs at interior doorways. They provide a level plaster line for installation of casing and other trim.

Grout. Mortar made of such consistency (by adding water) that it will just flow into the joints and cavities of the masonry work and fill them solid.

Gusset. A flat wood, plywood, or similar type member used to provide a connection at intersection of wood members. Most commonly used at joints of wood trusses. They are fastened by nails, screws, bolts, or adhesives.

Gutter or eave trough. A shallow channel or conduit of metal or wood set below and along the eaves of a house to catch and carry off rainwater from the roof.

Gypsum plaster. Gypsum formulated to be used with the addition of sand and water for base-coat plaster.

Header. (a) A beam placed perpendicular to joists and to which joists are nailed in framing for chimney, stairway, or other opening. (b) A wood lintel.

Hearth. The inner or outer floor of a fireplace, usually made of brick, tile, or stone.

Heartwood. The wood extending from the pith to the sapwood, the cells of which no longer participate in the life processes of the tree.

Hip. The external angle formed by the meeting of two sloping sides of a roof.

Hip roof. A roof that rises by inclined planes from all four sides of a building.

Humidifier. A device designed to increase the humidity within a room or a house by means of the discharge of water vapor. They may consist of individual room-size units or larger units attached to the heating plant to condition the entire house.

I-beam. A steel beam with a cross section resembling the letter *I*. It is used for long spans as basement beams or over wide wall openings, such as a double garage door, when wall and roof loads are imposed on the opening.

IIC. A new system utilized in the Federal Housing Administration recommended criteria for impact sound insulation.

INR (Impact Noise Rating). A single figure rating which provides an estimate of the impact sound-insulating performance of a floor-ceiling assembly.

Insulation board, rigid. A structural building board made of coarse wood or cane fiber in ½- and 25/32-inch thicknesses. It can be obtained in various size sheets, in various densities, and with several treatments.

Insulation, thermal. Any material high in resistance to heat transmission that, when placed in the walls, ceiling, or floors of a structure, will reduce the rate of heat flow.

Interior finish. Material used to cover the interior framed areas, or materials of walls and ceilings.

Jack rafter. A rafter that spans the distance from the wallplate to a hip, or from a valley to a ridge.

Jamb. The side and head lining of a doorway, window, or other opening.

Joint. The space between the adjacent surfaces of two members or components joined and held together by nails, glue, cement, mortar, or other means.

Joint cement. A powder that is usually mixed with water and used for joint treatment in gypsum-wallboard finish. Often called "spackle."

Joist. One of a series of parallel beams, usually 2 inches in thickness, used to support floor and ceiling loads, and supported in turn by larger beams, girders, or bearing walls.

Kiln dried lumber. Lumber that has been kiln dried often to a moisture content of 6 to 12 percent. Common varieties of softwood lumber, such as framing lumber are dried to a somewhat higher moisture content.

Knot. In lumber, the portion of a branch or limb of a tree that appears on the edge or face of the piece.

Landing. A platform between flights of stairs or at the termination of a flight of stairs.

Lath. A building material of wood, metal, gypsum, or insulating board that is fastened to the frame of a building to act as a plaster base.

Lattice. A framework of crossed wood or metal strips.

Leader. See **Downspout.**

Ledger strip. A strip of lumber nailed along the bottom of the side of a girder on which joists rest.

Light. Space in a window sash for a single pane of glass. Also, a pane of glass.

Lintel. A horizontal structural member that supports the load over an opening such as a door or window.

Lookout. A short wood bracket or cantilever to support an overhang portion of a roof or the like, usually concealed from view.

Louver. An opening with a series of horizontal slats so arranged as to permit ventilation but to exclude rain, sunlight, or vision. See also **Attic ventilators.**

Lumber. Lumber is the product of the sawmill and planing mill not further manufactured other than by sawing, resawing, and passing lengthwise through a standard planing machine, crosscutting to length, and matching.

Lumber, boards. Yard lumber less than 2 inches thick and 2 or more inches wide.

Lumber, dimension. Yard lumber from 2 inches to, but not including, 5 inches thick and 2 or more inches wide. Includes joists, rafters, studs, plank, and small timbers.

Lumber, dressed size. The dimension of lumber after shrinking from green dimension and after machining to size or pattern.

Lumber, matched. Lumber that is dressed and shaped on one edge in a grooved pattern and on the other in a tongued pattern.

Lumber, shiplap. Lumber that is edge-dressed to make a close rabbeted or lapped joint.

Lumber, timbers. Yard lumber 5 or more inches in least dimension. Includes beams, stringers, posts, caps, sills, girders, and purlins.

Lumber, yard. Lumber of those grades, sizes, and patterns which are generally intended for ordinary construction, such as framework and rough coverage of houses.

Mantel. The shelf above a fireplace. Also used in referring to the decorative trim around a fireplace opening.

Masonry. Stone, brick, concrete, hollow-tile, concrete-block, gypsum-block, or other similar building units or materials or a combination of the same, bonded together with mortar to form a wall, pier, buttress, or similar mass.

Mastic. A pasty material used as a cement (as for setting tile) or a protective coating (as for thermal insulation or waterproofing).

Metal lath. Sheets of metal that are slit and drawn out to form openings. Used as a plaster base for walls and ceilings and as reinforcing over other forms of plaster base.

Millwork. Generally all building materials made of finished wood and manufactured in millwork plants and planing mills are included under the term "millwork." It includes such items as inside and outside doors, window and doorframes, blinds, porchwork, mantels, panelwork, stairways, moldings, and interior trim. It normally does not include flooring, ceiling, or siding.

Miter joint. The joint of two pieces at an angle that bisects the joining angle. For example, the miter joint at the side and head casing at a door opening is made at a 45° angle.

Moisture content of wood. Weight of the water contained in the wood, usually expressed as a percentage of the weight of the ovendry wood.

Molding. A wood strip having a curved or projecting surface used for decorative purposes.

Mortise. A slot cut into a board, plank, or timber, usually edgewise, to receive tenon of another board, plank, or timber to form a joint.

Mullion. A vertical bar or divider in the frame between windows, doors, or other openings.

Muntin. A small member which divides the glass or openings of sash or doors.

Natural finish. A transparent finish which does not seriously alter the original color or grain of the natural wood. Natural finishes are usually provided by sealers, oils, varnishes, water-repellent preservatives, and other similar materials.

Newel. A post to which the end of a stair railing or balustrade is fastened. Also, *any* post to which a railing or balustrade is fastened.

Nonbearing wall. A wall supporting no load other than its own weight.

Nosing. The projecting edge of a molding or drip. Usually applied to the projecting molding on the edge of a stair tread.

Notch. A crosswise rabbet at the end of a board.

O. C., on center. The measurement of spacing for studs, rafters, joists, and the like in a building from the center of one member to the center of the next.

O. G., or ogee. A molding with a profile in the form of a letter *S*; having the outline of a reversed curve.

Outrigger. An extension of a rafter beyond the wall line. Usually a smaller member nailed to a larger rafter to form a cornice or roof overhang.

Paint. A combination of pigments with suitable thinners or oils to provide decorative and protective coatings.

Panel. In house construction, a thin flat piece of wood, plywood, or similar material, framed by stiles and rails as in a door or fitted into grooves of thicker material with molded edges for decorative wall treatment.

Paper, building. A general term for papers, felts, and similar sheet materials used in buildings without reference to their properties or uses.

Paper, sheathing. A building material, generally paper or felt, used in wall and roof construction as a protection against the passage of air and sometimes moisture.

Parting stop or strip. A small wood piece used in the side and head jambs of double-hung windows to separate upper and lower sash.

Partition. A wall that subdivides spaces within any story of a building.

Penny. As applied to nails, it originally indicated the price per hundred. The term now serves as a measure of nail length and is abbreviated by the letter *d.*

Perm. A measure of water vapor movement through a material (grains per square foot per hour per inch of mercury difference in vapor pressure).

Pier. A column of masonry, usually rectangular in horizontal cross section, used to support other structural members.

Pigment. A powdered solid in suitable degree of subdivision for use in paint or enamel.

Pitch. The incline slope of a roof or the ratio of the total rise to the total width of a house, i.e., an 8-foot rise and 24-foot width is a one-third pitch roof. Roof slope is expressed in the inches of rise per foot of run.

Pitch pocket. An opening extending parallel to the annual rings of growth, that usually contains, or has contained, either solid or liquid pitch.

Pith. The small, soft core at the original center of a tree around which wood formation takes place.

Plaster grounds. Strips of wood used as guides or strike-off edges around window and door openings and at base of walls.

Plate. Sill plate: a horizontal member anchored to a masonry wall. Sole plate: bottom horizontal member of a frame wall. Top plate: top horizontal member of a frame wall supporting ceiling joists, rafters, or other members.

Plough. To cut a lengthwise groove in a board or plank.

Plumb. Exactly perpendicular; vertical.

Ply. A term to denote the number of thicknesses or layers of roofing felt, veneer in plywood, or layers in built-up materials, in any finished piece of such material.

Plywood. A piece of wood made of three or more layers of veneer joined with glue, and usually laid with the grain of adjoining plies at right angles. Almost always an odd number of plies are used to provide balanced construction.

Pores. Wood cells of comparatively large diameter that have open ends and are set one above the other to form continuous tubes. The openings of the vessels on the surface of a piece of wood are referred to as pores.

Preservative. Any substance that, for a reasonable length of time, will prevent the action of wood-destroying fungi, borers of various kinds, and similar destructive agents when the wood has been properly coated or impregnated with it.

Primer. The first coat of paint in a paint job that consists of two or more coats; also the paint used for such a first coat.

Putty. A type of cement usually made of whiting and boiled linseed oil, beaten or kneaded to the consistency of dough, and used in sealing glass in sash, filling small holes and crevices in wood, and for similar purposes.

Quarter round. A small molding that has the cross section of a quarter circle.

Rabbet. A rectangular longitudinal groove cut in the corner edge of a board or plank.

Radiant heating. A method of heating, usually consisting of a forced hot water system with pipes placed in the floor, wall, or ceiling; or with electrically heated panels.

Rafter. One of a series of structural members of a roof designed to support roof loads. The rafters of a flat roof are sometimes called roof joists.

Rafter, hip. A rafter that forms the intersection of an external roof angle.

Rafter, valley. A rafter that forms the intersection of an internal roof angle. The valley rafter is normally made of double 2-inch-thick members.

Rail. Cross members of panel doors or of a sash. Also the upper and lower members of a balustrade or staircase extending from one vertical support, such as a post, to another.

Rake. Trim members that run parallel to the roof slope and form the finish between the wall and a gable roof extension.

Raw linseed oil. The crude product processed from flaxseed and usually without much subsequent treatment.

Reflective insulation. Sheet material with one or both surfaces of comparatively low heat emissivity, such as aluminum foil. When used in building construction the surfaces face air spaces, reducing the radiation across the air space.

Reinforcing. Steel rods or metal fabric placed in concrete slabs, beams, or columns to increase their strength.

Relative humidity. The amount of water vapor in the atmosphere, expressed as a percentage of the maximum quantity that could be present at a given temperature. (The actual amount of water vapor that can be held in space increases with the temperature.)

Resorcinol glue. A glue that is high in both wet and dry strength and resistant to high temperatures. It is used for gluing lumber or assembly joints that must withstand severe service conditions.

Ribbon (Girt). Normally a 1- by 4-inch board let into the studs horizontally to support ceiling or second-floor joists.

Ridge. The horizontal line at the junction of the top edges of two sloping roof surfaces.

Ridge board. The board placed on edge at the ridge of the roof into which the upper ends of the rafters are fastened.

Rise. In stairs, the vertical height of a step or flight of stairs.

Riser. Each of the vertical boards closing the spaces between the treads of stairways.

Roll roofing. Roofing material, composed of fiber and saturated with asphalt, that is supplied in 36-inch wide rolls with 108 square feet of material. Weights are generally 45 to 90 pounds per roll.

Roof sheathing. The boards or sheet material fastened to the roof rafters on which the shingle or other roof covering is laid.

Rubber-emulsion paint. Paint, the vehicle of which consists of rubber or synthetic rubber dispersed in fine droplets in water.

Run. In stairs, the net width of a step or the horizontal distance covered by a flight of stairs.

Saddle. Two sloping surfaces meeting in a horizontal ridge, used between the back side of a chimney, or other vertical surface, and a sloping roof.

Sand float finish. Lime mixed with sand, resulting in a textured finish.

Sapwood. The outer zone of wood, next to the bark. In the living tree it contains some living cells (the heartwood contains none), as well as dead and dying cells. In most species, it is lighter colored than the heartwood. In all species, it is lacking in decay resistance.

Sash. A single light frame containing one or more lights of glass.

Sash balance. A device, usually operated by a spring or tensioned weatherstripping designed to counterbalance double-hung window sash.

Saturated felt. A felt which is impregnated with tar or asphalt.

Scratch coat. The first coat of plaster, which is scratched to form a bond for the second coat.

Screed. A small strip of wood, usually the thickness of the plaster coat, used as a guide for plastering.

Scribing. Fitting woodwork to an irregular surface. In moldings, cutting the end of one piece to fit the molded face of the other at an interior angle to replace a miter joint.

Sealer. A finishing material, either clear or pigmented, that is usually applied directly over uncoated wood for the purpose of sealing the surface.

Seasoning. Removing moisture from green wood in order to improve its serviceability.

Semigloss paint or enamel. A paint or enamel made with a slight insufficiency of nonvolatile vehicle so that its coating, when dry, has some luster but is not very glossy.

Shake. A thick handsplit shingle, resawed to form two shakes; usually edge-grained.

Sheathing. The structural covering, usually wood boards or plywood, used over studs or rafters of a structure. Structural building board is normally used only as wall sheathing.

Sheathing paper. See **Paper, sheathing.**

Sheet metal work. All components of a house employing sheet metal, such as flashing, gutters, and downspouts.

Shellac. A transparent coating made by dissolving **lac,** a resinous secretion of the lac bug (a scale insect that thrives in tropical countries, especially India), in alcohol.

Shingles. Roof covering of asphalt, asbestos, wood, tile, slate, or other material cut to stock lengths, widths, and thicknesses.

Shingles, siding. Various kinds of shingles, such as wood shingles or shakes and nonwood shingles, that are used over sheathing for exterior sidewall covering of a structure.

Shiplap. See **Lumber, shiplap.**

Shutter. Usually lightweight louvered or flush wood or nonwood frames in the form of doors located at each side of a window. Some are made to close over the window for protection; others are fastened to the wall as a decorative device.

Siding. The finish covering of the outside wall of a frame building, whether made of horizontal weatherboards, vertical boards with battens, shingles, or other material.

Siding, bevel (lap siding). Wedge-shaped boards used as horizontal siding in a lapped pattern. This siding varies in butt thickness from ½ to ¾ inch and in widths up to 12 inches. Normally used over some type of sheathing.

Siding, Dolly Varden. Beveled wood siding which is rabbeted on the bottom edge.

Siding, drop. Usually ¾ inch thick and 6 and 8 inches wide with tongued-and-grooved or shiplap edges. Often used as siding without sheathing in secondary buildings.

Sill. The lowest member of the frame of a structure, resting on the foundation and supporting the floor joists or the uprights of the wall. The member forming the lower side of an opening, as a door sill, window sill, etc.

Sleeper. Usually, a wood member embedded in concrete, as in a floor, that serves to support and to fasten subfloor or flooring.

Soffit. Usually the underside of an overhanging cornice.

Soil cover (ground cover). A light covering of plastic film, roll roofing, or similar material used over the soil in crawl spaces of buildings to minimize moisture permeation of the area.

Soil stack. A general term for the vertical main of a system of soil, waste, or vent piping.

Sole or sole plate. See **Plate.**

Solid bridging. A solid member placed between adjacent floor joists near the center of the span to prevent joists from twisting.

Span. The distance between structural supports such as walls, columns, piers, beams, girders, and trusses.

Splash block. A small masonry block laid with the top close to the ground surface to receive roof drainage from downspouts and to carry it away from the building.

Square. A unit of measure—100 square feet—usually applied to roofing material. Sidewall coverings are sometimes packed to cover 100 square feet and are sold on that basis.

Stain, shingle. A form of oil paint, very thin in consistency, intended for coloring wood with rough surfaces, such as shingles, without forming a coating of significant thickness or gloss.

Stair carriage. Supporting member for stair treads. Usually a 2-inch plank notched to receive the treads; sometimes called a "rough horse."

Stair landing. See **Landing.**

Stair rise. See **Rise.**

STC. (Sound Transmission Class). A measure of sound stopping of ordinary noise.

Stile. An upright framing member in a panel door.

Stool. A flat molding fitted over the window sill between jambs and contacting the bottom rail of the lower sash.

Storm sash or storm window. An extra window usually placed on the outside of an existing one as additional protection against cold weather.

Story. That part of a building between any floor and the floor or roof next above.

Strip flooring. Wood flooring consisting of narrow, matched strips.

String, stringer. A timber or other support for cross members in floors or ceilings. In stairs, the support on which the stair treads rest; also stringboard.

Stucco. Most commonly refers to an outside plaster made with Portland cement as its base.

Stud. One of a series of slender wood or metal vertical structural members placed as supporting elements in walls and partitions. (Plural: studs or studding.)

Subfloor. Boards or plywood laid on joists over which a finish floor is to be laid.

Suspended ceiling. A ceiling system supported by hanging it from the overhead structural framing.

Tail beam. A relatively short beam or joist supported in a wall on one end and by a header at the other.

Termites. Insects that superficially resemble ants in size, general appearance, and habit of living in colonies; hence, they are frequently called "white ants." Subterranean termites establish themselves in buildings not by being carried in with lumber, but by entering from ground nests **after** the building has been constructed. If unmolested, they eat out the woodwork, leaving a shell of sound wood to conceal their activities, and damage may proceed so far as to cause collapse of parts of a structure before discovery. There are about 56 species of termites known in the United States; but the two major ones, classified by the manner in which they attack wood, are ground-inhabiting or subterranean termites (the most common) and dry-wood termites, which are found almost exclusively along the extreme southern border and the Gulf of Mexico in the United States.

Termite shield. A shield, usually of noncorrodible metal, placed in or on a foundation wall or other mass of masonry or around pipes to prevent passage of termites.

Terneplate. Sheet iron or steel coated with an alloy of lead and tin.

Threshold. A strip of wood or metal with beveled edges used over the finish floor and the sill of exterior doors.

Toenailing. To drive a nail at a slant with the initial surface in order to permit it to penetrate into a second member.

Tongued and grooved. See **Dressed and matched.**

Tread. The horizontal board in a stairway on which the foot is placed.

Trim. The finish materials in a building, such as moldings, applied around openings (window trim, door trim) or at the floor and ceiling of rooms (baseboard, cornice, and other moldings).

Trimmer. A beam or joist to which a header is nailed in framing for a chimney, stairway, or other opening.

Truss. A frame or jointed structure designed to act as a beam of long span, while each member is usually subjected to longitudinal stress only, either tension or compression.

Turpentine. A volatile oil used as a thinner in paints and as a solvent in varnishes. Chemically, it is a mixture of terpenes.

Undercoat. A coating applied prior to the finishing or top coats of a paint job. It may be the first of two or the second of three coats. In some usage of the word it may become synonymous with priming coat.

Under layment. A material placed under finish coverings, such as flooring, or shingles, to provide a smooth, even surface for applying the finish.

Valley. The internal angle formed by the junction of two sloping sides of a roof.

Vapor barrier. Material used to retard the movement of water vapor into walls and prevent condensation in them. Usually considered as having a perm value of less than 1.0. Applied separately over the warm side of exposed walls or as a part of batt or blanket insulation.

Varnish. A thickened preparation of drying oil or drying oil and resin suitable for spreading on surfaces to form continuous, transparent coatings, or for mixing with pigments to make enamels.

Vehicle. The liquid portion of a finishing material; it consists of the binder (nonvolatile) and volatile thinners.

Veneer. Thin sheets of wood made by rotary cutting or slicing of a log.

Vent. A pipe or duct which allows flow of air as an inlet or outlet.

Vermiculite. A mineral closely related to mica, with the faculty of expanding on heating to form lightweight material with insulation quality. Used as bulk insulation and also as aggregate in insulating and acoustical plaster and in insulating concrete floors.

Volatile thinner. A liquid that evaporates readily and is used to thin or reduce the consistency of finishes without altering the relative volumes of pigments and nonvolatile vehicles.

Wane. Bark, or lack of wood from any cause, on edge or corner of a piece of wood.

Water-repellent preservative. A liquid designed to penetrate into wood and impart water repellency and a moderate preservative protection. It is used for millwork, such as sash and frames, and is usually applied by dipping.

Weatherstrip. Narrow or jamb-width sections of thin metal or other material to prevent infiltration of air and moisture around windows and doors. Compression weather stripping prevents air infiltration, provides tension, and acts as a counter balance.

Wood rays. Strips of cells extending radially within a tree and varying in height from a few cells in some species to 4 inches or more in oak. The rays serve primarily to store food and to transport it horizontally in the tree.

INDEX